Numerical Methods for Evolutionary Differential Equations

COMPUTATIONAL SCIENCE & ENGINEERING

Computational Science and Engineering (CS&E) is widely accepted, along with theory and experiment, as a crucial third mode of scientific investigation and engineering design. This series publishes research monographs, advanced undergraduate- and graduate-level textbooks, and other volumes of interest to a wide segment of the community of computational scientists and engineers. The series also includes volumes addressed to users of CS&E methods by targeting specific groups of professionals whose work relies extensively on computational science and engineering.

Editor-in-Chief
Omar Ghattas
University of Texas at Austin

Editorial Board

David Keyes, Associate Editor
Columbia University

Ted Belytschko
Northwestern University

Clint Dawson
University of Texas at Austin

Lori Freitag Diachin
Lawrence Livermore National Laboratory

Charbel Farhat
Stanford University

James Glimm
Stony Brook University

Teresa Head-Gordon
University of California–Berkeley and
Lawrence Berkeley National Laboratory

Rolf Jeltsch
ETH Zurich

Chris Johnson
University of Utah

Laxmikant Kale
University of Illinois

Efthimios Kaxiras
Harvard University

Jelena Kovacevic
Carnegie Mellon University

Habib Najm
Sandia National Laboratory

Alex Pothen
Old Dominion University

Series Volumes

Ascher, Uri M., *Numerical Methods for Evolutionary Differential Equations*

Zohdi, T. I., *An Introduction to Modeling and Simulation of Particulate Flows*

Biegler, Lorenz T., Omar Ghattas, Matthias Heinkenschloss, David Keyes, and Bart van Bloemen Waanders, Editors, *Real-Time PDE-Constrained Optimization*

Chen, Zhangxin, Guanren Huan, and Yuanle Ma, *Computational Methods for Multiphase Flows in Porous Media*

Shapira, Yair, *Solving PDEs in C++: Numerical Methods in a Unified Object-Oriented Approach*

Numerical Methods for Evolutionary Differential Equations

Uri M. Ascher
University of British Columbia
Vancouver, British Columbia, Canada

Society for Industrial and Applied Mathematics
Philadelphia

Copyright © 2008 by the Society for Industrial and Applied Mathematics.

10 9 8 7 6 5 4 3 2 1

All rights reserved. Printed in the United States of America. No part of this book may be reproduced, stored, or transmitted in any manner without the written permission of the publisher. For information, write to the Society for Industrial and Applied Mathematics, 3600 Market Street, 6th Floor, Philadelphia, PA 19104-2688 USA.

Trademarked names may be used in this book without the inclusion of a trademark symbol. These names are used in an editorial context only; no infringement of trademark is intended.

MATLAB is a registered trademark of The MathWorks, Inc. For MATLAB product information, please contact The MathWorks, Inc., 3 Apple Hill Drive, Natick, MA 01760-2098 USA, 508-647-7000, Fax: 508-647-7101, info@mathworks.com, www.mathworks.com.

Figure 2.12 is reprinted with permission from E. Boxerman and U. Ascher, Decomposing cloth, Eurographics/*ACM SIGGRAPH Symposium on Computer Animation* (2004), 153-161.

Figures 9.4 and 9.5 are reprinted with permission from E. Haber, U. Ascher, and D. Oldenburg, Inversion of 3D electromagnetic data in frequency and time domain using an inexact all-at-once approach, *J. Geophysics*, 69 (2004), 1216-1228.

Figure 9.11 is reprinted with kind permission from Springer Science and Business Media from U. Ascher and E. Boxerman, On the modified conjugate gradient method in cloth simulation, *The Visual Computer*, 19 (2003), 526-531.

Figure 11.2 is reprinted with kind permission from Springer Science and Business Media from U. Ascher, H. Huang, and K. van den Doel, Artificial time integration, *BIT*, 47 (2007), 3-25.

The cover was produced from images created by and used with permission of the Scientific Computing and Imaging (SCI) Institute, University of Utah; J. Bielak, D. O'Hallaron, L. Ramirez-Guzman, and T. Tu, Carnegie Mellon University; O. Ghattas, University of Texas at Austin; K. Ma and H. Yu, University of California, Davis; and Mark R. Petersen, Los Alamos National Laboratory. More information about the images is available at http://www.siam.org/books/series/csecover.php

Library of Congress Cataloging-in-Publication Data

Ascher, U. M. (Uri M.), 1946-
  Numerical methods for evolutionary differential equations / Uri M. Ascher.
    p. cm. -- (Computational science and engineering ; 5)
  Includes bibliographical references and index.
  ISBN 978-0-898716-52-8
  1. Evolution equations--Numerical solutions. I. Title.
  QA377.A827 2008
  003'.5--dc22

                                 2008006667

 is a registered trademark.

To Nurit and Noam

Contents

Preface		xi
1	**Introduction**	**1**
	1.1 Well-Posed Initial Value Problems	4
	1.1.1 Simple model cases	7
	1.1.2 More general cases	10
	1.1.3 Initial-boundary value problems	11
	1.1.4 The solution operator	12
	1.2 A Taste of Finite Differences	12
	1.2.1 Stability ideas	17
	1.3 Reviews	24
	1.3.1 Taylor's theorem	24
	1.3.2 Matrix norms and eigenvalues	25
	1.3.3 Function spaces	28
	1.3.4 The continuous Fourier transform	29
	1.3.5 The matrix power and exponential	30
	1.3.6 Fourier transform for periodic functions	31
	1.4 Exercises	32
2	**Methods and Concepts for ODEs**	**37**
	2.1 Linear Multistep Methods	39
	2.2 Runge–Kutta Methods	42
	2.3 Convergence and 0-stability	48
	2.4 Error Control and Estimation	52
	2.5 Stability of ODE Methods	53
	2.6 Stiffness	55
	2.7 Solving Equations for Implicit Methods	59
	2.8 Differential-Algebraic Equations	64
	2.9 Symmetric and One-Sided Methods	66
	2.10 Highly Oscillatory Problems	66
	2.11 Boundary Value ODEs	71
	2.12 Reviews	72
	2.12.1 Gaussian elimination and matrix decompositions	73
	2.12.2 Polynomial interpolation and divided differences	74

		2.12.3 Orthogonal and trigonometric polynomials 77
		2.12.4 Basic quadrature rules 79
		2.12.5 Fixed point iteration and Newton's method 80
		2.12.6 Discrete and fast Fourier transforms 82
	2.13	Exercises 83

3 Finite Difference and Finite Volume Methods   91
     3.1   Semi-Discretization 92
           3.1.1   Accuracy and derivation of spatial discretizations 94
           3.1.2   Staggered meshes 98
           3.1.3   Boundary conditions 106
           3.1.4   The finite element method 110
           3.1.5   Nonuniform meshes 113
           3.1.6   Stability and convergence 120
     3.2   Full Discretization 120
           3.2.1   Order, stability, and convergence 122
           3.2.2   General linear stability 128
     3.3   Exercises 130

4 Stability for Constant Coefficient Problems   135
     4.1   Fourier Analysis 135
           4.1.1   Stability for scalar equations 137
           4.1.2   Stability for systems of equations 139
           4.1.3   Semi-discretization stability 142
           4.1.4   Fourier analysis and ODE absolute stability regions 143
     4.2   Eigenvalue Analysis 144
     4.3   Exercises 146

5 Variable Coefficient and Nonlinear Problems   151
     5.1   Freezing Coefficients and Dissipativity 153
     5.2   Schemes for Hyperbolic Systems in One Dimension 154
           5.2.1   Lax–Wendroff and variants for conservation laws 156
           5.2.2   Leapfrog and Lax–Friedrichs 158
           5.2.3   Upwind scheme and the modified PDE 162
           5.2.4   Box and Crank–Nicolson 165
     5.3   Nonlinear Stability and Energy Methods 168
           5.3.1   Energy method 169
           5.3.2   Runge–Kutta for skew-symmetric semi-discretizations 173
     5.4   Exercises 177

6 Hamiltonian Systems and Long Time Integration   181
     6.1   Hamiltonian Systems 182
     6.2   Symplectic and Other Relevant Methods 185
           6.2.1   Symplectic Runge–Kutta methods 188
           6.2.2   Splitting and composition methods 189
           6.2.3   Variational methods 194

		6.3	Properties of Symplectic Methods 195
		6.4	Pitfalls in Highly Oscillatory Hamiltonian Systems 198
		6.5	Exercises 205

7 Dispersion and Dissipation — 211

- 7.1 Dispersion 212
- 7.2 The Wave Equation 217
- 7.3 The KdV Equation 230
  - 7.3.1 Schemes based on a classical semi-discretization 232
  - 7.3.2 Box schemes 236
- 7.4 Spectral Methods 242
- 7.5 Lagrangian methods 246
- 7.6 Exercises 246

8 More on Handling Boundary Conditions — 253

- 8.1 Parabolic Problems 253
- 8.2 Hyperbolic Problems 257
  - 8.2.1 Boundary conditions for hyperbolic problems 257
  - 8.2.2 Boundary conditions for discretized hyperbolic problems . . . 261
  - 8.2.3 Order reduction for Runge–Kutta methods 268
- 8.3 Infinite or Large Domains 271
- 8.4 Exercises 272

9 Several Space Variables and Splitting Methods — 275

- 9.1 Extending the Methods We Already Know 276
- 9.2 Solving for Implicit Methods 279
  - 9.2.1 Implicit methods for parabolic equations 282
  - 9.2.2 Alternating direction implicit methods 290
  - 9.2.3 Nonlinear problems 292
- 9.3 Splitting Methods 292
  - 9.3.1 More general splitting 297
  - 9.3.2 Additive methods 306
  - 9.3.3 Exponential time differencing 310
- 9.4 Review: Iterative Methods for Linear Systems 312
  - 9.4.1 Simplest iterative methods 312
  - 9.4.2 Conjugate gradient and related methods 314
  - 9.4.3 Multigrid methods 317
- 9.5 Exercises 320

10 Discontinuities and Almost Discontinuities — 327

- 10.1 Scalar Conservation Laws in One Dimension 329
  - 10.1.1 Exact solution of the Riemann problem 333
- 10.2 First Order Schemes for Scalar Conservation Laws 334
  - 10.2.1 Godunov's scheme 338
- 10.3 Higher Order Schemes for Scalar Conservation Laws 340
  - 10.3.1 High-resolution schemes 341

		10.3.2 Semi-discretization and ENO schemes 342
		10.3.3 Strong stability preserving methods 346
		10.3.4 WENO schemes 349
	10.4	Systems of Conservation Laws 352
	10.5	Multidimensional Problems 356
	10.6	Problems with Sharp Layers 358
	10.7	Exercises 360

11 Additional Topics — 365

    11.1 What First: Optimize or Discretize? 365
        11.1.1 Symmetric matrices for nonuniform spatial meshes 366
        11.1.2 Efficient multigrid and Neumann BCs 366
        11.1.3 Optimal control 367
    11.2 Nonuniform Meshes 368
        11.2.1 Adaptive meshes for steady state problems 369
        11.2.2 Adaptive mesh refinement 371
        11.2.3 Moving meshes 371
    11.3 Level Set Methods 372

Bibliography   **375**

Index   **387**

Preface

Methods for the numerical simulation of dynamic mathematical models have been the focus of intensive research for well over 60 years. However, rather than reaching closure, there is a continuing demand today for better and more efficient methods, as the range of applications is increasing. Mathematical models involving evolutionary partial differential equations (PDEs) as well as ordinary differential equations (ODEs) arise in many diverse applications, such as fluid flow, image processing and computer vision, physics-based animation, mechanical systems, relativity, earth sciences, and mathematical finance. It is possible today to dream of, if not actually achieve, a realistic simulation of a clothed, animated figure in a video game, or of an accurate simulation of a fluid flowing in a complex geometry in three dimensions, or of simulating the dynamics of a large molecular structure for a realistic time interval without requiring several weeks of intense computing.

This text was developed from course notes written for graduate courses that I have taught repeatedly over the years. The students typically come from different disciplines, including Computer Science, Mathematics, Physics, Earth and Ocean Sciences, and a variety of Engineering disciplines. With the widening scope of practical applications comes a widened scope of an interested audience. This means not only varied background and expertise in a typical graduate class, but also that not all those who need to know how to simulate such PDE systems are or should be experts in fluid dynamics! The approach therefore chosen emphasizes the study of principles, properties, and usage of numerical methods from the point of view of general applicability. This text is not a collection of recipes, and basic analysis and understanding are emphasized throughout, yet a formal theorem–proof format is avoided and no topic is covered simply for its theoretical beauty. Moreover, while no one class of applications motivates the exposition, many examples from different application areas are discussed throughout.

In addition to not relying on strict fluid dynamics prerequisites, the other strategic decision made was to delay in each chapter as much as possible the separation of treatment of parabolic and hyperbolic equations. The other route, taken by many authors, is to devote separate chapters for the different PDE types. This often leads to a very neat presentation. In this text the approach is more concept oriented, however, and it is hoped that the differences necessarily highlighted by contrasting the treatments in this way shed more direct light on some issues. Moreover, questions about problems such as simulating a convection-dominated diffusion-convection process and about mixed hyperbolic-parabolic systems are more naturally addressed.

The introductory Chapter 1 is essential. First we develop a sense for the types of mathematical PDE models for which solving evolutionary problems makes sense, by studying well-posedness. Then we embark upon an introduction by example to numerical methods and issues that are developed more fully later on.

Several years ago L. Petzold and I wrote a book in a similar spirit about the numerical solution of ODEs. I still like that work, and thankfully so do others; however, there are many students and researchers in various disciplines who simply don't seem to have room in their program to accommodate a course or a text devoted solely to numerical ODEs. Therefore, I have included in the present text two chapters on this topic. Chapter 2 crams in all the material from our book [14] that is viewed as essential for a crash course on simulating ODEs, with an eye toward what is essential and relevant for PDEs. Chapter 6, not covered in [14], is more specialized and considers certain problems and methods in Geometric Integration, especially for Hamiltonian systems. These concepts and methods are relevant also for PDEs—see especially Chapter 7—and they allow a different and interesting look at numerical methods for differential equations. However, they are less essential in a way.

Chapter 3 develops in detail the basic concepts, issues, and discretization tools that arise in finite difference and finite volume methods for PDEs. It relates to and expands the material in the previous two chapters but they are not a strict prerequisite for reading it.

Stability is an essential concept when designing and analyzing methods for the numerical solution of time-dependent problems. Chapter 4 deals with constant coefficient PDEs where resulting criteria are relatively easy to check. Chapter 5 continues into variable coefficient and nonlinear problems, where stability criteria and approaches are defined and used, and where new methods for hyperbolic PDEs are introduced.

The first five chapters are the basic ones, and it is reasonable to teach in a semester course mainly these plus some forays into later chapters, e.g., Chapter 6 if one concentrates significantly on ODEs. On the other hand, many more delicate or involved issues in numerical PDEs, and much of the more recent research, are in the last five chapters. The topic of numerical dispersion in wave problems and of conservative vs. dissipative methods is considered in Chapter 7, while handling solutions with discontinuities is discussed at some length in Chapter 10. These chapters are both somewhat more topical and occasionally more advanced, and neither is a prerequisite for the other. The handling of boundary conditions is briefly considered in Chapter 3, but Chapter 8 discusses deeper and more specialized issues, particularly for hyperbolic PDEs. Two related but different topics are considered in Chapter 9. The first concerns handling additional issues that arise for problems in more than one space variable. There are many issues here, and they are necessarily covered occasionally in less depth than would be possible in a more specialized monograph. I have used this chapter also to introduce and discuss the interesting class of splitting methods, even though these arise not only in multidimensional problems. Finally, Chapter 11 quickly describes some highly interesting related topics that could require separate monographs if they were to be fully treated.

There are several reviews of background material collected in separate sections in Chapters 1 and 2. These are meant as refreshers to quickly help a reader who has been exposed to their contents beforehand, not to replace a proper introduction to their subject matters. The survey of iterative methods for linear systems of algebraic equations in Section 9.4 is somewhere between a review and a core material. At the end of each chapter there are also exercises, the first of which (numbered 0) consisting of more straightforward review questions. I have tried to indicate those exercises that have been found to be more difficult, or more time-consuming, among them.

No attempt has been made to make the bibliography complete: this would have been a vast and dangerous undertaking in itself. Wherever possible I have tried to refer to texts, monographs, and survey articles that contain a wealth of additional references. Several references to my own work are there simply because I've naturally drawn upon past work as a source for examples and exercises and because I am generally more familiar with them than with others.

Parts of Chapters 2, 6, and 7 were written for a short course that I gave at the Institute of Pure and Applied Mathematics (IMPA) in Rio de Janeiro. Here is an opportunity to thank my friends and colleagues there, especially D. Marchesin, A. Nachbin, M. Sarkis, L. Velho, and J. Zubelli, for their hospitality during several visits to the "marvelous city." I also gratefully acknowledge the hospitality of A. Iserles and the pleasant Isaac Newton Institute in Cambridge, where much of the material presented here was polished during a seven-week stay last year.

Many people have helped me in various ways to shape, reshape, refine, and debug this text. Several generations of our students had to endure its (and my) imperfections and their various comments have helped tremendously. In particular let me note Eddy Boxerman, Hui Huang, and Ewout van den Berg; there are many others not mentioned here. Colleagues who have read various versions of these notes and/or offered loads of advice and useful criticism include in particular Mark Ainsworth, Mihai Anitescu, Evaristo Biscaia Jr., Robert Bridson, Chris Budd, Philippe Chartier, Chen Greif, Eldad Haber, Ernst Hairer, Arieh Iserles, Robert McLachlan, Sarah Mitchell, Dinesh Pai, Linda Petzold, Ray Spiteri, David Tranah, and Jim Varah. I am indebted to you all!

Uri M. Ascher
Vancouver, December 2007

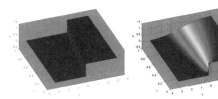

Chapter 1
Introduction

This text explores numerical methods based on finite difference and finite volume discretizations for solving evolutionary partial differential equations (PDEs). Mathematical models involving such differential equations arise in many diverse applications, such as fluid flow, image processing and computer vision, physics-based animation, mechanical systems, earth sciences, and mathematical finance. We will methodically develop tools for understanding and assessing methods and software for these problems.

We start at a relatively basic level, and rudimentary knowledge of differential equations and numerical methods is a sufficient prerequisite. In later chapters we discuss advanced techniques required for problems with nonlinearities, multidimensions, long-time integration, interfaces, and discontinuities such as shocks. Many examples are given to illustrate the various concepts and methods introduced.

In general, our typical PDE will depend on a few independent variables. Throughout this text we distinguish one of these variables as "*time*" while the other variables are treated as "*space*" variables, often not distinguished from one another.

To make things more concrete, let us consider a **motivating example**. We will not get into precise or full details as yet: these will come later. The *advection-diffusion equation* is given by the PDE

$$u_t + a u_x = \nu u_{xx}. \tag{1.1}$$

In this equation, the independent variables are time t and space x, and the unknown function sought is $u = u(t, x)$. The subscript notation corresponds to partial differentiation: $u_t = \frac{\partial u}{\partial t}$, $u_x = \frac{\partial u}{\partial x}$, and $u_{xx} = \frac{\partial^2 u}{\partial x^2}$.

There are two parameters in (1.1), assumed known, real, and constant. The parameter a corresponds to a fluid velocity in an advection process, while ν controls the amount of diffusion (or viscosity, depending on the application area) and is required to be nonnegative, $\nu \geq 0$. You don't have to be familiar with the physical interpretations of these quantities in order to proceed here.

Our differential equation is defined on a domain in space and time, $0 < x < b$ and $t \geq 0$; see Figure 1.1. For instance, set $b = 1$. The PDE is supplemented, in general, by initial conditions and boundary conditions.

1

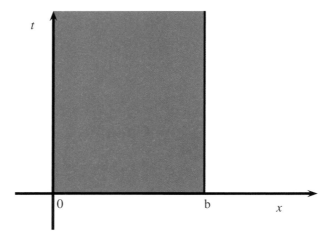

Figure 1.1. *Domain of definition for an initial-boundary value problem.*

At the initial time $t = 0$, our sought solution u must agree with a given function $u_0(x)$

$$u(0, x) = u_0(x), \qquad 0 \le x \le b. \tag{1.2}$$

The situation with boundary conditions is more complex and there is more to choose from. Conceptually, the simplest is to specify u for all t at both ends $x = 0$ and $x = b$,

$$u(t, 0) = g_0(t), \qquad u(t, b) = g_b(t), \tag{1.3a}$$

which makes sense if $\nu > 0$. These are called *Dirichlet boundary conditions*. For $\nu = 0$ only one condition, either at $x = 0$ or at $x = b$ (depending on the sign of a), may be specified: we shall return to this later, in Chapter 8.

Another favorite type, good for any $\nu \ge 0$, are *periodic boundary conditions*, which specify that $u(t, x) = u(t, b + x)$ for all relevant t and x. In particular

$$u(t, 0) = u(t, b). \tag{1.3b}$$

These boundary conditions say that what happens in the strip depicted in Figure 1.1 is replicated at similar strips to the right and to the left: $-b < x < 0$, $b < x < 2b$, etc. Thus, the problem is defined for all x real, as depicted in Figure 1.2. This corresponds to having no boundary conditions at all on an infinite domain in x, a case that is pursued more methodically in Section 1.1.

To better understand the expected behavior of solutions for the advection-diffusion equation (1.1) subject to the initial conditions (1.2), let us look at a solution that at each time t has the form

$$u(t, x) = \hat{u}(t, \xi) e^{i\xi x} \tag{1.4}$$

for some fixed, real value of ξ. We assume that the boundary conditions (1.3) allow us to do this (i.e., that they do not "interfere" with our intention to concentrate on the initial value problem for our PDE). The parameter ξ is called a **wave number**, and $|\hat{u}(t, \xi)|$ is the **amplitude** of the **mode** $e^{\iota\xi x}$. Here we use complex numbers, and this allows expressing things elegantly. We employ the standard notation

$$\iota = \sqrt{-1}$$

and recall that for any real angle θ in radians

$$e^{\iota\theta} = \cos\theta + \iota \sin\theta,$$
$$|e^{\iota\theta}| = \cos^2\theta + \sin^2\theta = 1.$$

Thus, for a complex number λ, $|e^\lambda| = e^{\mathcal{R}e\lambda}$. In particular, $|e^{\iota\xi x}| = e^0 = 1$ for any real ξ and x.

For our special form of the solution (1.4) we have

$$u_x = \iota\xi\, \hat{u}\, e^{\iota\xi x}, \quad u_{xx} = (\iota\xi)^2\, \hat{u}\, e^{\iota\xi x},$$

so, upon inserting these expressions into (1.1) and canceling the common exponent, we obtain for our advection-diffusion equation

$$\hat{u}_t(t, \xi) = -(\xi^2 \nu + \iota\xi a)\, \hat{u}(t, \xi).$$

For each wave number ξ this is an ordinary differential equation (ODE) in t that has the solution

$$\hat{u}(t, \xi) = e^{-(\xi^2 \nu + \iota\xi a)t} \hat{u}(0, \xi).$$

Therefore, the amplitude of this mode as it evolves in time is given by

$$|\hat{u}(t, \xi)| = |e^{-(\xi^2 \nu + \iota\xi a)t}|\, |\hat{u}(0, \xi)|$$
$$= e^{-\xi^2 \nu t}|\hat{u}(0, \xi)|.$$

For the special form of solution (1.4) we clearly have $|u(t, x)| = |\hat{u}(t, \xi)|$ at each time instance t, so

$$|u(t, x)| = e^{-\xi^2 \nu t}|u(0, x)|.$$

We see that

- If $\nu > 0$, then the amplitude of the solution mode decays in time. The larger the wave number ξ, the faster it decays. This is typical of a *parabolic* PDE.

- If $\nu = 0$, then the amplitude of the solution mode remains the same for all ξ: we have $|\hat{u}(t, \xi)| = |\hat{u}(0, \xi)|$. This is typical of a *hyperbolic* PDE.

The case of a very small but positive ν corresponds to a parabolic problem that is "almost hyperbolic."

This concludes our **motivating example**. In general, we may not expect the PDE to have constant coefficients, and even if it has we may not expect its solution to look like a single mode; and yet, some essential elements encountered above will keep appearing in what follows.

This introductory chapter has two essential parts. In Section 1.1 we develop a sense for the types of mathematical problems that are being considered here in a more methodical way than used in the preceding motivating example, still before any discretization or computation is applied to them. We consider this right away because without some basic feeling for the problem being solved it is hard to develop adequate numerical methods and to assess the quality of obtained computational results later on. Thus we elaborate upon the concept of well-posedness in example-driven fashion. Then in Section 1.2 we give a first taste of what most of the text is concerned with, by defining and following the performance of three simple discretizations for the simple advection equation. Various important properties are manifested in these simple examples. By the end of this section we will be able to give a rough guide to what follows in further chapters. There is also a section containing various reviews of relevant background topics, and of course some inevitable exercises.

1.1 Well-posed initial value problems

We want to investigate numerical methods that yield sensible results. For instance, if the physical problem being simulated has a bounded solution, then we want the numerical one to be bounded as well. For this we must make some assumptions on the PDE problem being discretized, essentially ensuring that it behaves sensibly *before* discretization. Thus, if a computer simulation of such a problem yields poor results, then we'll know that the fault is with the numerical method and seek to improve it. The book by Kreiss and Lorenz [111] covers well-posedness in depth. (See also Kreiss [110].)

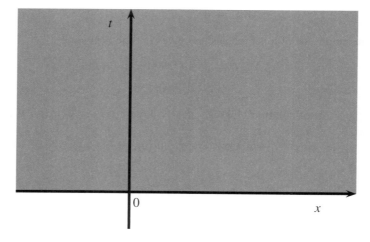

Figure 1.2. *Domain of definition for a Cauchy problem.*

Consider a linear PDE

$$u_t = P(\partial_x, t, x) u, \qquad -\infty < x < \infty, \, t > 0 \qquad (1.5a)$$

(see Figure 1.2). Here $u = u(t, x)$ is a scalar or vector unknown function in the time variable t and a space variable x, and P is an operator which is polynomial in ∂_x: $P =$

1.1. Well-Posed Initial Value Problems

$\sum_{j=0}^{m} p_j(t, x) \frac{\partial^j}{\partial x^j}$. For example, setting $m = 2$, $p_0 \equiv p_1 \equiv 0$, $p_2 \equiv 1$, gives the simple heat equation, $u_t = u_{xx}$. This in turn is a special case of (1.1) with $\nu = 1$ and $a = 0$.

The differential equations are supplemented by a given (square integrable) initial value function $u_0(x)$

$$u(0, x) = u_0(x), \qquad -\infty < x < \infty. \tag{1.5b}$$

The system (1.5) is then a *pure initial value problem* (IVP), otherwise known as a **Cauchy problem**.

> **Note:** For those weary of infinite domains, and even more so of integrals over infinite domains such as those appearing in (1.6) or (1.10), let us hasten to point out that we have no intention to program or compute solutions for any of these as such. They are here for the purpose of basic analysis, serving as a means to avoid introducing boundary conditions too early in the game: ∞ is our friend!

In general, a well-posed problem possesses the property that its solution depends continuously on the data: a small change in the data produces only a small change in the solution. This is obviously a desired feature for the subsequent numerical solution by a stable numerical process.

Let us define the $\mathcal{L}_2$-norm

$$\|u(t)\| \equiv \|u(t, \cdot)\| = \left(\int_{-\infty}^{\infty} |u(t, x)|^2 dx \right)^{1/2} \tag{1.6}$$

(see the review of function spaces in Section 1.3.3). We will say that the IVP (1.5) is **well-posed** (sometimes referred to as *properly posed*) if there are constants K and α such that for all $t \geq 0$

$$\|u(t)\| \leq K e^{\alpha t} \|u(0)\| = K e^{\alpha t} \|u_0\| \quad \forall \, u_0 \in \mathcal{L}_2. \tag{1.7}$$

Note that K and α may have definite values which can be retrieved for given problem instances.[1] Of particular importance is the value of α. It can possibly be zero, in which case the solution does not grow much in time, or even negative, in which case the solution decidedly decays.

More generally, we'll have a possibly finite, or semi-infinite, range for x, e.g., the domain depicted in Figure 1.1. Then the equations (1.5) (appropriately restricted in x) are supplemented by boundary conditions (BCs) at $(t, 0)$ and (t, b). We then have an *initial-boundary value problem*. The boundary conditions may be assumed homogeneous. A similar definition of well-posedness is then possible, with the integral range in the norm

---

[1] We regard the initial time as arbitrary here. Thus, strictly speaking we must require that

$$\|u(t)\| \leq K e^{\alpha(t - t_0)} \|u(t_0)\|$$

for any $0 \leq t_0 \leq t$. The same holds in the context of corresponding stability requirements for numerical methods which follow later on.

definition appropriately changed and all functions $u(t, x)$ considered satisfying the homogeneous boundary conditions.

Consider next the pure IVP (1.5) with *constant coefficients*

$$P = P(\partial_x) = \sum_{j=0}^{m} p_j \frac{\partial^j}{\partial x^j}. \tag{1.8}$$

Then we can get a good handle on well-posedness upon using a **Fourier transform** (see the review in Section 1.3.4). This gives both a justification and a generalization for our motivating treatment of the advection-diffusion equation (1.1) by (1.4).

The PDE (1.5a) is now written simply as

$$u_t = P(\partial_x)u, \qquad -\infty < x < \infty, \ t \geq 0,$$

and it is transformed in Fourier space with respect to x using (1.17)

$$\hat{u}(t, \xi) = \frac{1}{\sqrt{2\pi}} \int_{-\infty}^{\infty} e^{-\iota \xi x} u(t, x) dx.$$

This yields[2] for each *wave number* ξ an ordinary differential equation (ODE) in time

$$\hat{u}_t = P(\iota \xi)\hat{u}, \qquad t > 0. \tag{1.9}$$

The matrix $P(\iota \xi)$ (which is $s \times s$ for a system of s first order PDEs) is called the **symbol** of the PDE system. The ODE (1.9) has a given initial value $\hat{u}(0, \xi)$ which is the Fourier transform (1.17) of $u_0(x)$.

This initial value ODE is easy to solve:

$$\hat{u}(t, \xi) = e^{P(\iota \xi)t} \hat{u}(0, \xi);$$

hence for one PDE

$$\begin{aligned}
\|u(t, \cdot)\|^2 &= \|\hat{u}(t, \cdot)\|^2 = \int_{-\infty}^{\infty} |\hat{u}(t, \xi)|^2 d\xi \\
&= \int_{-\infty}^{\infty} |e^{P(\iota \xi)t} \hat{u}(0, \xi)|^2 d\xi \leq \sup_{\xi} |e^{P(\iota \xi)t}|^2 \|\hat{u}(0, \cdot)\|^2 \\
&= \sup_{\xi} |e^{P(\iota \xi)t}|^2 \|u_0\|^2.
\end{aligned} \tag{1.10}$$

Comparing this bound to (1.7) we see that a Cauchy problem with constant coefficients is well-posed if there are constants K and α independent of $t \geq 0$ such that

$$\sup_{-\infty < \xi < \infty} |e^{P(\iota \xi)t}| \leq K e^{\alpha t}. \tag{1.11}$$

For a PDE system with $s > 1$ a matrix norm replaces magnitude. The latter condition is easy to check in practice, even though ξ runs over the entire real line, so we proceed to consider special cases and verify their well-posedness.

---

[2]We call the Fourier variable corresponding to space a *wave number*, whereas the Fourier transform in time is traditionally described in terms of *frequencies*. Occasionally people mix the two, talking about "high frequencies," for instance, while meaning high (i.e., *large*) wave numbers. We will try to stay consistent about this terminology.

1.1. Well-Posed Initial Value Problems

1.1.1 Simple model cases

1. For the simple **heat equation**

$$u_t = u_{xx},$$

we have

$$P(\iota\xi) = (\iota\xi)^2 = -\xi^2.$$

Hence

$$|e^{P(\iota\xi)t}| = |e^{-\xi^2 t}| \leq 1 \quad \forall \xi.$$

The heat equation is the simplest instance of a *parabolic equation*. Thus, the solution of the parabolic initial value problem for $t > 0$ does not grow—a stronger property than what is required for a mere well-posedness. In fact, we see a decay in $\hat{u}(t, \xi)$ which is more pronounced for the higher wave numbers. The solution operator of the heat equation is thus a **smoother**.

On the other hand, for $u_t = -u_{xx}$ the problem is not well-posed. Indeed, it is not difficult to see from the equations in (1.10) that $\|u(t)\|$ grows faster than is allowed by (1.7) for any constant α. This PDE is just the heat equation for $t \leq 0$, i.e., backward in time, as can be seen by applying the change of variable $t \leftarrow -t$. Its ill-posedness corresponds to the fact from physics that heat flow is not reversible: the temperature distribution in the past cannot be determined from its present distribution.

2. Next, consider the **advection equation**

$$u_t + a u_x = 0,$$

where $a \neq 0$ is a real constant.[3] This is perhaps the simplest instance of a *hyperbolic equation*. We get

$$P(\iota\xi) = -\iota\xi a,$$

which is purely imaginary, hence

$$|e^{P(\iota\xi)t}| = 1$$

for each wave number ξ. The Cauchy problem is well-posed, with $\alpha = 0$ in (1.11), but unlike the heat equation there is no decay in $\hat{u}$ for any wave number.

The advection operator, therefore, is not a smoother. Consider for example the initial conditions

$$u_0(x) = \begin{cases} 1, & x < 0, \\ 0, & x \geq 0. \end{cases}$$

---

[3] We write the advection equation in this form because the PDE describes a substance (contaminant) with concentration $u(t, x)$ being carried downstream by a fluid flowing through a one-dimensional pipe with velocity a.

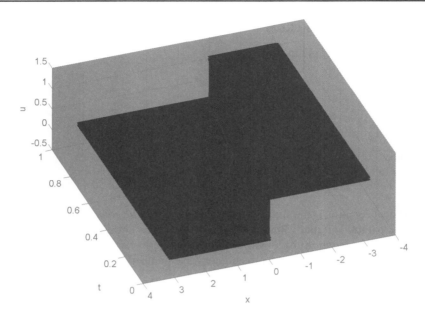

Figure 1.3. *Solution of the simple advection equation $u_t - u_x = 0$ starting from a step-function. The discontinuity simply moves along $x = -t$.*

The exact solution obviously has the form

$$u(t, x) = u_0(x - at),$$

so the initial discontinuity at $x = 0$ propagates undamped into the half-space $t > 0$: at a later time $t = \tilde{t}$ the solution is still a step function with the discontinuity located at $x = a\tilde{t}$. The discontinuity therefore propagates with wave speed $\frac{dx}{dt} = a$ along the **characteristic curve**

$$x - at = 0,$$

as can be seen in Figure 1.3.

On the other hand, the solution of the heat equation for the same initial values $u_0(x)$ is smooth for all $t > 0$; see Figure 1.4. This is a simple instance of a fundamental difference between parabolic problems (such as the heat equation) and hyperbolic problems (such as the advection equation). Parabolic problems are easier to solve numerically, in general, and methods which decouple the treatment of the time and space variables are natural for them, as we will see. Simple hyperbolic problems can be treated in a similar way, but in more complex situations the propagation of information along characteristics in hyperbolic problems must somehow be taken into account.

3. The **wave equation**

$$\phi_{tt} - c^2\phi_{xx} = 0,$$

1.1. Well-Posed Initial Value Problems

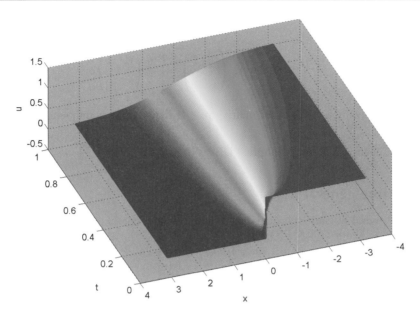

Figure 1.4. *Solution of the simple heat equation $u_t = u_{xx}$ starting from a step-function and $u(t, -\pi) = 1$, $u(t, \pi) = 0$. The discontinuity is diffused; in fact, $u(t, \cdot) \in C^\infty$ for $t > 0$.*

where c is a real constant, is also known as the *classical wave equation*, to distinguish it from other PDEs describing wave motion. It can be written as a first order system in the following way: define $u_1 = \phi_t$ and $u_2 = c\phi_x$, obtaining for $\mathbf{u} = (u_1, u_2)^T$

$$\mathbf{u}_t - \begin{pmatrix} 0 & c \\ c & 0 \end{pmatrix} \mathbf{u}_x = \mathbf{0}.$$

The eigenvalues of this matrix are $\pm c$. (See Section 1.3.2.) They are real, hence the wave equation is hyperbolic. The waves travel left and right along straight line characteristics at speeds $\pm c$.

Note that the Cauchy problem for $\mathbf{u}$ entails specification of ϕ_x and ϕ_t for all x at $t = 0$. The usual form of Cauchy problem for ϕ actually involves specifying $\phi(0, x)$ and $\phi_t(0, x)$ instead. However, obtaining $\phi_x(0, x)$ from $\phi(0, x)$ is straightforward.

4. The **Laplace equation** is the simplest instance of an *elliptic equation*. In the variables t and x it reads

$$\phi_{tt} + \phi_{xx} = 0.$$

This looks like the wave equation, but with imaginary "wave speeds" $c = \imath$. Thus, there are no real traveling waves here.

The symbol is

$$P(\iota\xi) = \iota\xi \begin{pmatrix} 0 & \iota \\ \iota & 0 \end{pmatrix} = -\xi \begin{pmatrix} 0 & 1 \\ 1 & 0 \end{pmatrix}.$$

The eigenvalues of the latter matrix are ± 1; hence there is unbounded growth like $e^{|\xi t|}$ in $\|e^{P(\iota\xi)t}\|$.

The Cauchy problem for elliptic PDEs is ill-posed!

1.1.2 More general cases

1. The general second order scalar PDE with constant coefficients

$$au_{xx} + bu_{xt} + cu_{tt} + du_x + eu_t + fu = 0$$

obviously has the simple heat, wave, and Laplace equations as special cases. Assuming for definiteness that $a \neq 0$, $c^2 + e^2 > 0$, it can be shown (see, e.g., [127] or the classic [47]) that

- if $b^2 - 4ac > 0$, then the PDE is hyperbolic;
- if $b^2 - 4ac = 0$, then the PDE is parabolic;
- if $b^2 - 4ac < 0$, then the PDE is elliptic.

This is the classical PDE classification. In case of variable coefficients the type determination is done locally in time and space. It can happen (and does so in gas dynamics) that a PDE would be hyperbolic in one part of the domain and elliptic in another.

2. For the second order **parabolic** system

$$\mathbf{u}_t = A\mathbf{u}_{xx},$$

where A is a symmetric positive definite matrix, we get

$$P(\iota\xi) = -\xi^2 A.$$

Since A is symmetric positive definite, its eigenvalues are real and positive, and moreover, we can use (1.24) with $\text{cond}(T) = 1$; see Section 1.3.5. Hence

$$\|e^{P(\iota\xi)t}\| = \|e^{-\xi^2 At}\| \leq 1 \quad \forall \xi.$$

We have obtained a direct extension of the simple heat equation. The essential properties turn out also to hold if A is allowed to depend on x and t, as will be discussed later.

3. Next, consider the first order system

$$\mathbf{u}_t + A\mathbf{u}_x = \mathbf{0}.$$

This system is **hyperbolic** if the matrix A is a diagonalizable matrix with real eigenvalues. Say $\Lambda = T^{-1}AT$ is a diagonal matrix, where T is the corresponding eigenvector matrix. By the Fourier transform we have the symbol

$$P(\iota\xi) = -\iota\xi A.$$

Hence, in (1.24) we get purely imaginary, scalar exponents whose modulus is 1. This yields

$$\|e^{P(\iota\xi)t}\| \leq \|T\|\|T^{-1}\|.$$

The point is that $K = \|T\|\|T^{-1}\|$ is a constant, independent of the wave number. We obtain a well-posed problem, but, as for the simple advection equation, there is no decay, even in high wave numbers. This translates into potential numerical trouble in more complex problems of this type.

1.1.3 Initial-boundary value problems

The preceding analysis is applied to the pure initial value problem. A key point is the diagonalization of the PDE system by the Fourier transform, which leads to a simple ODE for each wave number. However, on a finite interval in space a similar approach yields some additional terms relating to the boundary, and the elegance (at least) may be lost. Indeed, everything can potentially become much more complicated. An exception are *periodic boundary conditions*, and this is pursued further in Section 1.3.6, Example 1.1, and Exercise 2.

We will treat some issues regarding BCs in Section 3.1.3. A fuller treatment is delayed until later, in Chapter 8, although of course we will encounter BCs again much before that. Here, let us consider for a moment the heat equation in two space variables,

$$u_t = u_{xx} + u_{yy}.$$

For the pure initial value problem we apply a Fourier transform in both space variables x and y. A large initial value ODE system in time is obtained and well-posedness for all nonnegative time $t \geq 0$ is derived in exactly the same way as for the heat equation in one space variable. Specifically, if ξ and η are the wave numbers corresponding to x and y, respectively, then

$$P(\iota\xi, \iota\eta) = -(\xi^2 + \eta^2)$$

and

$$|e^{P(\iota\xi,\iota\eta)t}| = |e^{-(\xi^2+\eta^2)t}| \leq 1 \quad \forall \xi, \eta.$$

Now, if the domain of definition in space is a connected, bounded set $\Omega \subset \mathbb{R}^2$, then, as it turns out, boundary conditions are needed along the entire boundary of Ω, denoted $\partial\Omega$. Suppose that u is given on $\partial\Omega$—these are *Dirichlet boundary conditions*. The obtained initial-boundary value problem is well-posed [111].

Next, examine the **steady state** case of the heat equation in two space variables, where the solution does not depend on time. Then $u_t = 0$ and the Laplace equation is obtained. The boundary value problem for the Laplace equation is well-posed, even though we found earlier that the initial value problem is not!

1.1.4 The solution operator

Consider again the pure IVP with constant coefficients (1.8), (1.5b). Although not explicitly stated, we have obtained the solution using a Fourier transform in space:

$$\begin{aligned} u(t, x) &= \frac{1}{\sqrt{2\pi}} \int_{-\infty}^{\infty} e^{\imath \xi x} \hat{u}(t, \xi) d\xi \\ &= \frac{1}{\sqrt{2\pi}} \int_{-\infty}^{\infty} e^{\imath \xi x} e^{P(\imath \xi)t} \hat{u}(0, \xi) d\xi \\ &= \frac{1}{2\pi} \int_{-\infty}^{\infty} e^{\imath \xi x} e^{P(\imath \xi)t} \left[\int_{-\infty}^{\infty} e^{-\imath \xi \zeta} u_0(\zeta) d\zeta \right] d\xi. \end{aligned}$$

We can write this as

$$u(t, \cdot) = \mathcal{S}(t) u_0, \qquad (1.12)$$

where $\mathcal{S}(t)$ is the *solution operator*. It is easy to verify that the family of solution operators $\{\mathcal{S}(t), t \geq 0\}$ forms a *semi-group* on the solution space:

(a) $\mathcal{S}(0) = I$,

(b) $\mathcal{S}(t_1 + t_2) = \mathcal{S}(t_2)\mathcal{S}(t_1)$.

1.2 A taste of finite differences

Let us return to the pure IVP (1.5) defined on the domain depicted in Figure 1.2. We want to replace the derivatives by divided differences, thereby replacing the differential equation by algebraic equations which can be solved to yield an approximate solution. Consider a uniform mesh in t and x with **step sizes** $k = \Delta t$ and $h = \Delta x$. Thus, the mesh points are (t_n, x_j), where $x_j = jh$, $j = 0, \pm 1, \pm 2, \ldots$, and $t_n = nk$, $n = 0, 1, 2, \ldots$; see Figure 1.5. The solution $u(t, x)$ is approximated at mesh points as

$$v_j^n = v(t_n, x_j) \approx u(nk, jh).$$

For such a uniform mesh, under usual circumstances, the ratio

$$\mu = \frac{k}{h^m} \qquad (1.13)$$

plays an important role, as we will see below and in future chapters. Recall that m is the highest order of spatial differentiation that appears in (1.5a).

1.2. A Taste of Finite Differences 13

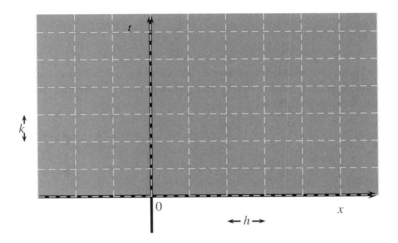

Figure 1.5. *Discretization mesh for a Cauchy problem.*

The function v is called a **mesh function**. Let us also denote all its values at a certain time level by $v^n = \{v_0^n, v_{\pm 1}^n, v_{\pm 2}^n, \ldots\}$. The superscript denotes the time level, not a power! Further, define the l_2-norm

$$\|v^n\| = h\sqrt{\sum_j (v_j^n)^2}. \tag{1.14}$$

Below we concentrate on a simple example, just to get a taste of the basic concepts. The general case will be considered in subsequent chapters.

For the advection equation
$$u_t + au_x = 0,$$
consider the following three simple schemes applied at each mesh point (t_n, x_j):

1.
$$\frac{1}{k}(v_j^{n+1} - v_j^n) + \frac{a}{h}(v_{j+1}^n - v_j^n) = 0.$$

This scheme is first order accurate in both t and x. Such accuracy order essentially means that if we substitute the exact solution $u(t_n, x_j)$ in place of v_j^n, etc., in the above difference equation, then the residual decreases by a factor M if both k and h are decreased by the same factor M. We will define order more precisely in the next two chapters; see in particular Section 3.2.1 and Example 3.9. Note that the difference in x is **one-sided** in that the point to the right, or east, of (t_n, x_j) participates but the point to the left, or west, does not.

2.
$$\frac{1}{k}(v_j^{n+1} - v_j^n) + \frac{a}{2h}(v_{j+1}^n - v_{j-1}^n) = 0.$$

This scheme is first order accurate in t and second order in x (see Section 3.2.1). The difference in x is **centered**.

3.
$$\frac{1}{2k}(v_j^{n+1} - v_j^{n-1}) + \frac{a}{2h}(v_{j+1}^n - v_{j-1}^n) = 0.$$

This is known as the **leapfrog** scheme. It is second order accurate, and the difference is *centered*, in both t and x. Note that this is a *two-step* scheme, whereas the other two are *one-step* in time.

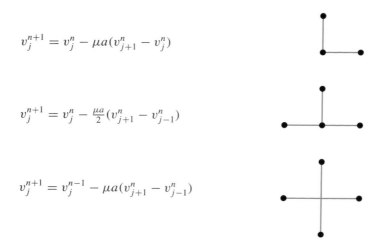

$$v_j^{n+1} = v_j^n - \mu a(v_{j+1}^n - v_j^n)$$

$$v_j^{n+1} = v_j^n - \frac{\mu a}{2}(v_{j+1}^n - v_{j-1}^n)$$

$$v_j^{n+1} = v_j^{n-1} - \mu a(v_{j+1}^n - v_{j-1}^n)$$

Figure 1.6. *Three simple schemes for the advection equation.*

It is worthwhile to keep in mind for each formula which locations of the mesh it involves. The "computational molecule" distinguishes the schemes from one another; see Figure 1.6. These schemes are all **explicit**. This means that the next unknown value, v_j^{n+1}, is defined by known values at the previous time levels n and $n-1$,

$$v_j^{n+1} = v_j^n - \mu a(v_{j+1}^n - v_j^n), \qquad (1.15a)$$
$$v_j^{n+1} = v_j^n - \frac{\mu a}{2}(v_{j+1}^n - v_{j-1}^n), \qquad (1.15b)$$
$$v_j^{n+1} = v_j^{n-1} - \mu a(v_{j+1}^n - v_{j-1}^n), \qquad (1.15c)$$

where $\mu = \frac{k}{h}$. Starting from $v_j^0 = u_0(x_j)$ we can march in time using (1.15a) or (1.15b), i.e., for $n = 0, 1, 2, \ldots$, compute v_j^{n+1} for all j in parallel based on the known v^n. In the case of (1.15c) we also need initial values at $n = 1$. These can be obtained by using the Taylor expansion (see Section 1.3.1),

$$u(k) = u(0) + ku_t(0) + \frac{k^2}{2}u_{tt}(0) + O(k^3) = u(0) - kau_x(0) + \frac{k^2 a^2}{2}u_{xx}(0) + O(k^3),$$

or simply a stable, second order accurate, one-step scheme.[4]

---
[4] A one-step scheme is occasionally referred to as a *two-level* method in the PDE literature. A two-step scheme is three-level, etc.

1.2. A Taste of Finite Differences

Below and throughout the text we occasionally refer to the mesh function v as if it is defined everywhere in the PDE problem's domain. Thus, we write $v(t_n, x_j) = v_j^n$ and refer more generally to $v(t, x)$. Although the mesh function is not defined, strictly speaking, at locations other than mesh points, we imagine that its domain of definition is extended by some appropriate local interpolation scheme. We do avoid taking derivatives of v on the continuum; thus the details of the extension of the mesh function are unimportant and may be left unspecified in what follows.

Example 1.1 The three schemes (1.15) all look reasonable and simple, and they approximate the differential equation well when k and h are small. So what can possibly go wrong?!

Consider the advection problem

$$u_t + u_x = 0,$$

$$u(0, x) = u_0(x) = \begin{cases} 1, & x \leq 0, \\ 0, & x > 0. \end{cases}$$

Here $a = 1$. Recall that the exact solution is

$$u(t, x) = u_0(x - t).$$

So, at $t = 1$

$$u(1, x) = \begin{cases} 1, & x \leq 1, \\ 0, & x > 1. \end{cases}$$

Now, consider applying the scheme (1.15a) for h and k as small as you wish (but in any case $h < 1$). If $x_0 = 0$, then $v_j^0 = 0 \ \forall \ j > 0$, which implies that $v_j^1 = 0 \ \forall \ j > 0$. This yields in a similar manner $v_j^2 = 0 \ \forall \ j > 0$, and so on. Thus, for N such that $Nk = 1$ we likewise obtain

$$v_j^N = 0 \quad \forall \ j > 0,$$

which has the error

$$|v_j^N - u(1, x_j)| = 1 \quad \text{for } 0 < x_j \leq 1.$$

Figure 1.7 depicts the situation. This is an unacceptable error, both quantitatively and qualitatively. ∎

The reason for the failure in Example 1.1 is that a fundamental restriction has been violated. The origin of this goes back to the ancient paper[5] by Courant, Friedrichs, and Lewy [52]. The condition is thus referred to as the **CFL condition**. It says that the domain of dependence of the PDE must be contained in that of the difference scheme, as in Figure 1.8. Indeed, the value of v_j^{n+1} according to (1.15a) depends on v_j^n and v_{j+1}^n,

---

[5] By "ancient" we mean, needless to say, preceding the modern computing era.

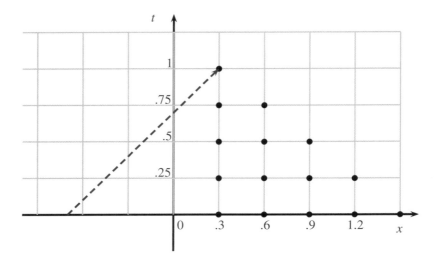

Figure 1.7. *Example* 1.1 *with* $k = .25$ *and* $h = .3$, *i.e.*, $\mu < 1$. *The exact solution for the PDE* $u_t + u_x = 0$ *at* $(t, x) = (1, .3)$ *is equal to the value at the foot of the characteristic curve leading to it (in dashed blue),* $u(1, .3) = u(0, -.7) = u_0(-.7) = 1$. *The approximate solution at the same location, on the other hand, is traced back to initial values to the right of* $x = .3$; *since all of these vanish,* $v(1, .3) = 0$.

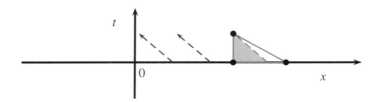

Figure 1.8. *Characteristic curves for* $u_t + au_x = 0$, $a < 0$. *The CFL condition implies that the domain of dependence of the PDE is contained in that of the difference scheme.*

whereas that of the exact solution $u(t_{n+1}, x_j)$ is simply transported along the characteristic curve, so that $u(t_{n+1}, x_j) = u(t_n, x_j - ka)$. Continuing interpreting such dependence back to the starting line we see that v_j^{n+1} depends only on the values of v_j^{n-1}, v_{j+1}^{n-1}, and v_{j+2}^{n-1}, then on v_j^{n-2}, v_{j+1}^{n-2}, v_{j+2}^{n-2}, and v_{j+3}^{n-2}, and so on backward in time until the initial values $u_0(x_j), \ldots, u_0(x_{j+n+1})$, whereas $u(t_{n+1}, x_j) = u_0(x_j - at_{n+1})$. Now, if $x_j > x_j - at_{n+1}$ or $x_{j+n+1} < x_j - at_{n+1}$, which would occur if $a > 0$ or $(-a)\mu > 1$, respectively, then it is possible to *arbitrarily* change the initial value function $u_0(x_j - at_{n+1})$ and thus the exact value $u(t_{n+1}, x_j)$ without affecting the value of the calculated approximate solution!

In particular, the scheme (1.15a) violates the CFL condition if $a > 0$, which explains the failure in Example 1.1. This is because the forward difference in x is commensurate with the characteristics traveling leftward in time, as in Figure 1.8, not rightward as in

1.2. A Taste of Finite Differences

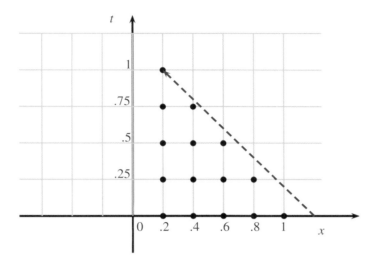

Figure 1.9. *Violation of the CFL condition when $|a|\mu > 1$. The exact solution for the PDE $u_t - u_x = 0$ at $(t, x) = (1, .2)$ is equal to the value at the foot of the characteristic curve leading to it (in dashed blue), $u(1, .2) = u(0, 1.2) = u_0(1.2)$. The approximate solution at the same location is traced back after N time steps, $Nk = 1$, to initial values in the range $[.2, .2 + Nh]$. If $\mu > 1$, then $.2 + Nh < .2 + Nk = 1.2$. Here $k = .25$, $h = .2$, hence $\mu = 1.2$, $N = 4$ and $.2 + Nh = 1$. Assuming no change in the initial values $u_0(x)$ over $[.2, 1]$, the same numerical solution is obtained at $t = 1$, $x = .2$ regardless of the exact value $u_0(1.2)$.*

Figure 1.7. Even if the characteristics travel leftward, though, i.e., $a < 0$, we must maintain $(-a)\mu \leq 1$: Figure 1.9 shows what happens if the latter condition is violated. Exercise 5 explores this further.

1.2.1 Stability ideas

Now, just as with well-posedness we want to make sure that for any h small enough $\|v^n\|$ does not increase too fast in n as $k \to 0$, $nk \leq t_f$, for a fixed maximum time t_f. In fact, we don't want $\|v^n\|$ to increase at all, if possible, in cases such as the advection equation where the exact solution does not increase at all ($\alpha = 0$ in (1.11)). This is the essence of *stability* of the difference scheme.

> **Note:** The stability requirement of a difference method corresponds to the well-posedness requirement of a differential problem.

Since the advection equation has constant coefficients, we have the same computational formula at all mesh points. This allows us to apply the Fourier analysis also in the discrete case. Writing

$$v(t, x) = \frac{1}{\sqrt{2\pi}} \int_{-\infty}^{\infty} e^{\iota \xi x} \hat{v}(t, \xi) d\xi,$$

we obtain for the first scheme (1.15a)

$$\int_{-\infty}^{\infty} e^{\iota \xi x} \hat{v}(t + k, \xi) d\xi = \int_{-\infty}^{\infty} \left[e^{\iota \xi x} \hat{v}(t, \xi) - \mu a (e^{\iota \xi (x+h)} - e^{\iota \xi x}) \hat{v}(t, \xi) \right] d\xi.$$

This must hold for any x, so the integrands must agree, yielding

$$\hat{v}(t + k, \xi) = \left[1 - \mu a (e^{\iota \xi h} - 1) \right] \hat{v}(t, \xi)$$
$$= \gamma_1(\zeta) \hat{v}(t, \xi),$$

where $\zeta = \xi h$ and

$$\gamma_1(\zeta) = 1 - \mu a (e^{\iota \zeta} - 1).$$

So, each Fourier mode is multiplied by a corresponding constant γ_1 over each time step.

The stability condition requires that $\|v^n\|$ remain bounded in terms of $\|v^0\|$ for all n. For a scheme which in Fourier space reads

$$\hat{v}(t + k, \xi) = \gamma_1(\zeta) \hat{v}(t, \xi),$$

this translates to requiring that

$$|\gamma_1(\zeta)| \leq 1 \;\forall \zeta. \tag{1.16}$$

The function $g(\zeta) = \gamma_1(\zeta)$ is called the **amplification factor**; see Section 4.1.

For the scheme (1.15a) note that $\gamma_1(\zeta)$ is a periodic function, so we need consider only $|\zeta| \leq \pi$. Writing $\gamma_1(\zeta) = 1 + \mu a - \mu a e^{\iota \zeta}$, the condition (1.16) is seen to hold (see Figure 1.10) iff[6] $a \leq 0$ and

$$\mu(-a) \leq 1.$$

This is a *stability restriction* which coincides with the CFL condition.

Applying the same analysis to the second scheme (1.15b) we readily obtain

$$\hat{v}(t + k, \xi) = \gamma_2(\zeta) \hat{v}(t, \xi),$$
$$\gamma_2(\zeta) = 1 - \frac{\mu a}{2} (e^{\iota \zeta} - e^{-\iota \zeta})$$
$$= 1 - \iota \mu a \sin \zeta.$$

Clearly, $\gamma_2(\zeta)$ does not satisfy the stability condition (1.16). We have stumbled upon a phenomenon which does not occur in ODEs at all, where an innocent-looking, consistent one-step scheme is nonetheless *unconditionally unstable*!

---

[6]**iff** stands for "if and only if."

1.2. A Taste of Finite Differences

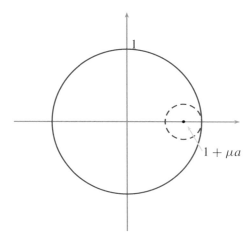

Figure 1.10. *Stability of the difference scheme* (1.15a) *for a < 0.*

For the two-step, leapfrog scheme (1.15c) we obtain

$$\hat{v}(t+k, \xi) = \hat{v}(t-k, \xi) - \mu a(e^{\iota\zeta} - e^{-\iota\zeta})\hat{v}(t, \xi) \text{ or}$$
$$\hat{v}^{n+1} = \hat{v}^{n-1} - 2(\iota\mu a \sin \zeta)\hat{v}^n.$$

We get a second order difference equation which we can solve by an *ansatz*: try $\hat{v}^n = \kappa^n$. Substituting and dividing through by κ^{n-1}, this is seen to work if κ satisfies the quadratic equation

$$\kappa^2 = 1 - 2(\iota\mu a \sin \zeta)\kappa.$$

Solving this equation we obtain the condition that $|\kappa| \leq 1$ (i.e., the stability condition (1.16) holds) provided again that the CFL condition holds. Indeed, here the condition is really

$$\mu |a| \leq 1,$$

i.e., the scheme is stable under the same condition regardless of the sign of a, unlike the one-sided scheme (1.15a).

The number $\mu |a|$ is often referred to in the literature as the **Courant number** or the **CFL number**.

Example 1.2 Let us carry out calculations using the three schemes (1.15) for the initial-value PDE

$$u_t - u_x = 0,$$
$$u_0(x) = \sin \eta x,$$

where η is a parameter. Periodic boundary conditions (BC) on the interval $[-\pi, \pi]$ are employed; thus, $u(t, -\pi + x) = u(t, \pi + x)$ for all relevant x and t, and in particular

$$u(t, -\pi) = u(t, \pi).$$

The exact solution is then
$$u(t,x) = \sin \eta(t+x).$$

With spatial and temporal step sizes h and k, respectively, let $J = \frac{2\pi}{h} - 1$ and seek solution values v_j^n for $j = 0, 1, \ldots, J$. The periodic BC give
$$v_{J+1}^n = v_0^n \quad \forall n.$$

Moreover, it is useful to imagine the spatial mesh points sitting on a ring, rather than on a straight line, with x_{J+1} identified with x_0. Thus, for $j = J$, $v_{j+1} = v_0$, and for $j = 0$, $v_{j-1} = v_J$, at any n; see Figure 1.11.

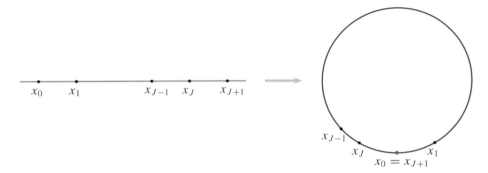

Figure 1.11. *For periodic boundary conditions, imagine the red mesh points sitting on a blue circle with $x_0 = x_{J+1}$ closing the ring. Thus, x_J is the left (clockwise) neighbor of x_0, x_1 is the right (counterclockwise) neighbor of x_{J+1}, and so on.*

A simple MATLAB® script for the leap-frog scheme follows:

```
% parameters
  eta = 1;
  h = .001*pi;
  k = 0.5*h; mu = k / h;

% spatial mesh
  xx = [-pi:h:pi];
  Jp2 = length(xx);
  Jp1 = Jp2 - 1; J = Jp1 - 1;
% final time tf = 1
  nsteps = floor(1/k);
  tf = k * nsteps;

% initial condition
  un = sin (eta * xx);
% exact solution at t=tf
  ue = sin (eta * (xx + tf));
```

1.2. A Taste of Finite Differences

```
%leap frog: need also initial values at t = k
  vo = un;
% cheat for simplicity: use exact solution for v at t=k
  vn = sin (eta * (k+xx));

% advance in time
  for n=2:nsteps

    voo = vo;
    vo = vn;
    % interior x-points
    vn(2:Jp1) = voo(2:Jp1) + mu * (vo(3:Jp2) - vo(1:J));
    % periodic BC: 1=Jp2 so left of 1 is 0=Jp1
    vn(1) = voo(1) + mu * (vo(2)-vo(Jp1));
    vn(Jp2) = vn(1);

  end

% plot results and calculate error at t=tf
  plot(xx,vn,'m')
  xlabel('x')
  ylabel('u')
  title('Leap-frog scheme')
  err3 = max(abs(vn-ue));
```

Table 1.1. *Maximum errors at $t = 1$ for the three simple difference schemes applied to the equation $u_t - u_x = 0$ with $u_0(x) = \sin \eta x$. Error blowup is denoted by *. Note that $k = \mu h$.*

η	h	μ	Error in (1.15a)	Error in (1.15b)	Error in (1.15c)
1	$.01\pi$	0.5	7.7e-3	7.8e-3	1.2e-4
	$.001\pi$	0.5	7.8e-4	*	1.2e-6
	$.001\pi$	1.1	*	*	*
10	$.01\pi$	0.5	5.4e-1	1.15	1.2e-1
	$.001\pi$	0.5	7.6e-2	*	1.2e-3
	$.0005\pi$	0.5	3.9e-2	*	3.1e-4

Table 1.1 displays the resulting maximum errors. Note that we avoid taking $k = h$, that is, $\mu = 1$—this would correspond to stepping exactly along characteristics, which is impossible in realistic mathematical models. (Our purpose here is to demonstrate typical phenomena, not to actually solve $u_t - u_x = 0$ numerically as an end in itself!)

For $\eta = 1$ we observe that when $\mu < 1$ is held fixed, the error decreases like h for (1.15a) and like h^2 for (1.15c). For $\mu > 1$ all schemes are unstable because the stability restriction that coincides with the CFL condition is violated. For $\eta = 10$ we observe the same first and second order convergence rates for the stable schemes, although in absolute value all the errors are larger than those for $\eta = 1$ by factors of about 100 using (1.15a) and 1000 using (1.15c). This corresponds to the second and third derivatives of the solution u, respectively, being involved in the error expressions of these two schemes, as we'll see in Chapter 3. ∎

Let us further investigate the unstable scheme (1.15b). In general, after n steps

$$\hat{v}^n = \gamma_2(\zeta)^n \hat{v}^0$$

(that's the amplification factor to the power n). Thus, with $nk = 1$ we have at wave number ξ the amplification

$$|\hat{v}^n| = |1 - \iota \frac{k}{h} a \sin(\xi h)|^{1/k} |\hat{u}_0(\xi)|.$$

The amplification factor is far from 1 when $\sin(\xi h)$ is not small, e.g., $\xi h \approx \pi/2$. For such a wave number ξ some basic calculus manipulations show that we get a blowup like $e^{\frac{\mu a^2}{2h}}$ as $k \to 0$. So, as $h \to 0$ the blowup is faster, which is apparent in Table 1.1, and it is particularly damaging for high wave-number components of the initial values. In fact, different initial value functions $u_0(x)$ may have high wave-number components with small or large amplitude, and thus exhibit different rates of blowup. The solution at $t = 1$ for $\eta = 1, h = .001\pi$, and $k = .5h$ is displayed in Figure 1.12.

For many physical problems it may be safely assumed that $\hat{u}_0(\xi) = 0$ for high wave numbers, say, $|\xi| > \xi_f$ for some value ξ_f. Then, as $h, k \to 0$ with the ratio μ kept fixed, *convergence* may be seen to occur even for an unstable scheme.[7] That is to say, convergence would have occurred in such cases had we worked in precise arithmetic! However, **roundoff errors** introduce components with high wave numbers into $\hat{u}_0(\xi)$, and these error components are magnified by an unstable scheme.

> **Note:** Generally, we need accuracy for low wave numbers and stability for high wave numbers of the error.

To close off this introductory chapter we next turn to a rough classification of the problems and numerical algorithms under consideration. The PDE problems considered in this text are well-posed as initial value problems. Attempting to classify them according to suitable numerical treatment, we can distinguish between smooth and nonsmooth in two different ways, namely, in the differential equation *operator* and in the *solution*, creating in effect three classes:

- For the class of parabolic-type PDEs, the differential operator itself is a smoother, as we have seen. The differential problem is diffusive. The solution of such an initial-

---

[7]We will soon make precise the meaning of stability, accuracy, consistency, and convergence. For now, we're appealing to the reader's intuition and related knowledge from elsewhere—e.g., a numerical ODE course. Please reread these paragraphs after you've read Chapters 2 and 3.

1.2. A Taste of Finite Differences

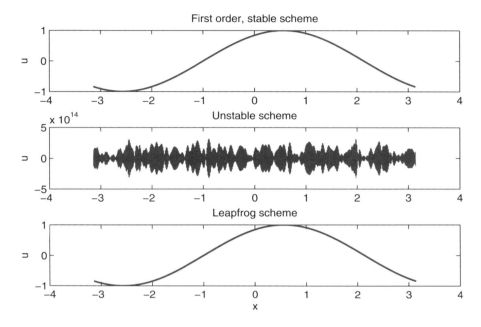

Figure 1.12. *Solution of $u_t - u_x = 0$ at $t = 1$ using the three schemes* (1.15) *with* $u_0 = \sin x$, $h = .001\pi$, $k = .5h$.

boundary value problem is smooth even if the initial or boundary value functions are not. Correspondingly, we can seek highly accurate, high order discrete approximations. The main computational challenge is often concerned with the mild **stiffness** that such problems exhibit, issues that will be made clearer in Sections 2.6, 2.7, 5.1, 9.2, and elsewhere.

- For the class of hyperbolic-type PDEs, the differential operator is not a smoother and the PDE is **conservative**. Still, in many situations concerning wave propagation the solution is smooth. For such problems efficient numerical methods with high order of accuracy are sought as well. The main issues here are stability, **dissipation**, and **dispersion**, elaborated upon in Section 1.2, Chapters 4, 5, and 7, and elsewhere.

- For the class of hyperbolic-type PDEs, nonsmooth solutions may occur. These may arise as a result of a discontinuity in the initial or boundary conditions, as in Figure 1.3. For nonlinear problems, solution **discontinuities** inside the domain in x and t may arise even if the solution is smooth along all boundaries. For such problems it is hard to construct accurate numerical methods near discontinuities. Methods are then sought that limit the effect of inevitable inaccuracies near discontinuities on the approximate solution elsewhere in the domain and produce physical-looking solutions with sharp profiles. They turn out to be rather different methods from those used for the smooth case. This is elaborated upon in Chapter 10, Sections 5.2 and 11.3, and elsewhere.

1.3 Reviews

In this section and in Section 2.12 we have collected short reviews on various mathematical and numerical topics that are used throughout the text but are not developed here. Thus, these topics are considered as background material. Of course the background of different readers may vary widely. For some, a quick review may serve as a reminder or refresher. Readers may find a review helpful to connect with other, related strands of knowledge. But there are certain to be readers who may find our short reviews insufficient. If you find that you still are clueless after reading one of these reviews, then please consult an appropriate introductory text.

1.3.1 Taylor's theorem

Let $w(\cdot)$ be a smooth function. **Taylor's theorem** expresses the value of w at locations $\hat{x} = x + h$ near a given argument value x in terms of w and its derivatives at x

$$w(\hat{x}) = w + h\frac{dw}{dx} + \frac{h^2}{2}\frac{d^2w}{dx^2} + \cdots + \frac{h^l}{l!}\frac{d^l w}{dx^l} + \cdots.$$

Thus, for instance, the first derivative of a function $u(t, x)$ with respect to x can be written as

$$u_x(t, x) = h^{-1}(u(t, x + h) - u(t, x)) - \frac{h}{2}u_{xx}(t, x) + O(h^2).$$

Here $O(h^p)$ is the customary notation for terms that tend to 0 at least as fast as h^p when $h \to 0$. Defining the *forward difference* for u_x as

$$D_+ u = u(t, x + h) - u(t, x), \quad @ (t, x),$$

Taylor's theorem yields an expression for the error

$$u_x - \frac{1}{h}D_+ u = -\frac{h}{2}u_{xx} + O(h^2), \quad @ (t, x).$$

Also, defining the *centered difference* for u_x as

$$D_0 u = u(t, x + h) - u(t, x - h), \quad @ (t, x),$$

then, by writing the Taylor expansions

$$u(x \pm h) = u \pm hu_x + \frac{h^2}{2}u_{xx} \pm \frac{h^3}{6}u_{xxx} + \frac{h^4}{4!}u_{xxxx} + \cdots,$$

we have

$$u_x - \frac{1}{2h}D_0 u = -\frac{h^2}{6}u_{xxx} + O(h^4), \quad @ (t, x).$$

Similarly, a centered difference for the second derivative u_{xx} is given by

$$D_+ D_- u = [u(t, x + h) - u(t, x)] - [u(t, x) - u(t, x - h)]$$
$$= u(t, x + h) - 2u(t, x) + u(t, x - h), \quad @ (t, x).$$

1.3. Reviews

By adding the Taylor expansions for $u(x \pm h)$ and subtracting $2u(x)$ we have

$$u_{xx} - \frac{1}{h^2}D_+D_-u = -\frac{h^2}{12}u_{xxxx} + O(h^4), \quad @ \ (t,x).$$

Recall from advanced calculus that **Taylor's theorem for a function of several variables** gives

$$w(t+k, x+h) = w + hw_x + kw_t + \frac{1}{2!}[h^2 w_{xx} + 2hk\, w_{xt} + k^2 w_{tt}]$$
$$+ \cdots + \frac{1}{l!}\ell^l w + \cdots,$$

where the functions on the right-hand side are evaluated at (t, x) and

$$\ell^l w = \sum_{j=0}^{l} \binom{l}{j} \left(\frac{\partial^l w}{\partial x^j \partial t^{l-j}}(x, t)\right) h^j k^{l-j}.$$

1.3.2 Matrix norms and eigenvalues

Let us first recall some matrix definitions. Many people are used to real-valued matrices, but in this book we occasionally encounter more general, complex-valued ones. The concepts extend smoothly from real to complex, and so does the associated vocabulary.

Consider a complex-valued $m \times m$ matrix A and denote by A^T the transpose and by A^* the conjugated transpose of A. Thus, $A^* = A^T$ if A is real, i.e., if all its elements are real.

The following definitions are collected in Table 1.2. The matrix A is **Hermitian** if $A^* = A$ and **unitary** if $A^* = A^{-1}$. Thus, for real matrices, A is Hermitian iff it is **symmetric** (i.e., $A^T = A$), and A is unitary iff it is **orthogonal** (i.e., $A^T A = I$). Further, A is **skew-Hermitian** if $A^* = -A$ and **skew-symmetric** if $A^T = -A$. The matrix A is **normal** if $AA^* = A^*A$. Hermitian, skew-Hermitian, and unitary matrices are all examples of normal matrices.

Table 1.2. *Some matrix definitions, for A complex- and real-valued.*

Real-valued A		Complex-valued A	
Symmetric	if $A^T = A$	Hermitian	if $A^* = A$
Skew-symmetric	if $A^T = -A$	Skew-Hermitian	if $A^* = -A$
Orthogonal	if $A^T A = I$	Unitary	if $A^* A = I$
Normal	if $A^T A = AA^T$	Normal	if $A^* A = AA^*$

Given an $l \times m$ complex matrix A and a vector norm $\|\cdot\|$, the corresponding **induced matrix norm** is defined by

$$\|A\| = \max_{\mathbf{x} \neq \mathbf{0}} \frac{\|A\mathbf{x}\|}{\|\mathbf{x}\|}.$$

- For the maximum vector norm, $\|\mathbf{x}\|_\infty = \max_{1 \le i \le m} |x_i|$, we have the induced matrix norm
$$\|A\|_\infty = \max_{1 \le i \le l} \sum_{j=1}^m |a_{ij}|.$$

- For the l_1 vector norm, $\|\mathbf{x}\|_1 = \sum_{i=1}^m |x_i|$, we have the matrix norm
$$\|A\|_1 = \max_{1 \le j \le m} \sum_{i=1}^l |a_{ij}|.$$

- For the l_2 vector norm, where there is an associated **inner product**, $\|\mathbf{x}\|_2 = \sqrt{\mathbf{x}^*\mathbf{x}} \equiv (\mathbf{x}, \mathbf{x})^{1/2}$, we have the matrix norm
$$\|A\|_2 = \sqrt{\rho(A^*A)},$$
where ρ is the *spectral radius*, defined below.

Note that if, for instance, $\mathbf{x} = (1, 1, \ldots, 1)^T$, then $\|\mathbf{x}\|_1 = m$ and $\|\mathbf{x}\|_2 = \sqrt{m}$. So, if m increases unboundedly, then the vector norm definition may require scaling, e.g., $\|\mathbf{x}\|_2 = \sqrt{\mathbf{x}^*\mathbf{x}/m}$. Indeed, in (1.14) we scale $\|v^n\|$ by the spatial step size h.

Let A be nonsingular. Then its **condition number** in a given norm is defined by
$$\text{cond}(A) = \|A\|\|A^{-1}\|.$$

The condition number features prominently in numerical linear algebra. It indicates a bound on the quality of the computed solution to a linear system of equations involving A: the smaller it is, the better is the guaranteed quality. The condition number is an even more central indicator for the iterative methods reviewed in Section 9.4. However, one rarely requires the exact value of $\text{cond}(A)$, as it is used just to indicate quantity sizes. Hence, any reasonable induced matrix norm may be used in its definition. Note that we always have
$$\text{cond}(A) = \|A\|\|A^{-1}\| \ge \|AA^{-1}\| = 1.$$

Any unitary or orthogonal matrix T is thus ideally conditioned because $\|T\|_2 = \|T^{-1}\|_2 = 1$, hence $\text{cond}(T) = 1$.

Given an $m \times m$, generally complex-valued matrix A, an **eigenvalue** λ is a scalar which satisfies
$$A\mathbf{x} = \lambda \mathbf{x}$$
for some vector $\mathbf{x} \ne \mathbf{0}$. In general, λ may be complex even if A is real. But if A is Hermitian, then λ is guaranteed to be real. If all eigenvalues of A are positive, then A is **positive definite**. The vector $\mathbf{x}$, which is clearly determined only up to a scaling factor, is called an **eigenvector**. Counting multiplicities, A has m eigenvalues, which we denote by $\lambda_1, \ldots, \lambda_m$.

The **spectral radius** of a matrix is defined by
$$\rho(A) = \max_{1 \le i \le m} |\lambda_i|.$$

1.3. Reviews

In general, in the l_2 matrix norm

$$\rho(A) \leq \|A\|_2$$

for any square matrix. But if A is normal, then

$$\rho(A) = \|A\|_2.$$

A **similarity transformation** is defined, for any nonsingular matrix T, by

$$B = T^{-1}AT.$$

The matrix B has the same eigenvalues as A and the two matrices are said to be similar. If B is diagonal, denoted $B = \text{diag}\{\lambda_1, \ldots, \lambda_m\}$, then the displayed λ_i are the eigenvalues of A, the corresponding eigenvectors are the columns of T, and A is said to be **diagonalizable**. Any normal matrix is diagonalizable, in fact, by a unitary transformation (i.e., T can be chosen to satisfy $T^* = T^{-1}$). This is nice because the similarity transformation is ideally conditioned. For a general matrix, however, a unitary similarity transformation can at best bring A only to a matrix B in upper triangular form. (This still features the eigenvalues on the main diagonal of B.)

For a general, square matrix A there is always a similarity transformation into a **Jordan canonical form**,

$$B = \begin{pmatrix} \Lambda_1 & & & & 0 \\ & \Lambda_2 & & & \\ & & \ddots & & \\ & & & \ddots & \\ 0 & & & & \Lambda_s \end{pmatrix} ; \quad \Lambda_i = \begin{pmatrix} \lambda_i & 1 & & & 0 \\ & \lambda_i & 1 & & \\ & & \ddots & \ddots & \\ & & & \lambda_i & 1 \\ 0 & & & & \lambda_i \end{pmatrix}, \; i = 1, \ldots, s.$$

The Jordan blocks Λ_i generally have different sizes.

A simple, sometimes useful, way to estimate matrix eigenvalues without computing them is to use **Gershgorin's theorem**. In its simplest form this theorem states that the eigenvalues of A are all located in the union $\cup_{i=1}^{m} D_i$ of disks (in the complex plane) centered at the diagonal elements of A,

$$D_i = \left\{ z \in \mathcal{C} : |z - a_{ii}| \leq \sum_{\substack{j=1 \\ j \neq i}}^{m} |a_{ij}| \right\}.$$

For example, if $a_{ii} > 0 \, \forall i$ and A is **diagonally dominant**, i.e., $\sum_{j=1, j \neq i}^{m} |a_{ij}| \leq a_{ii} \, \forall i$, then Gershgorin's theorem yields that $\mathcal{R}e \, \lambda_i \geq 0 \, \forall i$. Thus, if A is also Hermitian and nonsingular, then it is positive definite.

If A is normal and nonsingular, then, utilizing the existence of a unitary similarity transformation, it is easy to see that in the 2-norm

$$\text{cond}(A) = \frac{\max_i |\lambda_i|}{\min_i |\lambda_i|}.$$

For a positive definite matrix this simplifies into

$$\text{cond}(A) = \frac{\max_i \lambda_i}{\min_i \lambda_i}.$$

For a more general, possibly rectangular matrix A with a full column rank we have

$$\text{cond}(A) = \frac{\max_i \sigma_i}{\min_i \sigma_i},$$

where σ_i are the **singular values** of A. These are the square roots of the eigenvalues of $A^T A$.

A good reference for numerical linear algebra is Golub and van Loan [68]. There are many others, including the reader-friendly Trefethen and Bau [170] and Demmel [57]. See Saad [149] for iterative methods.

1.3.3 Function spaces

On the real line $\mathbb{R}$, define the $\mathcal{L}_2$ norm

$$\|w\| = \left(\int_{-\infty}^{\infty} |w(x)|^2 dx\right)^{1/2}.$$

The space $\mathcal{L}_2(\mathbb{R})$ consists of all square integrable functions, i.e., all functions whose $\mathcal{L}_2$ norm defined above is finite.

A similar definition applies more generally on a domain $\Omega \subset \mathbb{R}^d$ which may be bounded or unbounded: the space $\mathcal{L}_2(\Omega)$ consists of all square integrable functions over Ω, with the norm

$$\|w\| = \left(\int_{\Omega} |w(\mathbf{x})|^2 d\mathbf{x}\right)^{1/2}.$$

It is customary to omit the dependence on Ω when there is no room for confusion (at least in the author's mind).

Other norms that we will encounter are the sup (max), or the $\mathcal{L}_\infty$ norm

$$\|w\|_\infty = \sup_{\mathbf{x} \in \Omega} |w(\mathbf{x})|,$$

and the $\mathcal{L}_1$ norm

$$\|w\|_1 = \int_{\Omega} |w(\mathbf{x})| d\mathbf{x}.$$

Note that the corresponding spaces satisfy $\mathcal{L}_\infty \subset \mathcal{L}_2 \subset \mathcal{L}_1$.

Estimates that depend on function derivatives are also very common in PDE analysis. These lead to **Sobolev spaces**. Based, say, on the $\mathcal{L}_2$ norm, a Sobolev space $\mathcal{H}_s$ is the space of functions for which all the derivatives of order j, for all $0 \leq j \leq s$, are square integrable, and the corresponding Sobolev norm with which this space is equipped is just the sum of the norms of these derivatives. Thus, in particular, $\mathcal{H}_0 = \mathcal{L}_2$ (where we identify the zeroth derivative with the function itself) and $\mathcal{H}_{s+1} \subset \mathcal{H}_s$.

1.3. Reviews

For example, let Ω be the unit circle defined by $x^2 + y^2 \leq 1$, so its boundary $\partial\Omega$ is defined by $x^2 + y^2 = 1$, and let

$$q(\mathbf{x}) = q(x, y) = \begin{cases} 1 & x < 0 \\ 0 & x \geq 0 \end{cases}.$$

Then q is square integrable but q_x is not. Thus, $q \in \mathcal{H}_0(\Omega)$ but $q \notin \mathcal{H}_1(\Omega)$. Now, if the function $\phi(x, y)$ satisfies Poisson's equation $\phi_{xx} + \phi_{yy} = q$ in Ω and $\phi = 0$ on $\partial\Omega$, then ϕ and its derivatives up to order 2 are square integrable. Hence $\phi \in \mathcal{H}_2(\Omega)$, and this higher Sobolev index expresses the additional smoothness that ϕ has over q.

1.3.4 The continuous Fourier transform

For a square-integrable function $v \in \mathcal{L}_2(\mathbb{R})$ define the Fourier transform

$$\hat{v}(\xi) = \frac{1}{\sqrt{2\pi}} \int_{-\infty}^{\infty} e^{-\iota\xi x} v(x) dx, \tag{1.17}$$

where $\iota = \sqrt{-1}$. It then follows that the *inverse transform* is given by

$$v(x) = \frac{1}{\sqrt{2\pi}} \int_{-\infty}^{\infty} e^{\iota\xi x} \hat{v}(\xi) d\xi. \tag{1.18}$$

The variable ξ is traditionally referred to as the **wave number** when x is a space variable and as the **frequency** when x is time.

The Fourier transform has many beautiful properties, among which are the following two that can be easily verified:

- The $\mathcal{L}_2$ norm of the transform is equal to (and not just bounded by) the norm of the original function

$$\|v\| = \|\hat{v}\| \tag{1.19}$$

  (this is called *Parseval's equality*).

- A differentiation of $v(x)$ is transformed into a multiplication of $\hat{v}(\xi)$ by $\iota\xi$,

$$v_x \Longleftrightarrow \iota\xi\hat{v}(\xi). \tag{1.20}$$

In d dimensions the Fourier transform readily generalizes to

$$\hat{v}(\boldsymbol{\xi}) = \frac{1}{(2\pi)^{d/2}} \int_{\mathbb{R}^d} e^{-\iota\boldsymbol{\xi}\cdot\mathbf{x}} v(\mathbf{x}) d\mathbf{x}, \tag{1.21}$$

where $\boldsymbol{\xi} \cdot \mathbf{x} = \sum_{i=1}^{d} \xi_i x_i$.

Let us demonstrate the use of (1.20) for the PDE

$$u_t = u_{xx} - 3u_x, \quad -\infty < x < \infty, \ t \geq 0. \tag{1.22a}$$

This is the advection-diffusion equation (1.1) with $\nu = 1$ and $a = 3$. Applying the Fourier transform in x we have

$$u(t, x) = \frac{1}{\sqrt{2\pi}} \int_{-\infty}^{\infty} e^{\imath \xi x} \hat{u}(t, \xi) d\xi.$$

There is only one place under the integral sign where either t or x appears. Taking derivatives we obtain

$$u_t(t, x) = \frac{1}{\sqrt{2\pi}} \int_{-\infty}^{\infty} e^{\imath \xi x} \hat{u}_t(t, \xi) d\xi,$$

$$u_x(t, x) = \frac{1}{\sqrt{2\pi}} \int_{-\infty}^{\infty} e^{\imath \xi x} (\imath \xi) \hat{u}(t, \xi) d\xi,$$

$$u_{xx}(t, x) = \frac{1}{\sqrt{2\pi}} \int_{-\infty}^{\infty} e^{\imath \xi x} (\imath \xi)^2 \hat{u}(t, \xi) d\xi.$$

Substituting in (1.22a) written as $u_t - u_{xx} + 3u_x = 0$ we therefore have

$$\frac{1}{\sqrt{2\pi}} \int_{-\infty}^{\infty} e^{\imath \xi x} \left[\hat{u}_t + \xi^2 \hat{u} + 3\imath \xi \hat{u} \right] (t, \xi) d\xi = 0.$$

This equality is to hold for any real x, hence the expression in square parentheses must vanish. Thus, for each real wave number ξ the following ODE in time,

$$\hat{u}_t = -\left(\xi^2 + 3\imath \xi \right) \hat{u}, \quad t \geq 0, \tag{1.22b}$$

must hold. The PDE (1.22a) has thus been transformed into a set of scalar, constant coefficient ODEs (1.22b) with the wave number ξ as a parameter. The ODE (1.22b) can be solved exactly for each ξ.

> **Note:** The following identities hold for any scalar θ:
>
> $$e^{\imath \theta} = \cos \theta + \imath \sin \theta,$$
>
> $$\cos \theta = \frac{e^{\imath \theta} + e^{-\imath \theta}}{2},$$
>
> $$\sin \theta = \frac{e^{\imath \theta} - e^{-\imath \theta}}{2\imath},$$
>
> $$\cos \theta - 1 = -2 \sin^2(\theta/2).$$

1.3.5 The matrix power and exponential

We will often want to bound the powers A^n (i.e., A times itself n times), where A is an $m \times m$ matrix. If $A = T \Lambda T^{-1}$, where Λ is a diagonal matrix, then

$$A^n = (T \Lambda T^{-1})(T \Lambda T^{-1}) \cdots (T \Lambda T^{-1})$$
$$= T \Lambda (T^{-1} T) \Lambda (T^{-1} T) \cdots \Lambda T^{-1} = T \Lambda^n T^{-1}.$$

1.3. Reviews

Thus, in any consistent matrix norm

$$\|A^n\| \leq \|\Lambda^n\|\mathrm{cond}(T) = [\rho(A)]^n \mathrm{cond}(T), \qquad (1.23)$$

where $\mathrm{cond}(T) = \|T\|\|T^{-1}\|$ is the condition number of the transformation.

The matrix exponential is defined via a power series expansion by

$$e^{At} = \sum_{n=0}^{\infty} \frac{t^n A^n}{n!} = I + tA + \frac{t^2 A^2}{2} + \frac{t^3 A^3}{6} + \cdots.$$

If $A = T\Lambda T^{-1}$, where Λ is a diagonal matrix, then it is easy to see that $e^{At} = Te^{\Lambda t}T^{-1}$, where $e^{\Lambda t} = \mathrm{diag}\{e^{\lambda_1 t}, \ldots, e^{\lambda_m t}\}$. Thus,

$$\|e^{At}\| \leq e^{\max_i \mathcal{R}e\lambda_i t} \mathrm{cond}(T). \qquad (1.24)$$

The calculation of e^{At} for a general square matrix A, and even more so of $e^{At}\mathbf{v}$ for a given vector $\mathbf{v}$ in case that A is very large, can be surprisingly exciting [135].

1.3.6 Fourier transform for periodic functions

Consider the function space $\mathcal{L}_2[0, 2\pi]$. Define the **inner product**

$$(v, w) = \frac{1}{2\pi} \int_0^{2\pi} v(x) w^*(x) dx, \qquad (1.25)$$

which also holds for complex-valued vector functions. Then $\|w\| = (w, w)^{1/2}$. Let

$$\phi_j(x) = e^{\iota j x}, \qquad j = 0, \pm 1, \pm 2, \ldots, \qquad (1.26a)$$

be the **trigonometric basis functions**. Then

$$(\phi_j, \phi_l) = \delta_{jl} = \begin{cases} 0, & j \neq l, \\ 1, & j = l, \end{cases}$$

and any function $v \in \mathcal{L}_2[0, 2\pi]$ can be written as

$$v(x) = \sum_{j=-\infty}^{\infty} \hat{v}_j \phi_j(x), \quad \text{where} \qquad (1.26b)$$

$$\hat{v}_j = (v, \phi_j). \qquad (1.26c)$$

Thus, the trigonometric functions form a complete, orthonormal basis for $\mathcal{L}_2[0, 2\pi]$. Parseval's equality (1.19) now reads (due to the basis orthonormality)

$$\|v\| = \sqrt{\sum_{j=-\infty}^{\infty} |\hat{v}_j|^2}.$$

Now, for a *periodic* function v (i.e., $v(2\pi) = v(0)$) we have the same diagonalization effect on derivatives as in the infinite case (Section 1.3.4), because

$$(v_x, \phi_j) = \frac{1}{2\pi} \int_0^{2\pi} v_x(x) e^{-\iota j x} dx = \frac{\iota j}{2\pi} \int_0^{2\pi} v(x) e^{-\iota j x} dx = \iota j \hat{v}_j.$$

Above, the crucial step follows from integration by parts using the periodicity of v to avoid unwanted boundary terms.

1.4 Exercises

0. **Review questions**

    (a) What is a Cauchy problem?

    (b) Let $\hat{u}(t, \xi)$ be the Fourier transform of $u(t, x)$, where t acts as a parameter. What is the Fourier transform of $u_x(t, x)$?

    (c) Is the heat equation operator a smoother? Is the advection operator a smoother?

    (d) For the PDE $u_t = u_x$, what is a characteristic curve? Does x grow or decrease as t grows along this curve?

    (e) Write down the wave equation and the Laplace equation. For which is the pure IVP well-posed? For which is the boundary value problem well-posed?

    (f) Define a mesh function.

    (g) What is the leapfrog scheme for the advection equation?

    (h) State the CFL condition. Is it equivalent to stability?

    (i) What is an amplification factor? What happens if it equals 1.2 for all ζ?

1. For the Cauchy problem with constant coefficients (1.8) show that if $\iota^m p_m$ has an eigenvalue with a positive real part, then the problem cannot be well-posed.

2. For the initial-boundary value, constant-coefficient problem

    $$u_t = P(\partial_x)u, \qquad 0 < x < 2\pi, \ t \geq 0,$$
    $$u(0, x) = u_0(x), \quad u(t, 2\pi) = u(t, 0),$$

    derive a well-posedness condition analogous to (1.11).

3. The celebrated Black–Scholes model [26] for the pricing of stock options is central in mathematical finance; see, e.g., [183]. The PDE is given by

    $$u_t + \frac{1}{2}\sigma^2 x^2 u_{xx} + r x u_x - r u = 0, \quad 0 < x < \infty, \ t \leq T. \qquad (1.27)$$

    For the sake of completeness let us add that u is the sought value of the option under consideration, t is time, x is the current value of the underlying asset, r is the interest rate, σ the volatility of the underlying asset, T the expiry date, and E the exercise

1.4. Exercises

price. In general, r and σ may vary, but here they are assumed to be known constants, as are E and T.

For the European call option we have the terminal condition

$$u(T, x) = \max(x - E, 0) \tag{1.28a}$$

and the boundary conditions

$$u(t, 0) = 0, \quad u(t, x) \sim x - Ee^{-r(T-t)} \text{ as } x \to \infty. \tag{1.28b}$$

(a) Show that the transformation

$$x = Ee^y, \quad t = T - \frac{2s}{\sigma^2}, \quad u = Ev(s, y)$$

results in the initial value PDE

$$v_s = v_{yy} + (\kappa - 1)v_y - \kappa v, \quad -\infty < y < \infty, \tag{1.29}$$
$$v(0, y) = \max(e^y - 1, 0),$$

where $\kappa = \frac{2r}{\sigma^2}$.

(b) Show further that transforming

$$v = e^{\gamma y + \beta s} w(s, y), \quad \text{where}$$
$$\gamma = (1 - \kappa)/2, \quad \beta = -(\kappa + 1)^2/4,$$

yields the PDE problem

$$w_s = w_{yy}, \quad -\infty < y < \infty, \; s \geq 0, \tag{1.30}$$
$$w(0, y) = \max(e^{\frac{1}{2}(\kappa+1)y} - e^{\frac{1}{2}(\kappa-1)y}, 0).$$

(c) Prove that the terminal-value PDE (1.27)–(1.28) is well-posed.

(Note that the solution of (1.30), and therefore also of (1.29) and (1.27)–(1.28), can be specified exactly in terms of the integral

$$\frac{1}{\sqrt{2\pi}} \int_{-\infty}^{z} e^{-\zeta^2/2} d\zeta.$$

However, you don't need this for the purpose of the present exercise.)

4. Consider the *advection-diffusion* equation

$$u_t + au_x = \nu u_{xx}$$

with constant coefficients $\nu > 0$ and a. Show that the Cauchy problem is well-posed. What happens as $\nu \to 0$?

5. (a) For the advection equation $u_t + au_x = 0$ with $a > 0$, show that the one-sided scheme (1.15a) is unconditionally unstable.

   (b) On the other hand, show that the scheme
   $$v_j^{n+1} = v_j^n - \mu a(v_j^n - v_{j-1}^n)$$
   is stable for $a > 0$, provided that $\mu a \leq 1$, but unstable for any $\mu \geq 0$ if $a < 0$. Explain.

   (c) An **upwind** (also known as *upstream*) difference scheme for the advection equation can be defined by
   $$v_j^{n+1} = v_j^n - \mu a \begin{cases} (v_{j+1}^n - v_j^n), & a < 0, \\ (v_j^n - v_{j-1}^n), & a \geq 0. \end{cases}$$
   Show that this scheme is stable if $\mu|a| \leq 1$, just like the leapfrog scheme, regardless of the sign of a.

   (d) Without programming, construct an additional column in Table 1.1 corresponding to the upwind scheme.

6. For the advection equation
   $$u_t + au_x = 0,$$
   with $a < 0$, we know that the exact solution satisfies
   $$u(t + k, x) = u(t, x - ak).$$
   This suggests a *Lagrangian* numerical method: set
   $$v_j^{n+1} = \hat{v}_j^n,$$
   where $\hat{v}_j^n$ is an interpolation of nearby mesh values at the known time level t_n for a value at the point $(t_n, x_j - ak)$.

   (a) Show that the CFL condition implies that the point $(t_n, x_j - ak)$ must lie between (t_n, x_j) and (t_n, x_{j+1}).

   (b) Suppose that $\hat{v}_j^n$ is defined as the linear interpolation of (x_j, v_j^n) and (x_{j+1}, v_{j+1}^n) at $x_j - ak$. Show that the obtained scheme is the one-sided (1.15a).

7. The nonlinear Schrödinger equation (NLS) is given by
   $$\imath \psi_t + \psi_{xx} = V'(|\psi|)\psi,$$
   where $V : \mathbb{R} \to \mathbb{R}$ is a smooth function (e.g., the cubic NLS $V'(w) = \pm w^2$) and ψ is complex-valued.

   Linearizing by setting
   $$V' = c,$$

1.4. Exercises

where c is a real constant, write the resulting equation as a system for a real $\mathbf{u} = (u_1, u_2)^T$, where
$$\psi = u_1 + \iota u_2,$$
and show that the corresponding Cauchy problem is well-posed. Do you expect the solution for discontinuous initial data to become smooth as in Figure 1.4 or stay discontinuous for $t > 0$ as in Figure 1.3?

8. The so-called *logarithmic norm* of a square, complex-valued matrix A based on a given matrix norm $\|\cdot\|$, is defined by
$$\nu(A) = \lim_{k \downarrow 0} \frac{\|I + kA\| - 1}{k}. \tag{1.31}$$

    This is a very useful concept because of item (d) below.

    (a) Show that for any $k \geq 0$, $-\|A\| \leq k^{-1}(\|I + kA\| - 1) \leq \|A\|$.

    (b) Show that $k^{-1}(\|I + kA\| - 1)$ is monotonically nondecreasing in k. Hence the limit as $k \downarrow 0$ exists.

    (c) Explain why $\nu(A)$ is *not* a matrix norm.

    (d) Show that for any scalar $K \in \mathbb{R}$
    $$\nu(A) \leq K \quad \text{iff} \quad \|e^A\| \leq e^K. \tag{1.32}$$

    (Note that K can be negative.)

9. Continuing the previous exercise, show that the logarithmic norm satisfies the following properties:

    (a)
    $$\nu(sI + tA) = s + t\nu(A) \quad \forall\, s \in \mathbb{R}, t \geq 0,$$
    $$\nu(A + B) \leq \nu(A) + \nu(B),$$
    $$|\nu(A) - \nu(B)| \leq \|A - B\|,$$
    $$\nu(A) \geq -\|A\mathbf{x}\|/\|\mathbf{x}\| \quad \forall\, \mathbf{x} \neq \mathbf{0}.$$

    (b) For the usual 1, 2, and ∞-norms
    $$\nu_\infty(A) = \max_i \left(\mathcal{R}e(a_{ii}) + \sum_{j \neq i} |a_{ij}| \right),$$
    $$\nu_1(A) = \max_j \left(\mathcal{R}e(a_{jj}) + \sum_{i \neq j} |a_{ij}| \right),$$
    $$\nu_2(A) = \max\{\lambda|\ \lambda \text{ eigenvalue of } (A + A^*)/2\}.$$

10. Repeat the experiments of Example 1.2 for $\eta = 2$, $\mu = 0.5$, and the three spatial step sizes $h = .1\pi$, $.01\pi$, and $.001\pi$. Record maximum errors at $t = 1$ using the three schemes (1.15). What are your observations?

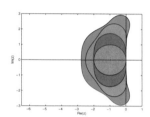

Chapter 2
Methods and Concepts for ODEs

The numerical solution of ODEs is a subject that has received a lot of attention in past decades. Many applications, for example, the motion of a system of rigid bodies obeying Newton's second law, give rise to mathematical models that involve dynamical behavior in time but do not involve spatial derivatives. Several important concepts and methods required for time-dependent PDEs arise also in the simpler ODE setting.

Moreover, a common approach for discretizing an initial value PDE problem is to first discretize the space variables, obtaining an initial value problem for a set of ODEs in time. This is called a **semi-discretization**. For instance, a semi-discretization for the advection equation
$$u_t + au_x = 0$$
considered in Section 1.2 could be the one-sided spatial discretization
$$\frac{dv_j}{dt} = -\frac{a}{h}(v_{j+1} - v_j) \qquad (2.1a)$$
or the centered discretization
$$\frac{dv_j}{dt} = -\frac{a}{2h}(v_{j+1} - v_{j-1}). \qquad (2.1b)$$
In (2.1) the index j ranges over the spatial mesh, yielding an ODE system for a mesh function $v(t)$ that is closed using some boundary conditions that are not our concern here. The full discretization (1.15a) is a particular discretization in time, called *forward Euler*, of the one-sided semi-discretization (2.1a), whereas the other two schemes (1.15b) and (1.15c) are two different time discretizations of the centered semi-discretization (2.1b).

We consider semi-discretizations more methodically in future chapters; see especially Section 3.1. Here our interest is to describe methods and concepts relevant once an ODE system is obtained. In fact, Chapter 1 is *not a prerequisite* to the current chapter. Note that ODEs commonly arise also in other situations when solving PDEs, e.g., from a Fourier analysis or when applying splitting methods (see Section 9.3). And yet, many PDE researchers appear to be peculiarly ignorant of methods and concepts for solving ODEs. (Many ODE researchers, for their part, seem to have developed methods and theory in a manner often suited only for small problems.) We aim to attain essential ODE expertise by introducing the present basic chapter and the more advanced Chapter 6. No attempt is made here at achieving completeness of ODE expositions.

The numerical solution of initial value problems for ODEs has been treated in many fine books [82, 85, 44, 154, 113, 66]. A MATLAB-oriented treatment is found in [155]. An ODE-enriched treatment of reaction-convection-diffusion PDEs is given by Hundsdorfer and Verwer [98]. LeVeque's [117] is a recent textbook covering basic methods for both ODEs and PDEs. But several other PDE texts, such as [137, 159], hardly address ODEs directly at all. The present chapter is partly extracted from Ascher and Petzold [14]. The discussion of stability concepts is particularly important.

> **Note:** Following is a synopsis of the present chapter. The "must know" topics are in Sections 2.1, 2.2, 2.5, 2.6, 2.7, and 2.9.

There are two popular families of methods in practice: linear multistep and Runge–Kutta. They are introduced in Sections 2.1 and 2.2, respectively. Basic concepts of *local truncation error*, *consistency*, and *order of accuracy* are introduced. These are extended to PDEs in Section 3.2.

The relative simplicity of ODEs allows us to discuss various issues in a direct and concise manner. Thus, convergence is considered in Section 2.3 and the practical issue of estimating and controlling the discretization error is briefly exposed in Section 2.4. The central issue of (absolute) *stability* is considered in Section 2.5 and related questions of *stiffness* are found in Section 2.6.

Implicit methods for ODE systems require the solution of a system of *algebraic equations* at each time step. This is considered in Section 2.7. The mixing of *differential and algebraic equations* is briefly described in Section 2.8. Two quick sections follow, 2.9 and 2.10. They introduce *symmetric schemes* and a glimpse at *highly oscillatory problems*. Another brief section summarizing a large topic is Section 2.11, where methods for *boundary value ODEs* are considered. This is followed by several background reviews and exercises.

Turning to basic methods, consider a possibly nonlinear ODE system

$$\frac{dy}{dt} \equiv y' = f(t, y), \qquad t > 0, \tag{2.2a}$$

subject to initial conditions

$$y(0) = y_0. \tag{2.2b}$$

In general, y and f have m components each, $m \geq 1$. It may be convenient at first to think of a scalar ODE, $m = 1$: the numerical methods described below all readily generalize to ODE systems. When we want to emphasize that $m > 1$ and that y and f are vector functions, e.g., in Section 2.7 and in particular examples, we use boldface notation.

> **Note:** Regarding notation, in the chapters concerning PDE discretizations we use subscripts to denote location in space and superscripts to denote location in time (e.g., v_j^n). Here, there is only one independent variable, normally time. But using superscripts without subscripts may appear confusing, so we use subscripts alone (e.g., y_n to approximate $y(t_n)$). The time step is consistently denoted by $k = \Delta t$.

2.1 Linear multistep methods

An s-step method uses at time t_n the known solution values

$$y_{n+1-s}, \ldots, y_{n-1}, y_n$$

to define the next unknown value, y_{n+1}, that approximates $y(t_{n+1})$. The iteration process increments n, starting at $n = s - 1$. Thus, the initial values $y_{s-1}, \ldots, y_0$, and not only the given y_0, are required to start the algorithm. An instance with $s = 2$ is provided by the leapfrog scheme (1.15c) given in its semi-discretized form (2.1b); see Example 1.2. Other instances will follow shortly.

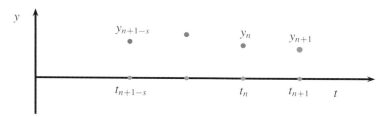

Figure 2.1. *The setup for a linear multistep method: The known solution values y_n at the current location t_n and $y_{n-1}, \ldots, y_{n+1-s}$ at previous mesh points, together with the values of f there, are used to form an approximation for the next unknown y_{n+1} at t_{n+1}, employing a linear formula.*

Assuming for simplicity a uniform mesh, $t_n = nk$, $n = 0, 1, \ldots$, where k is the **step size** in time, a general linear s-step method for the ODE (2.2a) reads

$$\sum_{j=0}^{s} \alpha_j y_{n+1-j} = k \sum_{j=0}^{s} \beta_j f_{n+1-j}. \tag{2.3}$$

Here, $f_{n+1-j} = f(t_{n+1-j}, y_{n+1-j})$ and α_j, β_j are coefficients. See Figure 2.1. Since (2.3) can be scaled arbitrarily, we set $\alpha_0 = 1$ for definiteness. The method is **explicit** if $\beta_0 = 0$ and **implicit** otherwise. Note that for an explicit method the unknown y_{n+1} appears only on the left-hand side, so the method can be written as

$$y_{n+1} = \sum_{j=1}^{s} \left[-\alpha_j y_{n+1-j} + k \beta_j f_{n+1-j} \right],$$

with the right-hand side thus involving only known quantities. An implicit method, on the other hand, involves the unknown also through $f_{n+1} = f(t_{n+1}, y_{n+1})$, so carrying out the time stepping iteration may be more involved; see Section 2.7. The method is called *linear* because, unlike general Runge–Kutta, for instance, the expression in (2.3) is linear in f.

Let us define the **local truncation error** for the method (2.3) as

$$d_n = k^{-1} \sum_{j=0}^{s} \alpha_j y(t_{n+1-j}) - \sum_{j=0}^{s} \beta_j y'(t_{n+1-j}). \tag{2.4}$$

This is the amount by which the exact solution fails to satisfy the difference equations (2.3) divided by k. The method is said to have **order of accuracy** (or *order* for short) p if

$$d_n = O(k^p)$$

for all problems with sufficiently smooth exact solutions $y(t)$. If $p \geq 1$, as it practically always is, then the method is said to be **consistent**.

The most popular families of linear multistep methods are the **Adams** family and the **backward differentiation formula** (BDF) family.

The Adams methods are derived by considering the integration of (2.2a)

$$y(t_{n+1}) = y(t_n) + \int_{t_n}^{t_{n+1}} f(t, y(t)) dt$$

and approximating the integrand $f(t, y)$ by passing an interpolating polynomial through previously computed values of $f(t_l, y_l)$. (See reviews of polynomial interpolation in Section 2.12.2 and of numerical integration in Section 2.12.4.) In the general form (2.3) we therefore set

$$\alpha_0 = 1, \ \alpha_1 = -1, \quad \text{and } \alpha_j = 0 \ \forall \ j > 1,$$

for all Adams methods.

The s-step *explicit Adams method*, also called the **Adams–Bashforth** method, is obtained by interpolating f through the previous points $t = t_n, t_{n-1}, \ldots, t_{n+1-s}$, giving an explicit method of accuracy $p = s$. Table 2.1 displays the coefficients of the Adams–Bashforth methods for s up to 6. For $s = 1$ we obtain the **forward Euler** method

$$y_{n+1} = y_n + k f_n. \tag{2.5}$$

The two-step Adams–Bashforth method reads

$$y_{n+1} = y_n + \frac{k}{2}(3 f_n - f_{n-1}), \tag{2.6}$$

and so on.

Example 2.1 Recalling briefly our semi-discretizations (2.1) for the advection equation, it is easy to verify that the scheme (1.15a) is obtained by applying forward Euler to (2.1a) while the unstable scheme (1.15b) is obtained by applying forward Euler to (2.1b). The

2.1. Linear Multistep Methods

Table 2.1. *Coefficients of Adams–Bashforth methods up to order 6.*

p	s	j →	1	2	3	4	5	6
1	1	β_j	1					
2	2	$2\beta_j$	3	−1				
3	3	$12\beta_j$	23	−16	5			
4	4	$24\beta_j$	55	−59	37	−9		
5	5	$720\beta_j$	1901	−2774	2616	−1274	251	
6	6	$1440\beta_j$	4277	−7923	9982	−7298	2877	−475

Table 2.2. *Coefficients of Adams–Moulton methods up to order 6.*

p	s	j →	0	1	2	3	4	5
1	1	β_j	1					
2	1	$2\beta_j$	1	1				
3	2	$12\beta_j$	5	8	−1			
4	3	$24\beta_j$	9	19	−5	1		
5	4	$720\beta_j$	251	646	−264	106	−19	
6	5	$1440\beta_j$	475	1427	−798	482	−173	27

leapfrog scheme (1.15c) is obtained by applying to the semi-discretization (2.1b) a linear multistep time discretization which does not belong to the Adams family. ∎

The s-step *implicit Adams method*, also called the **Adams–Moulton** method, is obtained by passing an interpolating polynomial of f through the previous points plus the next one, $t = t_{n+1}, t_n, t_{n-1}, \ldots, t_{n+1-s}$, giving an implicit method of accuracy $p = s + 1$. Table 2.2 displays the coefficients of the Adams–Moulton methods for s up to 5. Note that there are two one-step methods here: **backward Euler** ($s = p = 1$),

$$y_{n+1} = y_n + kf_{n+1}, \tag{2.7}$$

and **trapezoidal** ($s = 1$, $p = 2$; sometimes referred to as *Crank–Nicolson*),

$$y_{n+1} = y_n + \frac{k}{2}(f_n + f_{n+1}). \tag{2.8}$$

The four-step Adams–Moulton method (AM4) reads

$$y_{n+1} = y_n + \frac{k}{720}\left(251 f_{n+1} + 646 f_n - 264 f_{n-1} + 106 f_{n-2} - 19 f_{n-3}\right). \tag{2.9}$$

This method is fifth order accurate. Some of these methods and their order of accuracy are demonstrated in Example 2.5, after we connect order of accuracy to convergence.

Table 2.3. *Coefficients of BDF methods up to order 6.*

p	s	β_0	α_0	α_1	α_2	α_3	α_4	α_5	α_6
1	1	1	1	-1					
2	2	$\frac{2}{3}$	1	$-\frac{4}{3}$	$\frac{1}{3}$				
3	3	$\frac{6}{11}$	1	$-\frac{18}{11}$	$\frac{9}{11}$	$-\frac{2}{11}$			
4	4	$\frac{12}{25}$	1	$-\frac{48}{25}$	$\frac{36}{25}$	$-\frac{16}{25}$	$\frac{3}{25}$		
5	5	$\frac{60}{137}$	1	$-\frac{300}{137}$	$\frac{300}{137}$	$-\frac{200}{137}$	$\frac{75}{137}$	$-\frac{12}{137}$	
6	6	$\frac{60}{147}$	1	$-\frac{360}{147}$	$\frac{450}{147}$	$-\frac{400}{147}$	$\frac{225}{147}$	$-\frac{72}{147}$	$\frac{10}{147}$

The s-step *BDF method* is obtained by evaluating f only at the right end of the current step, (t_{n+1}, y_{n+1}), driving an interpolating polynomial of y through the previous points plus the next one, $t = t_{n+1}, t_n, t_{n-1}, \ldots, t_{n+1-s}$, and differentiating the resulting interpolant. This gives an implicit method of accuracy order $p = s$. Table 2.3 displays the coefficients of the BDF methods for s up to 6. For $s = 1$ we again obtain the backward Euler method (2.7). These methods are particularly useful for *stiff problems*, as discussed in Section 2.6.

2.2 Runge–Kutta methods

Runge–Kutta are one-step methods in which f is repeatedly evaluated within one mesh interval to obtain a higher order method. For instance, an "explicit trapezoidal method" is obtained upon using forward Euler

$$Y_1 = y_n, \quad Y_2 = y_n + kf(t_n, Y_1),$$

followed by setting

$$y_{n+1} = y_n + \frac{k}{2}(f(t_n, Y_1) + f(t_{n+1}, Y_2)).$$

This is a two-stage Runge–Kutta method of order 2.

In general, an *s-stage Runge–Kutta* method for the ODE (2.2a) can be written in the form

$$Y_i = y_n + k \sum_{j=1}^{s} a_{ij} f(t_n + c_j k, Y_j), \quad 1 \leq i \leq s, \tag{2.10a}$$

$$y_{n+1} = y_n + k \sum_{i=1}^{s} b_i f(t_n + c_i k, Y_i). \tag{2.10b}$$

The Y_i's are intermediate approximations to the solution at times $t_n + c_i k$, which may be correct to a lower order of accuracy than the order for the solution y_{n+1} at the end of the step. We shall always require

$$1 = \sum_{j=1}^{s} b_j, \quad c_i = \sum_{j=1}^{s} a_{ij}, \quad i = 1, \ldots, s.$$

2.2. Runge–Kutta Methods

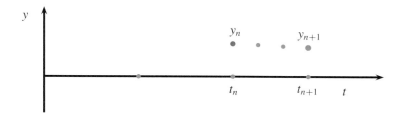

Figure 2.2. *The setup for a Runge–Kutta method. Repeated evaluations of the function f in the one mesh subinterval $[t_n, t_{n+1}]$ are used to obtain a higher order approximation y_{n+1} at t_{n+1}. Note that $y'(t)$ is allowed to be discontinuous at t_n without loss of accuracy, and no values before the current y_n must be recorded to apply the method.*

Note that Y_i are local to the step from t_n to t_{n+1}, and the only approximation that the next step "sees" is y_{n+1}. A Runge–Kutta step is depicted in Figure 2.2.

A summary of the relative advantages and disadvantages of linear multistep and RK methods is given at the end of Section 2.7.

The coefficients of the RK method are chosen in part to make error terms cancel in such a way that y_{n+1} is more accurate. Thus, whereas in an s-step multistep method the solution must be smooth over the several subintervals forming $[t_{n+1-s}, t_{n+1}]$ to realize its high order, here the exact solution may be merely continuous at t_n.

The method can be conveniently represented in a shorthand notation:

$$
\begin{array}{c|cccc}
c_1 & a_{11} & a_{12} & \cdots & a_{1s} \\
c_2 & a_{21} & a_{22} & \cdots & a_{2s} \\
\vdots & \vdots & \vdots & \ddots & \vdots \\
c_s & a_{s1} & a_{s2} & \cdots & a_{ss} \\
\hline
 & b_1 & b_2 & \cdots & b_s
\end{array}
$$

As for multistep methods, the **local truncation error** d_n is defined as the amount by which the exact solution fails to satisfy the numerical formula per step (2.10b), divided by k. The **order** of the method is then p if $d_n = O(k^p)$. A method whose order is at least 1 is said to be **consistent**. In addition, the **stage order** is defined similarly, as the minimum order of the stages, i.e., the amount by which the exact solution fails to satisfy (2.10a) divided by k. It becomes occasionally relevant, although more rarely than the order p itself.

The method is **explicit** if

$$a_{ij} = 0 \ \forall i \leq j.$$

Examples of explicit Runge–Kutta methods are the forward Euler method (2.5), given in tableau notation by

$$
\begin{array}{c|c}
0 & 0 \\
\hline
 & 1
\end{array},
$$

the explicit trapezoidal

$$\begin{array}{c|cc} 0 & 0 & 0 \\ 1 & 1 & 0 \\ \hline & \frac{1}{2} & \frac{1}{2} \end{array},$$

and the classical fourth order method (RK4)

$$\begin{array}{c|cccc} 0 & 0 & 0 & 0 & 0 \\ \frac{1}{2} & \frac{1}{2} & 0 & 0 & 0 \\ \frac{1}{2} & 0 & \frac{1}{2} & 0 & 0 \\ 1 & 0 & 0 & 1 & 0 \\ \hline & \frac{1}{6} & \frac{1}{3} & \frac{1}{3} & \frac{1}{6} \end{array} \qquad (2.11)$$

To show that (2.11) actually has order 4 is surprisingly nontrivial; see [14, 82]. In general, the order p of an explicit Runge–Kutta method satisfies $p \leq s$, with $p = s$ for both forward Euler (2.5) and the fourth order (2.11) above. There are no such methods with $p = s > 4$. Some of these explicit Runge–Kutta variants and their order of accuracy are demonstrated in Example 2.5 later on. But first, let us demonstrate how to work with such a method.

Example 2.2 The famous Lorenz equations are given by

$$\begin{aligned} y_1' &= \sigma(y_2 - y_1), \\ y_2' &= ry_1 - y_2 - y_1 y_3, \\ y_3' &= y_1 y_2 - by_3, \end{aligned}$$

where σ, r, b are positive parameters. We can obviously write these equations as an autonomous ODE system $y' = f(y)$, i.e., f in (2.2a) does not depend explicitly on t, with $m = 3$. This provides a simple example of a *chaotic system*; see [160, 161].

Here we use these equations merely to demonstrate the application of the explicit Runge–Kutta method RK4, using a constant step size. Following Lorenz we set $\sigma = 10$, $b = 8/3$, $r = 28$, integrate starting from $y(0) = (0, 1, 0)^T$, and plot y_3 vs. y_1 to obtain the famous "butterfly"; see Figure 2.3. The following MATLAB script should be easy to follow:

```
% Simulating the Lorenz butterfly using the simple ODE solver
    rk4.

% parameters
k = .01;                      % constant step size
tf = 100;                     % final time

% given initial conditions
y0(1) = 0; y0(2) = 1; y0(3) = 0;
```

2.2. Runge–Kutta Methods

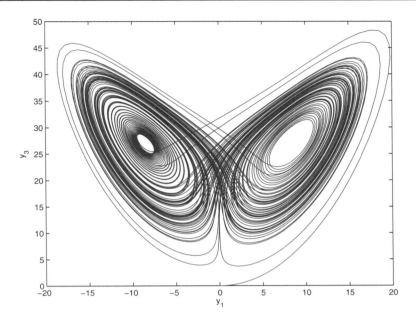

Figure 2.3. *Lorenz "butterfly" in the $y_1 \times y_3$ plane.*

```
 y0 = shiftdim(y0);

% integrate
 [tout,yout] = rk4(tf,k,y0);

% plot solution in phase space
 plot(yout(1,:),yout(3,:))
 xlabel('y_1')
 ylabel('y_3')

function [tout,yout] = rk4 (tf,k,y0)
%
% [tout,yout] = rk4 (tf,k,y0)
% solves non-stiff IVODEs based on classical 4-stage Runge-
   Kutta
%
%     y' = ffun (t,y),   y(0) = y0,   0 < t < tf
%     ffun returns a vector of m components for a given
%     t and y of m components.
%
% Input:
%   tf  - final time
%   k   - uniform step size
```

```
%   y0 - initial values y(0)   (size (m,1))
%
% Output:
%   tout - times where solution is computed
%   yout - solution vector at times tout.

% prepare
  N = tf / k;
  tout = [0];
  fo = ffun (tout(1),y0);
  yout(:,1) = y0;

% loop in time

  for n=1:N
     t = tout(n);
     y = yout(:,n);
     Y1 = y;
     K1 = ffun(t,Y1);
     Y2 = y + k/2*K1;
     t = t + k/2;
     K2 = ffun(t,Y2);
     Y3 = y + k/2*K2;
     K3 = ffun(t,Y3);
     Y4 = y + k*K3;
     t = t + k/2;
     K4 = ffun(t,Y4);
     tout(n+1) = t;
     yout(:,n+1) = y + k/6*(K1+2*K2+2*K3+K4);
  end

function f = ffun(t,y)

% function defining ODE

sigma = 10; b = 8/3; r = 28;

f = zeros(3,1);
f(1) = sigma *(y(2)-y(1));
f(2) = r * y(1) - y(2) - y(1)*y(3);
f(3) = y(1)*y(2) - b*y(3);
```

Again we emphasize that our purpose here is to show just how the Runge–Kutta loop (in function rk4) may look. We do not intend this to replace a good, general-purpose ODE software such as ode45; see also Section 2.4. ∎

2.2. Runge–Kutta Methods

In general, one should of course avoid evaluating the function f more than is absolutely necessary. The formulation (2.10) can alternatively be written as (2.42).

Examples of **implicit** Runge–Kutta methods include the backward Euler method (2.7)

$$\begin{array}{c|c} 1 & 1 \\ \hline & 1 \end{array}$$

and the trapezoidal method (2.8)

$$\begin{array}{c|cc} 0 & 0 & 0 \\ 1 & \frac{1}{2} & \frac{1}{2} \\ \hline & \frac{1}{2} & \frac{1}{2} \end{array}$$

An important variant of the trapezoidal method called the **midpoint** method is given by

$$\begin{array}{c|c} \frac{1}{2} & \frac{1}{2} \\ \hline & 1 \end{array},$$

i.e.,

$$y_{n+1} = y_n + kf\left(t_{n+1/2}, \frac{y_n + y_{n+1}}{2}\right). \qquad (2.12)$$

This method satisfies $s = 1$, $p = 2$. There are explicit Runge–Kutta variants of both the midpoint and the trapezoidal method, with $s = p = 2$.

Families of methods derived by **collocation** at *Gaussian points* and at *Radau points* yield high-order implicit Runge–Kutta methods. Such a collocation method constructs a *piecewise polynomial* approximate solution $y_\Delta(t)$ that is continuous over the entire integration interval in t, reduces to a polynomial on each mesh subinterval $[t_n, t_{n+1}]$, and satisfies the ODE at the collocation points $t_n + c_i k$ appearing also in (2.10). The resulting method is equivalent to a particular implicit Runge–Kutta one. See Sections 2.12.2–2.12.4 for a quick recap on piecewise polynomials, Gaussian points, and basic quadrature rules, and see [14, 85] for derivation and formulas of the collocation-implicit Runge–Kutta methods. Here we simply list the first few members of these families:

- **Gauss methods.** These are the maximum order Runge–Kutta methods; an s-stage Gauss method has order $p = 2s$. For $s = 1$ the midpoint method (2.12) is obtained. For $s = 2$ we have

$$\begin{array}{c|cc} \frac{3-\sqrt{3}}{6} & \frac{1}{4} & \frac{3-2\sqrt{3}}{12} \\ \frac{3+\sqrt{3}}{6} & \frac{3+2\sqrt{3}}{12} & \frac{1}{4} \\ \hline & \frac{1}{2} & \frac{1}{2} \end{array} \qquad s = 2, \quad p = 4.$$

- **Radau methods.** These correspond to quadrature rules where one end of the interval is included ($c_1 = 0$ or $c_s = 1$), and they attain order $p = 2s - 1$. We consider only the case $c_s = 1$. For $s = 1$ the backward Euler method is obtained. For $s = 2$ we have

$$\begin{array}{c|cc} \frac{1}{3} & \frac{5}{12} & -\frac{1}{12} \\ 1 & \frac{3}{4} & \frac{1}{4} \\ \hline & \frac{3}{4} & \frac{1}{4} \end{array} \qquad s = 2, \quad p = 3.$$

The important question regarding carrying out a step of implicit Runge–Kutta is discussed in Section 2.7.

The task of determining the *order* of a high order Runge–Kutta method[8] is much harder than the corresponding task for linear multistep methods. Denoting $C = \mathrm{diag}\{c_1, \ldots, c_s\}$, $\mathbf{b}^T = (b_1, b_2, \ldots, b_s)$, and $\mathbf{1} = (1, 1, \ldots, 1)^T$, the following conditions are *necessary* for the method to have order p:

$$\mathbf{b}^T A^i C^{l-1} \mathbf{1} = \frac{(l-1)!}{(l+i)!} = \frac{1}{l(l+1)\cdots(l+i)}, \qquad 1 \le l + i \le p. \tag{2.13}$$

(The indices run as follows: for each l, $1 \le l \le p$, we have order conditions for $i = 0, 1, \ldots, p - l$. Note also that superscripts denote powers in (2.13) and that $C^{l-1}\mathbf{1} = (c_1^{l-1}, \ldots, c_s^{l-1})^T$.) These conditions are easy to verify. Exercise 6 encourages you to work out what they imply for $p = 1, 2$, and 3.

The good news is that for orders up to 3 the conditions (2.13) are also sufficient. But for higher orders they are not! The complete picture regarding determination of order for Runge–Kutta methods is given, e.g., in [44, 82].

2.3 Convergence and 0-stability

Let us consider an initial value problem for an ODE system, (2.2). As done for PDEs in Chapter 1, our approach when studying essential concepts is to ask what is required for the ODE problem to be well-posed and then to require that the numerical method behave similarly. For ODEs, however, the requirement for well-posedness can be easily stated in a more general way, requiring only Lipschitz continuity of f in terms of y. Here is the fundamental theorem for initial value ODEs.

---

[8] For implicit Runte–Kutta methods based on piecewise polynomial collocation, the order of accuracy can be derived in ways different from, and simpler than, what is required for other Runge–Kutta methods; see [14].

2.3. Convergence and 0-stability

Theorem 2.3.
Let $f(t, y)$ be continuous for all (t, y) in a region $\mathcal{D} = \{0 \leq t \leq b, |y| < \infty\}$. Moreover, assume Lipschitz continuity in y: there exists a constant L such that for all (t, y) and $(t, \hat{y})$ in $\mathcal{D}$,

$$|f(t, y) - f(t, \hat{y})| \leq L|y - \hat{y}|. \tag{2.14}$$

Then

1. *For any initial value vector $y_0 \in \mathbb{R}^m$ there exists a unique solution $y(t)$ for the IVP throughout the interval $[0, b]$. This solution is differentiable.*

2. *The solution y depends continuously on the initial data: if $\hat{y}$ also satisfies the ODE (but not the same initial values), then*

$$|y(t) - \hat{y}(t)| \leq e^{Lt}|y(0) - \hat{y}(0)|. \tag{2.15}$$

3. *If $\hat{y}$ satisfies, more generally, a perturbed ODE*

$$\hat{y}' = f(t, \hat{y}) + r(t, \hat{y}),$$

where r is bounded on $\mathcal{D}$, $\|r\| \leq M$, then

$$|y(t) - \hat{y}(t)| \leq e^{Lt}|y(0) - \hat{y}(0)| + \frac{M}{L}(e^{Lt} - 1). \tag{2.16}$$

Thus we have solution existence, uniqueness, and continuous dependence on the data—in other words, a *well-posed problem*—provided that the conditions of the theorem hold. Let us check these conditions: if f is differentiable in y (we shall automatically assume this throughout), then the constant L can be taken as a bound on the first derivatives of f with respect to y. Denote by f_y the *Jacobian matrix*

$$(f_y)_{ij} = \frac{\partial f_i}{\partial y_j}, \qquad 1 \leq i, j \leq m.$$

Writing

$$f(t, y) - f(t, \hat{y}) = \int_0^1 \frac{d}{ds} f(t, \hat{y} + s(y - \hat{y})) \, ds$$
$$= \int_0^1 f_y(t, \hat{y} + s(y - \hat{y})) (y - \hat{y}) \, ds,$$

we can therefore choose $L = \sup_{(t,y) \in \mathcal{D}} \|f_y(t, y)\|$.

In many cases we must restrict the region $\mathcal{D}$ in order to be assured of the existence of such a (finite) bound L. For instance, if we restrict $\mathcal{D}$ to include bounded y such that

$|y - y_0| \leq \gamma$, and on this $\mathcal{D}$ both the Lipschitz bound (2.14) holds and $|f(t, y)| \leq M$, then a unique existence of the solution is guaranteed for $0 \leq t \leq \min(b, \gamma/M)$, giving the basic existence result a more local flavor.

Considering next the numerical method approximating the ODE problem, let y_π be a mesh function which takes on the value y_n at each t_n, i.e., $y_\pi(t_n) = y_n$, $n = 0, 1, \ldots, N$. (You can think of $y_\pi(t)$ as a function defined for all t, but all that matters is what it does at the mesh points.) Then the numerical method is given by

$$\mathcal{N}_\pi y_\pi(t_n) = 0,$$

and for the exact solution $\mathcal{N}_\pi y(t_n) = d_n$, the local truncation error. For example, backward Euler yields

$$\mathcal{N}_\pi y_\pi(t_n) = \frac{y_\pi(t_n) - y_\pi(t_{n-1})}{k} - f(t_n, y_\pi(t_n)), \quad n = 1, \ldots, N.$$

Under the conditions of the fundamental theorem, Theorem 2.3, we define the method to be **0-stable** if there are positive constants k_0 and K such that for any mesh functions x_π and z_π with $k \leq k_0$

$$|x_n - z_n| \leq K \left\{ |x_0 - z_0| + \max_{1 \leq j \leq N} |\mathcal{N}_\pi x_\pi(t_j) - \mathcal{N}_\pi z_\pi(t_j)| \right\}, \quad 1 \leq n \leq N. \quad (2.17)$$

What this bound says in effect is that the difference operator $\mathcal{N}_\pi$ is invertible and that its inverse is bounded by K.

The definition (2.17) is for one-step methods, but the extension to multistep methods is straightforward. With 0-stability it is an easy exercise to show that a method that has accuracy order p converges with the same order. In fact, we have the next theorem.

Theorem 2.4.

$$\text{consistency} + \text{0-stability} \Longrightarrow \text{convergence}.$$

Indeed, if the method is consistent of order p and 0-stable, then it is convergent of order p and

$$|y_n - y(t_n)| \equiv |e_n| \leq K \max_j |d_j| = O(k^p). \quad (2.18)$$

Example 2.5 Let us check the convergence error, both absolute value and order, for various methods previously encountered in this chapter. The problem

$$u'' + e^{u+1} = 0,$$
$$u(0) = 0, \quad u'(0) = \theta \tanh(\theta/4)$$

2.3. Convergence and 0-stability

has a solution that satisfies $u(1) = 0$ if θ satisfies the nonlinear algebraic equation

$$\theta = \sqrt{2e}\,\cosh(\theta/4).$$

See [14]. One such value is $\theta = 3.03623184819656$. We therefore rewrite the ODE in first order form for $y_1 = u$ and $y_2 = u'$,

$$y_1' = y_2,$$
$$y_2' = -e^{y_1+1},$$

and apply forward Euler, explicit trapezoidal (RK2), RK4, the two-step Adams–Bashforth (AB2), and the four-step Adams–Moulton (AM4) methods for this simple problem, measuring absolute errors at $t = 1$. We use $k = 10^{-l}$, $l = 1, 2, 3, 4$. The resulting errors are recorded in Figure 2.4.

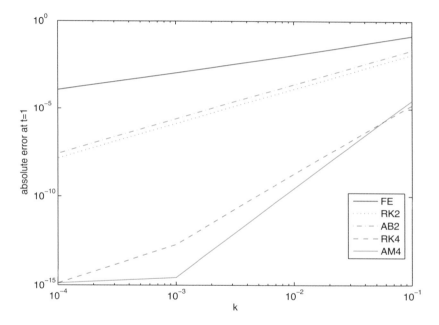

Figure 2.4. *Errors for different methods at different step sizes k plotted on a log-log scale. Before roundoff error enters the picture, the errors are all along straight lines with the slope equal to the method's order.*

It is clear from Figure 2.4 that the error is $O(k)$ in forward Euler, $O(k^2)$ in RK2 and AB2, $O(k^4)$ in RK4, and $O(k^5)$ in AM4. Since the plot is on a log-log scale we obtain straight lines with the slope p for a method of order p. For the two highest order methods the error is no longer on a straight line when it is so small in absolute value that roundoff error takes over. ∎

It is important to distinguish between the stability concepts arising in PDEs and those arising in ODEs. They are related but are not quite the same. The numerical ODE concept which, in the sense of mimicking well-posedness, most closely mirrors the PDE stability condition (3.33a) is 0-stability as defined above. However, whereas in the PDE the *ratio* of the discretization steps k in time and h in space as $k, h \to 0$ is of paramount importance, here there is no such ratio: basically, we first determine the ODE problem and thus its size, and only then we let k shrink to 0 in the difference method. Indeed, all ODE methods specifically mentioned in this chapter are unconditionally 0-stable.

2.4 Error control and estimation

For the sake of automating the integration process and robustly obtaining high-quality results at low cost, one wishes to write a mathematical software routine where a user would specify $f(t, y)$, the integration interval, y_0 and a tolerance Tol, and the routine would subsequently calculate $\{y_n\}_{n=0}^N$ accurate to within Tol. However, to control the **global error** $e_n = y_n - y(t_n)$ is difficult, so one is typically content to control the **local error**,

$$l_{n+1} = \bar{y}(t_{n+1}) - y_{n+1},$$

where $\bar{y}(t)$ is the solution of the ODE $y' = f(t, y)$ which satisfies $\bar{y}(t_n) = y_n$ (but $\bar{y}(0) \neq y_0$ in general).

Thus, we consider the nth mesh subinterval as if there is no error accumulated at its left end t_n and wonder what might result at the right end t_{n+1}. We basically assume that

$$|l_{n+1}| \simeq k|e_{n+1}|.$$

To control the local error we keep the step size small enough, so, in particular, $k = k_n$ is no longer the same for all steps n.

Suppose now that we use a pair of schemes, one of order q and the other of order $q + 1$, to calculate two approximations y_{n+1} and $\hat{y}_{n+1}$, respectively, at t_{n+1}, both starting from y_n at t_n. Then we can estimate

$$|l_{n+1}| \approx |\hat{y}_{n+1} - y_{n+1}|.$$

So, at the nth step we calculate these two approximations and compare: the step is accepted if

$$|\hat{y}_{n+1} - y_{n+1}| \leq k \, Tol.$$

We then set $y_{n+1} \leftarrow \hat{y}_{n+1}$ (the more accurate of the two values calculated) and increment n.

If the step is not accepted, then we decrease k to $\tilde{k}$ and repeat the step. This is done as follows: since the local error in the less accurate method is $l_{n+1} = ck^{q+1}$, upon decreasing k to a satisfactory $\tilde{k}$ its absolute value will become $|c|\tilde{k}^{q+1} \approx 0.9\tilde{k} \, Tol$, where the factor 0.9 is for safety. Dividing, we get

$$\left(\frac{\tilde{k}}{k}\right)^{q+1} \approx \frac{0.9\tilde{k} \, Tol}{|\hat{y}_{n+1} - y_{n+1}|}.$$

Thus, set

$$\tilde{k} = k \left(\frac{0.9k \, Tol}{|\hat{y}_{n+1} - y_{n+1}|}\right)^{\frac{1}{q}}. \tag{2.19}$$

2.5. Stability of ODE Methods

The only question left is how to choose the pair of formulas of orders q and $q + 1$ wisely. For Runge–Kutta methods this is achieved by searching for a pair of formulas which *share the internal stages Y_j* as much as possible. A good Runge–Kutta pair of orders 4 and 5 would use only 6 (rather than 9 or 10) function evaluations per step. Such pairs of formulas are implemented in MATLAB's `ode45`, for instance, as well as in several publicly available codes [14, 82].

One chief advantage of Runge–Kutta methods in the general context is that the step size $k = k_n$ need not be the same in consecutive time steps. General-purpose codes usually simulate a given initial value ODE attempting to keep the *local* error below a user-given tolerance. This is done by estimating the local error and adjusting the step size k_n for each step n. For more details see Chapter 4 of [14].

For multistep methods, a method for determining $\tilde{k}$ is also possible, indeed is simpler than Runge–Kutta—for example, using (2.19) with a pair of Adams formulas; see Section 2.7. However, there is more overhead in implementing a step size change, as the formulas involved must be updated as well; see Chapter 5 of [14]. In the present exposition we continue to assume that k is constant for the sake of simplicity.

2.5 Stability of ODE methods

We next ask what happens when the step size k is not very small. This corresponds to a finite ratio of time and space discretization steps k and h in the PDE case, because the ODE problem (2.2) may correspond to a semi-discretization such as (2.1) (see also (3.2) in Section 3.1) where the spatial step size h directly affects the size and contents of the ODE. We will discuss this connection in the next chapter.

Here, the fact that (3.2) is a large system and not a scalar ODE is crucial; and yet it is possible to learn a lot by considering the **test equation**

$$y' = \lambda y, \qquad t > 0, \tag{2.20}$$

where λ is a constant, complex scalar standing, e.g., for an eigenvalue of (3.2).

The solution of the test equation, $y(t) = e^{\lambda t} y(0)$, satisfies

$$|y(t)| \leq |y(0)| \iff \mathcal{R}e\, \lambda \leq 0.$$

Correspondingly, for a numerical method we define the **region of absolute stability** as that region of the complex z-plane containing the origin where

$$|y_{n+1}| \leq |y_n|, \qquad n = 0, 1, 2, \ldots, \tag{2.21}$$

when applying the method for the test equation (2.20), with $z = k\lambda$ from within this region. Note that 0-stability for the test equation corresponds to having the origin $z = 0$ belong to the absolute stability region.

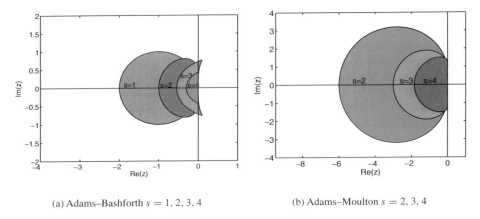

(a) Adams–Bashforth $s = 1, 2, 3, 4$ (b) Adams–Moulton $s = 2, 3, 4$

Figure 2.5. *Absolute stability regions of Adams methods.*

For a Runge–Kutta method in the general form (2.10) applied to the test equation we have

$$Y_i = y_n + z \sum_{j=1}^{s} a_{ij} Y_j, \quad y_{n+1} = y_n + z \sum_{j=1}^{s} b_j Y_j.$$

Eliminating the internal stages $Y_1, \ldots, Y_s$ and substituting in the expression for y_{n+1} using the same notation as in (2.13) we obtain

$$y_{n+1} = R(z) y_n, \quad R(z) = 1 + z \mathbf{b}^T (I - zA)^{-1} \mathbf{1}. \qquad (2.22)$$

For example, you can directly verify that $R(z) = 1 + z$ for forward Euler, $R(z) = (1-z)^{-1}$ for backward Euler, and $R(z) = (1 - z/2)^{-1}(1 + z/2)$ for the implicit trapezoidal method. The absolute stability region is then the connected set, containing the origin, of all complex valued numbers z for which $|R(z)| \leq 1$.

Plotting absolute stability regions is a favorite pastime for some; see, e.g., [85, 14]. Figure 2.5 displays these regions for the first few Adams methods. For instance, the forward Euler method is absolutely stable for a real, negative λ only if $|y_{n+1}| = |y_n + k f_n| = |y_n + k \lambda y_n| \leq |y_n|$, so the condition is $|1 + k\lambda| \leq 1$, which gives

$$k|\lambda| \leq 2.$$

Observe also that the stability regions of the explicit Adams–Bashforth methods shrink rapidly as s increases. High-order methods of this sort typically yield stability restrictions which are deemed too unwieldy in practice.

The first two Adams–Moulton methods contain the entire left half-plane in their absolute stability region. Such methods are called **A-stable**.

Explicit methods of the type we have exposed cannot be A-stable. The stability regions for explicit Runge–Kutta methods of the first few orders are depicted in Figure 2.6. Here, unlike for the Adams families, the absolute stability regions grow with s. However, they grow by not much: if we were to shrink the stability region for the s-stage method by a factor s (to yield a fairer comparison of step size allowed per function evaluation of f), then

2.6. Stiffness

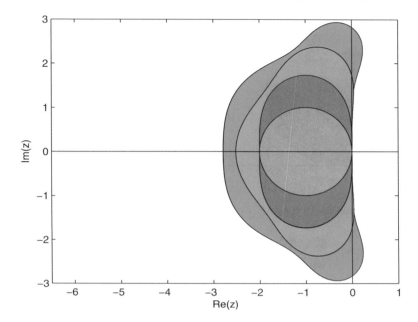

Figure 2.6. *Stability regions for p-stage explicit Runge–Kutta methods of order p, $p = 1, 2, 3, 4$. The inner circle corresponds to forward Euler, $p = 1$. The larger p is, the larger the stability region. Note the "ear lobes" of the fourth order method protruding into the right half-plane.*

none of the methods depicted in Figure 2.6 has a longer segment of the negative real line contained in their stability region than forward Euler.[9]

It is important to note that whereas the regions for the Runge–Kutta methods of order 1 and 2 have no intersection with the imaginary axis, those for the higher order methods do. The observation in Figure 2.6 for the classical four-stage Runge–Kutta method of order 4 regarding the intersection of its absolute stability region with the imaginary axis is a major reason for the popularity of this method in time discretizations of hyperbolic PDEs.

The absolute stability regions of the BDF methods have the "kidney shape" depicted in Figure 2.7. These methods (for $s \leq 6$) all have unbounded absolute stability regions.

2.6 Stiffness

An important concept in numerical ODE theory is stiffness. Loosely speaking, the initial value ODE is **stiff** if the absolute stability requirement for explicit Runge–Kutta methods dictates a much smaller step size than is needed to satisfy approximation requirements

---

[9]A class of s-stage explicit Runge–Kutta methods called **Runge–Kutta–Chebyshev** is particularly good in this regard, containing a stable segment of the negative real line that is proportional in length to s^2; see [98, 85] and further references therein. A quick review of Chebyshev polynomials and points is found in Section 2.12.3. These methods are particularly useful for mildly stiff problems with real eigenvalues, a situation yielded by semi-discretizations of certain parabolic PDEs in several space dimensions!

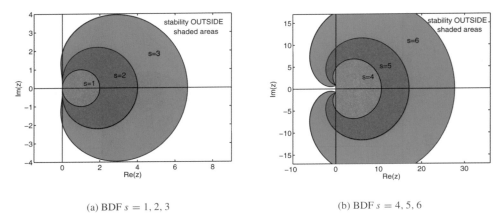

(a) BDF $s = 1, 2, 3$ (b) BDF $s = 4, 5, 6$

Figure 2.7. *BDF absolute stability regions. The stability regions are outside the shaded area for each method.*

alone; see [14, 85]. In this case other methods, which do not have such a restrictive absolute stability requirement, should be considered. Thus, explicit methods are suitable for nonstiff problems, whereas for stiff ones implicit methods such as BDF or collocation at Radau and Gauss points are appropriate.

L-stability and stiff decay

Let us generalize the test equation a bit to include an inhomogeneity,

$$y' = \lambda(y - g(t)), \quad 0 < t < b, \tag{2.23}$$

where $g(t)$ is a bounded, but otherwise arbitrary, function. As $\mathcal{R}e(\lambda) \to -\infty$, the (bounded) exact solution of (2.23) satisfies

$$y(t) \longrightarrow g(t), \quad 0 < t < b,$$

regardless of the initial value y_0. This is a case of extreme stiffness. Correspondingly, we say that a method has *stiff decay* if for $0 < t_n < b$ fixed

$$|y_n - g(t_n)| \to 0 \quad \text{as} \quad k\mathcal{R}e(\lambda) \to -\infty. \tag{2.24}$$

The trapezoidal and midpoint methods (which are A-stable) do not have stiff decay. However, the BDF methods and collocation at Radau points do possess stiff decay! These methods are therefore particularly suitable for very stiff problems. They are also an obvious choice for the direct discretization of differential-algebraic equations (DAEs); see Section 2.8 and [14].

People often call a method which has stiff decay *L-stable*. Indeed, the latter term may sound a bit less awkward, being an adjective, and we occasionally use it too. But note that L-stability was originally defined for implicit, A-stable Runge–Kutta methods, restrictions

2.6. Stiffness

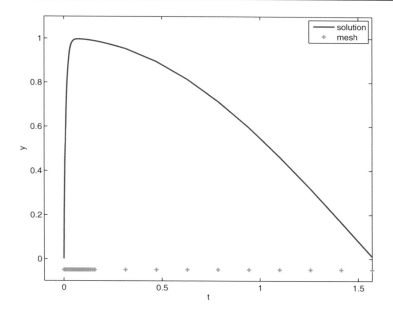

Figure 2.8. *Approximate solution and plausible mesh, Example* 2.6.

that we do not necessarily intend to honor when checking for stiff decay. Specifically, an A-stable Runge–Kutta method is said to be *L-stable* if

$$\lim_{\mathcal{R}e(z) \to -\infty} R(z) = 0. \quad (2.25)$$

So, readers should be careful to avoid confusion.

It can be easily seen that the implicit Runge–Kutta methods that are equivalent to collocation methods based on Radau points are all *L*-stable. For these methods the matrix A in (2.22) is nonsingular and the row vector $\mathbf{b}^T$ coincides with the last row of A, hence $\mathbf{b}^T A^{-1} \mathbf{1} = 1$. For backward Euler clearly $R(-\infty) = 0$, while for the trapezoidal method $R(-\infty) = -1$.

Example 2.6 The scalar problem

$$y' = -100(y - \cos t), \quad t \geq 0, \quad y(0) = 0,$$

has a solution which starts at the given initial value and grows rapidly. But after a short while, say, for $t \geq .03$, $y(t)$ varies much more slowly, satisfying $y(t) \approx \cos t$; see Figure 2.8. For the initial small interval of rapid change (commonly referred to as an initial **layer** or **transient**) we expect to use small step sizes, so that $100k_n \leq 1$, say. This is within the absolute stability region of the forward Euler method. But when $y(t) \approx \cos t$, accuracy considerations alone allow a much larger step size, so we want $100k_n \gg 2$. A reasonable mesh is plotted using markers on the t-axis in Figure 2.8. Obviously, however, the plotted solution in this figure was not found using the forward Euler method with this mesh, because the absolute stability restriction of the forward Euler method is severely violated here.

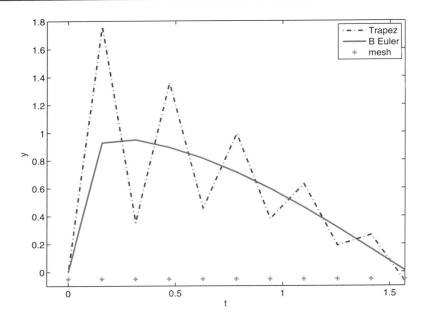

Figure 2.9. *Approximate solution on a coarse uniform mesh for Example* 2.6, *using backward Euler (the smoother curve) and trapezoidal (the oscillating one).*

Indeed, the solution plotted in Figure 2.8 was obtained using the A-stable trapezoidal method (2.8). The mesh was constructed as follows: for $t < .05\pi$ it was obtained by the nonstiff code `ode45` with default tolerances; for larger t a uniform step size $.05\pi$ was used, which is much sparser than what `ode45` comes up with. The trapezoidal method was then run on the combined mesh.

Suppose next that we are not interested in the details of the initial layer! This is a common occurrence in computational fluid dynamics (CFD), for instance. Can we then use, say, a uniform step size k which skips these details but is small enough to reproduce the smooth part of the solution?

This can be done using a method with stiff decay. For instance, employing backward Euler (2.7) with a fixed step $k = .05\pi$ to integrate this problem, the initial layer is poorly approximated, and still the solution is qualitatively recovered where it varies slowly; see Figure 2.9. On the other hand, the trapezoidal method, which does not have stiff decay, produces disagreeable oscillations in these circumstances and cannot be used to skip layers.

Layer skipping means that there is a region in time where the discrete dynamical system produced by the discretization no longer mimics closely the behavior of the continuous dynamical system. Herein lies a great potential for efficient use, as well as a great danger of misuse, of discretization methods such as backward Euler. ■

Returning to the generalized test equation (2.23) we can clearly generalize the observation regarding its solution for any limit $|\lambda| \to \infty$, and we can also consider the ODE subject to *boundary conditions*, as discussed in Section 2.11, instead of initial conditions.

2.7 Solving equations for implicit methods

Correspondingly we can also change the definitions of stiff decay (2.24) and L-stability (2.25) to hold for $|z| \to \infty$. Such a modification does not alter the list of our methods that qualify. For example, the backward Euler method damps the solution also when $z \gg 1$, i.e., when the exact solution grows rapidly; see Figure 2.7. Indeed, it is easy to see that for any Runge–Kutta method, $R(-\infty) = 0$ iff $R(\infty) = 0$. However, the interpretation of this sort of damping as *stability* is problematic, so we have stuck to the definitions (2.24) and (2.25).

2.7 Solving equations for implicit methods

We have seen that implicit methods have superior stability and accuracy properties, compared to explicit ones. However, carrying out a step using an implicit method is also more complicated and potentially much more expensive. We can write an s-step scheme (2.3) at a fixed step n as

$$y_{n+1} = k\beta_0 f(t_{n+1}, y_{n+1}) + \sum_{j=1}^{s} \left[-\alpha_j y_{n+1-j} + k\beta_j f_{n+1-j} \right].$$

If $\beta_0 \neq 0$, i.e., the method is implicit, then the unknown y_{n+1} appears on the right-hand side. If f is nonlinear in y (*not* in t: the dependence on t is unimportant), then a nonlinear algebraic problem must be approximately solved at each time step. Even if f were linear there is a difficulty to reckon with in case that (2.2) stands for a large ODE system. To emphasize that m may be large and to correspond to the notation in Section 2.12.5, let us use boldface notation: the ODE system considered is

$$\mathbf{y}' = \mathbf{f}(t, \mathbf{y})$$

and an implicit multistep scheme reads

$$\mathbf{y}_{n+1} = k\beta_0 \mathbf{f}(t_{n+1}, \mathbf{y}_{n+1}) + \sum_{j=1}^{s} \left[-\alpha_j \mathbf{y}_{n+1-j} + k\beta_j \mathbf{f}_{n+1-j} \right]. \tag{2.26}$$

A simple iterative method for solving (2.26) for $\mathbf{y}_{n+1}$ is to write the right-hand side as $\boldsymbol{\phi}(\mathbf{y}_{n+1})$ and apply **fixed point iteration**

$$\mathbf{y}_{n+1}^{\nu+1} = \boldsymbol{\phi}(\mathbf{y}_{n+1}^{\nu}), \quad \nu = 0, 1, \ldots.$$

Note that ν is an iteration counter, not a power. An excellent initial guess $\mathbf{y}_{n+1}^0$ can usually be obtained from the previous solution values as discussed below. Before that, however, we must ask whether this iteration converges. That will happen (see Section 2.12.5) if there is a constant ρ such that

$$\left\| k\beta_0 \frac{\partial \mathbf{f}}{\partial \mathbf{y}} \right\| = \left\| \frac{\partial \boldsymbol{\phi}}{\partial \mathbf{y}} \right\| \leq \rho < 1.$$

Thus, the simple fixed point iteration can be used only if the problem is nonstiff!

For nonstiff problems we must first explain why one would use implicit methods at all. The answer lies in the unacceptably small absolute stability regions of higher order Adams–Bashforth methods (see Figure 2.5) coupled with the high utility of a **predictor-corrector** approach in obtaining an accurate approximation as well as an error estimate. By contrast, implicit Runge–Kutta methods are rarely used for nonstiff problems. Upon using an $(s+1)$-step Adams–Bashforth method to predict $\mathbf{y}_{n+1}^0$, essentially one fixed point iteration for the Adams–Moulton method (2.26) suffices to obtain an error $ck^{s+1} + O(k^{s+2})$ whose leading term coefficient c, and not only the order $s+1$, are the same as those of the implicit method. Thus, the general fixed point iteration is abandoned and just one iteration is applied. The favorite variant is PECE: Predict $\mathbf{y}_{n+1}^0$ using Adams–Bashforth, Evaluate $\mathbf{f}(t_{n+1}, \mathbf{y}_{n+1}^0)$, Correct using Adams–Moulton as above obtaining $\mathbf{y}_{n+1} = \mathbf{y}_{n+1}^1$, and Evaluate $\mathbf{f}_{n+1} = \mathbf{f}(t_{n+1}, \mathbf{y}_{n+1})$ in preparation for the next step. See [85] for the absolute stability regions of the PECE variant.

The advantages of the predictor-corrector approach are that (i) a higher order method is cheaply obtained for $s > 2$, because only two evaluations of $\mathbf{f}$ are employed per step (think what would be required by an explicit higher order Runge–Kutta method!); and (ii) the pair $\mathbf{y}_{n+1}$ and $\mathbf{y}_{n+1}^0$ can be used to obtain an estimate for the local error so that error control is enabled according to Section 2.4. See Chapter 5 of [14] for fuller details.

For stiff problems the situation is more complex. Writing (2.26) as $\mathbf{g}(\mathbf{y}_{n+1}) = \mathbf{0}$, which is done by simply moving all terms to the left side, we have

$$J = \frac{\partial \mathbf{g}}{\partial \mathbf{y}_{n+1}} = I - k\beta_0 \frac{\partial \mathbf{f}}{\partial \mathbf{y}}, \qquad (2.27)$$

a matrix that in the stiff case is no longer dominated by the identity I. The simple fixed point iteration and its predictor-corrector truncations no longer work, and an Adams–Bashforth method is not suitable for predicting $\mathbf{y}_{n+1}^0$. Instead we set $\mathbf{y}_{n+1}^0 = \mathbf{y}_n$, which is still $O(k)$-accurate, and employ some variant of **Newton's method**.

Newton's method is known to converge very rapidly under certain conditions, provided that the initial guess is close to the solution. Here we do have such an initial guess (indeed, one can reduce the step size k and repeat the step if the method does not converge). The main challenge is to keep the cost of the iteration that now involves the solution of a possibly large linear system of equations manageable: such a system must be solved at each nonlinear iteration, at each time step. One thing people do is to freeze J locally and use its LU-decomposition (see Section 2.12.1) for several time steps. Another idea is to apply just one Newton iteration per time step, obtaining a **semi-implicit** variant which yields a first order method overall (see Exercise 10) and is thus particularly suitable in conjunction with the backward Euler method.

An extension of the latter idea for higher order Runge–Kutta methods is a **Rosenbrock method**; see [85, 98]. For the autonomous ODE system $\mathbf{y}' = \mathbf{f}(\mathbf{y})$, and with $A = \frac{\partial \mathbf{f}}{\partial \mathbf{y}}(\mathbf{y}_n)$, an s-stage Rosenbrock method reads

$$\mathbf{K}_i = k\mathbf{f}(\mathbf{y}_n + \sum_{j=1}^{i-1} \alpha_{ij} \mathbf{K}_j) + kA \sum_{j=1}^{i} \gamma_{ij} \mathbf{K}_j, \qquad 1 \le i \le s,$$

$$\mathbf{y}_{n+1} = \mathbf{y}_n + \sum_{i=1}^{s} b_i \mathbf{K}_i. \qquad (2.28)$$

2.7. Solving Equations for Implicit Methods

If all $\gamma_{ij} = 0$, then this is a usual explicit Runge–Kutta method; see Exercise 5. More interestingly, if γ_{ii} are all nonzero, then s linear systems involving A must be solved per time step; yet if $\gamma_{ii} = \gamma$ are all equal, then the same matrix $J = I - k\gamma A$ can be *LU*-decomposed once, and only a forward-backward substitution is subsequently required for each stage.

Also, depending on the application, one may seek approximations to J by dropping less necessary but varying terms from it. Note that the size of J in (2.27) or (2.28) is $m \times m$; it has the same size and sparsity pattern as that of the Jacobian matrix of the ODE system.

A major practical impediment to using fully implicit Runge–Kutta methods such as those from the Gauss and Radau families is that at each step a large system of equations must be solved. Whereas for implicit multistep methods such as BDF the size of the Jacobian matrix is $m \times m$ as we have seen, here the method's Jacobian matrix has size $sm \times sm$. In addition to (2.28) there are other ways to ease the pain, but not completely; see [14, 85]. Still, the general-purpose code RADAU5 of Hairer and Wanner [85] that is based on Radau collocation is competitive against any known alternative and better than most.

A popular approach is to seek **diagonally implicit Runge–Kutta methods** (DIRK). Here one requires
$$a_{ij} = 0 \ \forall i < j.$$
Then only (at most s) $m \times m$ systems are encountered, as the stages are resolved one at a time; see Exercise 3. These methods are close to the Rosenbrock methods.

Example 2.7 The following ODE system models a chemical reaction system and has been used extensively as a test problem for stiff solvers [154, 85, 14]:
$$\begin{aligned} y_1' &= -\alpha y_1 + \beta y_2 y_3, \\ y_2' &= \alpha y_1 - \beta y_2 y_3 - \gamma y_2^2, \\ y_3' &= \gamma y_2^2. \end{aligned}$$

Here $\alpha = 0.04$, $\beta = 1.e+4$, and $\gamma = 3.e+7$ are slow, fast, and very fast reaction rates. The starting point is $\mathbf{y}(0) = (1, 0, 0)^T$.

It is easy to find the steady state (where $\mathbf{y}' = \mathbf{0}$) for this ODE. First, note that $\sum_{i=1}^{3} y_i' = 0$, hence $\sum_{i=1}^{3} y_i(t) = \sum_{i=1}^{3} y_i(0) = 1$, for all t. Next, the third ODE yields at steady state $y_2 = 0$, then the second yields $y_1 = 0$, so given the invariant sum the steady state is $\mathbf{y}(\infty) = (0, 0, 1)^T$. However, it is known that the steady state is reached very slowly for this problem, so let us investigate this by integrating up to $b = 10^6$.

We use this opportunity to display a simple MATLAB script that employs the backward Euler method with a constant step size.

```
% program to simulate Robertson's equations
% using BDF with a constant step size

% initial data
  y0 = [1,0,0]';

% run parameters
  tf = 1.e+6;              % final time
```

```matlab
  k = input('enter constant time step : ');
  N = tf / k;
  tol = 1.e-4;           % tolerance for Newton iteration
  niter = 10;            % no more than niter Newton itns per
                         step
  tp = 0:k:tf;           % output times
  yp = zeros(3,N+1);     % initialize output solution array
  yp(:,1) = y0;

% Backward Euler

  for n = 1:N
      yo = yp(:,n);      % old y value (=y_n)
      yn = yo;           % initial Newton iterate
      iter = 0;          % Newton iteration counter
      goon = 1;          % iteration control variable
                           (= 'go on')

      while goon    % Newton iteration
          yc = yn;
          [f,J] = robertsonf (tp(n+1),yc);
          A = eye(3) - k*J;
          b = yo - yc + k*f;
          dy = A \ b;
          yn = yc + dy;
          iter = iter + 1;
          if norm(dy) < tol * norm(yn) | iter == niter
              goon = 0;
          end
      end
      yp(:,n+1) = yn;
  end

% plot solution
  semilogx (tp,yp(1,:),'b',tp,yp(2,:),'g--',tp,yp(3,:),'r-.')
  axis([0 1.e+6 -.1 1.1])
  legend('y_1','y_2','y_3','Location','East')
  xlabel('t')
  ylabel('y')

function [f,J] = robertsonf(t,y)
%   function [f,J] = robertsonf(t,y)
%   this function defines the robertson eqns plus Jacobian

  a = 0.04; b = 1.e+4; c = 3.e+7;
```

2.7. Solving Equations for Implicit Methods

```
f = [-a*y(1) + b*y(2)*y(3); ...
     a*y(1) - b*y(2)*y(3) - c*y(2)*y(2); ...
     c*y(2)*y(2)];

if nargout > 1
   J = zeros(3,3);
   J(1,1) = -a;  J(1,2) = b*y(3);  J(1,3) = b*y(2);
   J(2,1) = a;   J(2,2) = -b*y(3) - 2*c*y(2); J(2,3) =
                                    -b*y(2);
   J(3,2) = 2*c*y(2);
end
```

Results using this program with time step $k = 1$ are displayed in Figure 2.10(a). Figure 2.10(c) has been obtained by the MATLAB function `ode23s`. This routine is adaptive and performs much faster than our simple program with unit time step, taking only

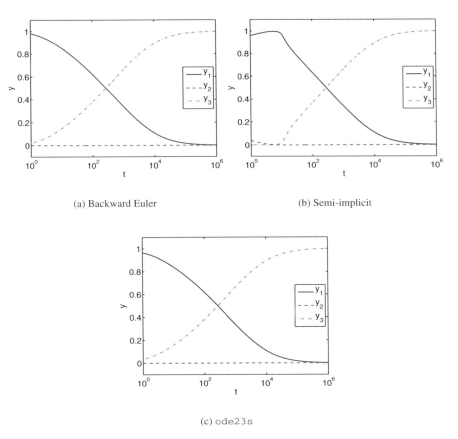

Figure 2.10. *Solving a very stiff problem using Backward Euler with $k = 1$, semi-implicit Backward Euler with $k = 1$, and a MATLAB stiff solver.*

60 steps in total. The results agree qualitatively with backward Euler, although `ode23s` is more accurate. Figure 2.10(b) is obtained upon setting niter = 1 in our program, so only one Newton iteration is taken at each time step. This semi-implicit variant does not perform well here, especially for smaller times, even though 10^6 steps are taken. This would normally indicate that our choice of fixed time step is not much smaller than necessary for using backward Euler, even for qualitative results. However, as it turns out, here a much larger step size can be taken for backward Euler, still achieving qualitatively correct results: it is the semi-implicit version that fails spectacularly!

This example therefore demonstrates, in addition to a simple implementation of an implicit method for a stiff problem, both the advantages of adaptive time stepping using a higher order method and the need for caution when using the semi-implicit variant. ∎

Note: Now we can finally summarize the relative advantages and disadvantages of linear multistep and Runge–Kutta methods.

- Runge–Kutta has the advantage of flexibility, for both stiff and nonstiff problems:
    - no other, additional method is required to find extra initial values;
    - there are fewer difficulties in locating events (which are specific times, such as collisions in a particle system, when nonsmooth changes occur);
    - affecting a step size change is straightforward.
- For nonstiff problems, PECE variants of Adams method pairs can prove significantly less expensive than Runge–Kutta if high accuracy is desired, especially if the evaluations of **f** dominate the cost.
- The linear system that must be solved for stiff problems has minimal size when using BDF methods, making them a relatively cheap and popular choice in practice.

2.8 Differential-algebraic equations

Mathematical models often involve various relationships between dependent variables. It often happens that some such relationships involve derivatives while others express interdependence algebraically. Classical examples involve chemical reactions and electrical circuits, among others; see [14, 85, 34]. Such a combination of differential and algebraic equations (DAEs) can be written as a system,

$$F(t, y, y') = 0, \qquad t > 0, \tag{2.29}$$

2.8. Differential-Algebraic Equations

which obviously generalizes the ODE form (2.2a) in that y' appears implicitly. The more significant change from ODE, though, is when the Jacobian matrix $\frac{\partial F}{\partial y'}$ is *singular*.

Often DAEs arise in **semi-explicit** form, where differential and algebraic relations are partially separated. For instance, consider the motion of a collection of rigid bodies. Newton's law relates accelerations, which are second derivatives of positions, to masses and forces. If in addition there are constraints on the relative motion of these bodies, e.g., that two wheels in a car must maintain a constant separation distance, or that a door must remain connected to a wall through its hinges, then we have a system of the form

$$M(x)x'' = f(x, x') - G^T(x)z, \qquad (2.30a)$$
$$0 = g(x), \qquad (2.30b)$$

where $x = x(t)$ are body positions, $v(t) = x'(t)$ are corresponding velocities, and $z = z(t)$ are Lagrange multiplier functions. Also, M is a symmetric positive definite (SPD) mass matrix, f are forces (having nothing to do with f in (2.2)), g are the algebraic (holonomic) constraints, and $G = \frac{\partial g}{\partial x}$. Assume that $GM^{-1}G^T$ is SPD as well. It is not difficult to see that (2.30) is a special case of (2.29), where $y = (x, v, z)$; see Exercise 12.

An important concept in DAEs is the **index**. It expresses a sense of how far the DAE is from being an ODE in disguise. The higher the index, the further the DAE is from an explicit ODE, so intuitively also more surprises may be encountered upon applying numerical ODE methods to it. For general DAE systems (2.29), the index along a solution $y(t)$ is the minimum number of differentiations of the system required to solve for y' uniquely in terms of y and t (i.e., to define an ODE of the form (2.2) for y).

In case of the multibody system (2.30) we see that upon two differentiations of the constraints we get $Gx'' + L(x, v) = 0$. Multiplying the ODE part by GM^{-1} and substituting we can express z at each t as

$$z = (GM^{-1}G^T)^{-1}\left(GM^{-1}f(x, v) + L(x, v)\right).$$

Differentiating this once more finally gives an explicit expression for y', hence the index equals 3.

There are two general approaches for solving DAE systems numerically. The first is to discretize directly. Note that if we replace the constraint in (2.30b) by $\epsilon Sx' = g(x)$, where $0 < \epsilon \ll 1$ and S is a matrix of signs ensuring model stability, then a very stiff ODE system is obtained. Therefore, seeing the DAE as a limit case with $\epsilon \to 0$, an L-stable method such as BDF or Radau collocation must be employed in a direct discretization approach. This turns out to work well for general index-1 DAEs and certain index-2 DAEs that are common in practice, especially if the system is stiff anyway. The general-purpose codes `DASSL` written by Petzold [34] and `RADAU5` are good both for these types of DAEs and for stiff ODEs.

The other approach is to transform the system, reducing its index prior to discretization. This is called **index reduction**. For instance, given a constrained multibody system (2.30) we can actually carry out two differentiations and obtain a nonstiff ODE by eliminating z. The possibility of applying a nonstiff discretization method to the resulting ODE is certainly a powerful incentive to at least investigate this route. However, note that the constraints may have to be stabilized, the approach is less automatic, and some sparsity may be lost in the transformation as well; see [14, 34, 85].

2.9 Symmetric and one-sided methods

The trapezoidal method (2.8) and the midpoint method (2.12) are examples of *symmetric* methods: a change of variable $\tau = -t$ on $[t_n, t_{n+1}]$, i.e., integrating from right to left, leaves them unchanged. Like backward Euler they are implicit and A-stable, and their cost of evaluation per step is similar as well. But the symmetry yields the higher order of accuracy 2, whereas Euler methods are only first order accurate. Therefore, symmetric time integration methods were preferred for parabolic PDEs in the early days of scientific computing [54].

To define this notion more precisely and generally, consider for notational simplicity the ODE $y' = f(y)$. A discretization method given by $y_{n+1} = y_n + k\psi(y_n, y_{n+1}, k)$ is **symmetric** if
$$\psi(u, v, k) = \psi(v, u, -k).$$

For a symmetric method, then, by letting $z_{n+1} \leftarrow y_n$, $z_n \leftarrow y_{n+1}$, and $k \leftarrow -k$, we get the same method for z_n as for y_n.

However, symmetric methods cannot have stiff decay: the information damping property required for stiff decay yields increases when integrating in the opposite direction. For instance, backward Euler becomes forward Euler, and no infinite absolute stability region is possible in both directions. Methods such as backward Euler are called **one-sided**. Higher order symmetric Runge–Kutta methods are given by collocation at Gauss points or at Lobatto points, whereas collocation at Radau points provides higher order one-sided methods. BDF methods are one-sided as well.

The realization of the power of one-sided methods in both space and time is commonly considered as the most important advance in CFD in modern times. However, they have their serious limitations, too, and we will see many situations where the symmetric methods offer an added edge; see especially Chapters 6 and 7.

2.10 Highly oscillatory problems

Consider yet again the test equation (2.20), this time with λ purely imaginary. The exact solution $y(t) = e^{\lambda t} y(0)$ now oscillates with $|y(t)| = |y(0)|$ $\forall t$. In particular, there is no decay in the exact solution. Numerical methods which cause significant decay in $|y_n|$ may therefore be missing important information.

In general, the local error of a method of the type we have seen of order p is $\approx c|y^{(p+1)}|k^{p+1}$, where $y^{(p+1)} = \frac{d^{p+1}y}{dt^{p+1}}$ and c is a constant. For the test equation, $y^{(p+1)}(t) = \lambda^{p+1} y(t)$. Thus, we cannot expect decent pointwise accuracy anywhere (not only in an initial layer as in Section 2.6) unless $k|\lambda| < 1$.

Example 2.8 The harmonic oscillator
$$u'' + \omega^2 u = 0, \qquad 0 < t < b,$$
$$u(0) = 1, \qquad u'(0) = 0,$$

where ω is a real scalar has the solution
$$u(t) = \cos \omega t.$$

2.10. Highly Oscillatory Problems

If the frequency ω is high, $\omega \gg 1$, then the derivatives grow larger and larger, because

$$\|u^{(p)}\| = \omega^p.$$

The local error of a discretization method of order p is

$$O(k^{p+1}\omega^{p+1}).$$

This means that to recover the highly oscillatory solution $u(t)$ accurately we must restrict

$$k < 1/\omega$$

regardless of the order of the method. In fact, for $k > 1/\omega$ increasing the order of the method as such is useless. ∎

Even if high-frequency solutions are not of interest in themselves, the damping produced by a backward Euler discretization may be too much for some problems. Examples are provided by symplectic maps and by hyperbolic PDEs with smooth solutions, studied in future chapters. See especially Chapters 6 and 7.

Example 2.9 (Cloth simulation) This example is longer and more intensive than the previous ones and is better regarded as a case study. We first give some background to the problem and then concentrate on the numerical issues. However, as we will see, the tasks of deriving a mathematical model and of solving it numerically end up being somewhat intertwined. Our exposition is based on [31, 6], whose work has its roots in Baraff and Witkin [20] and Choi and Ko [48].

The problem of animating cloth (see Figure 2.11) is of direct relevance for virtual character animation, garment design, and textile engineering. But these different purposes give rise to different treatments. Textile engineers require a relatively accurate, *quantitative* analysis and therefore must somehow solve complex sets of PDEs describing deformation of elasticity continua. The spatial discretization usually employs *finite elements*, but we do not dwell on this further here. The essential problem with applying such modeling techniques to cloth animation in computer graphics is their computational expense, which is considered prohibitive for interactive, real-time use. For virtual character animation the quest is then for cheaper methods which still yield believable cloth motion, as far as our eye and our perception are concerned.

With the latter goal in mind one is typically led to a *particle* method (also referred to as a *mass-spring* model in the present context): the cloth is considered as a mesh of particles (see Figure 2.12) which is more amenable to handling strange geometries and is intuitively easier to render as well, and equations of motion are derived and solved for this system. We proceed to discuss the latter approach, noting that there also exist various hybrids of the two.

Thus, consider a uniform mesh in an underlying plane (u, v). For instance, this mesh can consist of square or triangular cells where each square cell is bisected diagonally: all this mesh does is maintain neighborhood relationships between the particles. Each node (u_i, v_i), $1 \leq i \leq J$, has an associated particle $\mathbf{x}_i$ consisting of three cartesian components. Denote also $\mathbf{e}_{ij} = \mathbf{x}_i - \mathbf{x}_j$. In the absence of constraints our unknown vector of particle positions $\mathbf{x} = \mathbf{x}(t) \in \mathbb{R}^{3J}$ is then $\mathbf{x} = (\mathbf{x}_1^T, \mathbf{x}_2^T, \ldots, \mathbf{x}_J^T)^T$.

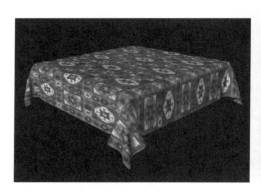

(a) Spread over a table (b) Draped over a sphere

Figure 2.11. *A square piece of fabric in two positions.*

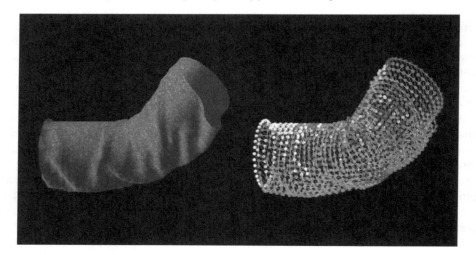

Figure 2.12. *A particle mesh representation of a three-dimensional body surface.*

Each particle has an associated mass $m_i > 0$.[10] The particle system then obeys Newton's second law of motion (which states that force equals mass times acceleration),

$$\mathbf{x}' = \mathbf{v}, \tag{2.31a}$$

$$M\mathbf{v}' = \mathbf{f}(\mathbf{x}, \mathbf{v}). \tag{2.31b}$$

---

[10]This mass can be calculated as follows [20]: triangulate the mesh and assign to each triangle a mass which is the product of the cloth's density and the triangle's fixed area in the (u, v) system. Then sum up the masses of all triangles having $\mathbf{x}_i$ as a node and let m_i be the third of that total.

2.10. Highly Oscillatory Problems

Here **v** are the particle velocities (each $\mathbf{v}_i$ having three components, of course) and M is the diagonal mass matrix

$$M = \text{diag}(m_1, m_1, m_1, m_2, \ldots, m_J, m_J, m_J).$$

This then is our ODE system, but its specification is not complete until the construction of the forces **f** is discussed, which we do next.

The forces acting on the system are of two types, $\mathbf{f} = \mathbf{f}_I + \mathbf{f}_{II}$. External and damping forces, $\mathbf{f}_I$, which may explicitly depend on both positions **x** and velocities **v**, include damping, gravity, air drag, contact, and constraint forces. The remaining forces, $\mathbf{f}_{II}$, on which we concentrate next, can all be written as the negative gradient of an internal energy function in the position variables **x** only. This is where the modeling details of the mass-spring system come in. There are three such forces—*stretching, shearing* and *bending*—and we use the connectivity diagram depicted in Figure 2.13 to model these using springs, following [48].

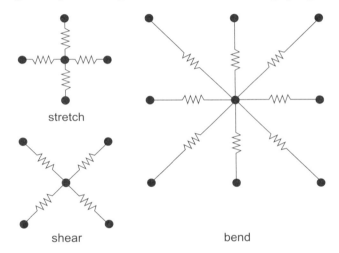

Figure 2.13. *Connectivity structure through the underlying (u, v)-mesh for stretch, shear, and bend springs.*

Stretch and shear springs are formulated identically, though they apply to different neighbors of a particle i and possess rather different material constants. They are linear springs that act only in extension and not in compression. Consider two particles i and j connected by one of these springs. The elastic force acting on particle i due to the deformation between it and particle j is

$$\mathbf{f}_i^j = \begin{cases} \kappa(|\mathbf{e}_{ij}| - L)\frac{\mathbf{e}_{ij}}{|\mathbf{e}_{ij}|}, & |\mathbf{e}_{ij}| \geq L, \\ \mathbf{0}, & |\mathbf{e}_{ij}| < L, \end{cases} \quad (2.32a)$$

where κ is the spring stiffness ($\kappa = \kappa_{st}$ for stretching and $\kappa = \kappa_{sh}$ for shearing) and L is the spring's rest length.

Another feature of the model [48] is its nonlinear handling of bending resistance. The equilibrium shape of buckled cloth is approximated as a circular arc. The curvature is thus determined as a function of the axial spring strain $\frac{|\mathbf{e}_{ij}|}{L}$, and the corresponding restorative force is expressed as

$$\mathbf{f}_i^j = \begin{cases} \mathbf{0}, & |\mathbf{e}_{ij}| \geq L, \\ f_{bend}(\frac{|\mathbf{e}_{ij}|}{L})\kappa_{bend}\frac{\mathbf{e}_{ij}}{|\mathbf{e}_{ij}|}, & |\mathbf{e}_{ij}| < L, \end{cases} \qquad (2.32b)$$

where f_{bend} is approximated as a fifth order polynomial function of the axial strain.

Let us not dwell here further on the modeling of damping or the very interesting collision and constraint forces and concentrate instead on the purpose of this case study in the present chapter's context. For this we note first that if there were no forces $\mathbf{f}_I$, the system (2.31) would have been of the highly oscillatory type. Indeed, since the stretching force is rather stiff as compared to bending and shearing forces (try applying such forces to your shirt or towel!), which translates to a much larger spring constant κ_{st}, we have a generalization of the simple Example 2.8. In the early days of cloth simulation explicit integration methods (that is, forward Euler; and also the leapfrog method considered in Chapter 6) were almost exclusively used.[11] But the resulting step size k is prohibitively small because essentially k must be proportional to $1/\sqrt{\kappa_{st}}$.

A major breakthrough was achieved in [20], where a variant of the backward Euler method was proposed for the time discretization of (2.31). Their claim was that this eliminates stability restrictions and therefore allows using large time steps. To be precise, the proposed method was a *semi-implicit* variant, where only one Newton iteration per time step is used (see Section 2.7), which amounts to solving one large, sparse linear system at each time step. Since the convergence of Newton's method is guaranteed only locally, this means that surprises, including instabilities, can happen with very large time steps, as has indeed been subsequently reported by disenchanted experimentalists. Nonetheless, in practice much larger steps than when using an explicit time discretization are often possible.

However, the reader who has really understood the material in this chapter must wonder at this point how it is possible to use the highly damping semi-implicit backward Euler discretization for a problem that contains no damping! The answer is that the problem of interest, to recall, does contain damping, included in the forces $\mathbf{f}_I$. Remember that (2.31) is merely a model for computation, not one describing a precise physical reality. Thus, the parameter controlling the damping in $\mathbf{f}_I$ can be reduced in view of the numerical damping introduced by the time discretization. For this application it is acceptable to numerically damp the in-surface, stretching oscillations, which also happen to be the problematic highly oscillatory ones. The out-of-surface, bending modes are moderate and well approximated. The result is a more efficient computation for a somewhat different mathematical model than we have advertised, but still a computation that produces pleasant looking results at a good price.

An implicit time step, even a semi-implicit one, results in the necessity to solve a system of linear equations. Several interesting issues arise in this context [20, 6, 31]. Some knowledge of the material covered in Sections 9.3 and 9.4 is necessary to pursue this further, however, so we postpone additional discussion of this case study until Example 9.11. ■

---

[11]Jumping the gun on Chapter 6 we note that (2.31) with $\mathbf{f} = \mathbf{f}_{II}$ is a *Hamiltonian* system and that the leapfrog scheme is *symplectic*. So is the explicit method obtained by applying forward Euler to (2.31b) followed by backward Euler to (2.31a). Such methods do not introduce damping.

2.11 Boundary value ODEs

Like our DAE description, the present section gives more of a synopsis than an in-depth treatment of boundary value ODE (BVODE) problems. See [14] for a serious treatment of the subject and Ascher, Mattheij, and Russell [10] for much more.

Considering an initial-boundary value problem such as (1.1)–(1.3) for a parabolic PDE system in one space variable, the usual numerical approach is to discretize in space first, obtaining an initial value ODE problem in time (IVP). However, if we discretize in time first, using, say, one of the methods that have been discussed in this chapter, then a BVODE system in the spatial variable results. This approach is called the *transversal method of lines* (transversal MOL). Here the solution is not specified in its entirety at any location of the independent variable x. Rather, a combination of the solution unknowns u and u_x is specified at $x = 0$, and possibly another combination of these variables is specified at $x = b$.

The general formulation of a BVODE in first order form can be written as

$$y' = f(t, y), \qquad (2.33\text{a})$$
$$0 = g(y(0), y(b)). \qquad (2.33\text{b})$$

Usually the BC are linear

$$g(y(0), y(b)) = B_0 y(0) + B_b y(b) - \beta, \qquad (2.33\text{c})$$

where B_0 and B_b are $m \times m$ matrices and β is an inhomogeneity vector. Note that the initial value problem discussed hitherto is a special case of (2.33) with

$$g(y(0), y(b)) = y(0) - \beta,$$

i.e., $B_0 = I$, $B_b = 0$.

The difference between the general boundary value problem (2.33) and the initial value ODE problem is fundamental, however. For the initial value problem we know the entire solution at one point $t = 0$, so we can proceed to integrate the ODE *locally*, marching along in t with the entire solution information available for making adaptive decisions. For the BVODE, on the other hand, we cannot proceed locally because the solution in its entirety is not known anywhere. So we must integrate the ODE simultaneously, or *globally*, on the entire interval $[0, b]$, and numerically on the entire mesh approximating it. This can be much harder to carry out, and indeed the transversal MOL approach is rarely used in practice.

But there are many other applications where BVODE systems (2.33) naturally arise. Some problems require a solution that is periodic in time and this involves specifying the solution at more than one point. Another common situation is where a phenomenon is described by means of an IVP, but the challenge is to find or identify some property such as a diffusion coefficient or control variable from measurements of the solution at various times. Such an **inverse problem** obviously has a global rather than a local nature.

A simple approach for solving BVODEs numerically is the **shooting method**. Denoting by $y(t; c)$ the solution of the ODE (2.33a) satisfying $y(0) = c$, we try to find c for which the boundary conditions are satisfied. This gives a possibly nonlinear system of equations

$$g(c, y(b; c)) = 0. \qquad (2.34)$$

The algebraic equations (2.34) are solved for c and the problem reduces to an initial value one. This approach therefore involves solving several initial value problems numerically for one BVODE system.

If we perturb the initial values of an ODE system, then the resulting trajectories may differ more and more as the interval of integration grows longer. It is not surprising that in such situations the solution of (2.34) as well as the subsequent initial value integration for the chosen initial value vector can become highly sensitive and the shooting method may run into trouble. The **multiple shooting** method seeks to limit the potential damage by controlling the size of the subintervals over which initial value problems are integrated. The given interval is thus divided into subintervals by a mesh of **shooting points**,

$$0 = t_0 < t_1 < \cdots < t_N = b,$$

IVP integrations are performed on each subinterval $[t_n, t_{n+1}]$, and the resulting solution segments are patched together to form a continuous approximate solution over the entire interval $[0, b]$ that satisfies the boundary conditions (2.33b).

Of course the IVP integration on a subinterval $[t_n, t_{n+1}]$ also employs a discretization; call its points $t_n = t_{n_0} < t_{n_1} < \cdots t_{n_{J_n}} = t_{n+1}$. There are therefore two meshes here, really. The coarse one comprises the shooting points and the fine one includes all the points where y is calculated along the way. In the extreme we use only one integration step between each two consecutive shooting points, whence the two meshes coincide and the method becomes simply a **one-step finite difference** method.

If the given problem is linear and as usual we denote the approximate solution at t_n by y_n, then the resulting multiple shooting or finite difference system can be written as

$$R_n y_{n+1} + S_n y_n = r_n, \quad n = 0, \ldots, N-1, \tag{2.35a}$$

$$B_0 y_0 + B_b y_N = \beta. \tag{2.35b}$$

This can further be written as one linear system for the global solution $\mathbf{y} = (y_0^T, y_1^T, \ldots, y_N^T)^T$. The resulting matrix has a specific sparsity structure that allows construction of fast direct solvers. This is true especially if the boundary conditions are **separated**, which means that for each j either the jth row of B_0 or the jth row of B_b vanishes. This allows organizing the matrix as banded; see Section 2.12.1.

Finally, the multiple shooting setup (2.35) naturally lends itself to a **parallel** implementation, both in forming the blocks S_n and R_n and in solving (2.35a) for y_{n+1} if this is deemed useful. This procedure is in marked contrast to the usual IVP algorithms that march in time. A natural direction to seek parallel algorithms for IVP integration is therefore to explore BVP methods such as multiple shooting. Of course this makes sense especially if the ODE problem to be solved is very large, such as it would be if arising as a semi-discretization of an evolutionary PDE in several space variables. In such a case iterative methods for solving (2.35) are typically sought.

2.12 Reviews

The present section is a continuation of Section 1.3, and we refer to Section 1.3 for an explanation of what these reviews are all about. Here the topics concern basic numerical

2.12.1 Gaussian elimination and matrix decompositions

Consider a linear system of m equations

$$A\mathbf{x} = \mathbf{b},$$

where A is real, square, and nonsingular, $\mathbf{b}$ is given, and $\mathbf{x}$ is a solution vector to be found. The solution is given by

$$\mathbf{x} = A^{-1}\mathbf{b}.$$

However, it is usually bad practice to attempt to form A^{-1}.

The well-known algorithm of *Gaussian elimination* (without pivoting) is equivalent to forming an **LU-decomposition** of A, followed by forward and backward substitutions.

The decomposition stage is independent of the right-hand side $\mathbf{b}$. It can be carried out without knowing $\mathbf{b}$ and subsequently can be used for more than one right-hand side. We deompose A as

$$A = LU,$$

where L is a unit lower triangular matrix (i.e., $l_{ij} = 0$, $i < j$, and $l_{ii} = 1$) and U is upper triangular (i.e., $u_{ij} = 0$, $i > j$). The LU-decomposition requires $\frac{1}{3}m^3 + O(m^2)$ flops (i.e., elementary floating point operations).

Given a data vector $\mathbf{b}$ we can now find $\mathbf{x}$ by writing

$$L(U\mathbf{x}) = A\mathbf{x} = \mathbf{b}.$$

Solving $L\mathbf{z} = \mathbf{b}$ for $\mathbf{z}$ involves **forward substitution** and costs $O(m^2)$ flops. Subsequently solving $U\mathbf{x} = \mathbf{z}$ completes the solution process using a **back substitution** and another $O(m^2)$ flops. The solution process is therefore much cheaper, when m is large, than the cost of the decomposition, at least in terms of sequential operation count and ignoring questions of fast memory access. Assuming that the LU-decomposition exists and is stable, the algorithm for *Gaussian elimination without pivoting* proceeds as described in the framed box below.

Not every nonsingular matrix has an LU-decomposition, and even if there exists such a decomposition the numerical process may become unstable. Thus, partial pivoting must be applied (unless the matrix has some special properties, for instance, if it is *symmetric positive definite*). A row-partial pivoting involves permuting rows of A to enhance stability and results in the decomposition

$$A = PLU,$$

where P is a permutation matrix (i.e., the columns of P are the m unit vectors, in some permuted order). We will refer to an LU-decomposition, assuming that partial pivoting has been applied as necessary.

If A is large and **sparse** (i.e., most of its elements are zero), then the LU-decomposition approach may or may not remain suitable. This depends on the sparsity pattern. For instance, if all the nonzero elements of A are contained in a narrow band, i.e., in a few diagonals

> **Algorithm:** LU-**decomposition and Gaussian elimination.**
>
> 1. *LU-decomposition*
>   For $j = 1 : m - 1$
>   (a) for $i = j + 1 : m$
>   i. $l_{ij} = a_{ij}/a_{jj}$
>   ii. $A[i, j+1:m] = A[i, j+1:m] - l_{ij} A[j, j+1:m]$
>
>   Comment: At this stage U is stored in the upper part of A. We could also store L in the lower part of A.
>
>   Next, for each given right-hand side **b**:
>
> 2. *Forward substitution*
>   For $i = 2 : m$
>   (a) $b_i = b_i - \sum_{j=1}^{i-1} l_{ij} b_j$.
>
> 3. *Back substitution*
>   For $i = m : -1 : 1$
>   (a) $x_i = [b_i - \sum_{j=i+1}^{m} a_{ij} x_j]/a_{i,i}$.

neighboring the main diagonal (in which case A is called **banded**), then the Gaussian elimination algorithm can be easily adjusted to avoid doing any work outside the band. For finite differences with one spatial variable, this typically leads to a reduction in the algorithm's complexity from cubic to linear in the matrix dimension. But inside the band the sparsity is usually lost, and other, iterative algorithms become more attractive. The latter is typically the case for discretizations of PDEs in more than one spatial variable, and it is further considered in Sections 9.2 and 9.4.

A special case of banded matrix is a **tridiagonal matrix**, corresponding to (3.3) with $l = r = 1$, and arising in compact difference schemes. If the matrix is also positive definite, then no pivoting is needed, and the above simplifies into the framed algorithm below.

This algorithm costs about $3m$ flops, which is comparable in its order to applying a simple explicit scheme at m mesh points, at least so long as we may pretend that fast memory access and parallelism are not an issue.

2.12.2 Polynomial interpolation and divided differences

Polynomial interpolation is a central building block for many computational tasks rooted in calculus, such as integration, differentiation, and solution of ODEs and PDEs.

2.12. Reviews

Algorithm: Tridiagonal systems.

1. *LU-decomposition*

   For $i = 1 : m - 1$

   (a) $l_{i+1,i} = a_{i+1,i}/a_{ii}$

   (b) $a_{i+1,i+1} = a_{i+1,i+1} - l_{i+1,i}a_{i,i+1}$

   Comment: At this stage U is stored in the superdiagonal of A. We could also store L in the subdiagonal of A.

   Next, for each given right-hand side **b**:

2. *Forward substitution*

   For $i = 2 : m$

   (a) $b_i = b_i - l_{i,i-1}b_{i-1}$.

3. *Back substitution*

   For $i = m : -1 : 1$

   (a) $x_i = [b_i - a_{i,i+1}x_{i+1}]/a_{i,i}$.

Suppose $g(x)$ is a function to be interpolated at the $q+1$ distinct points $x_0, x_1, x_2, \ldots, x_q$ by the unique polynomial $\phi(x)$ of degree $\leq q$ which satisfies the relations

$$\phi(x_l) = g(x_l), \quad l = 0, 1, 2, \ldots, q.$$

The polynomial can be written explicitly in Lagrangian form,

$$\phi(x) = \sum_{j=0}^{q} g(x_j) L_j(x), \qquad (2.36a)$$

where L_j are the **Lagrange polynomials**,

$$L_j(x) = \Pi_{i=0, i \neq j}^{q} \frac{(x - x_i)}{(x_j - x_i)}. \qquad (2.36b)$$

These are polynomials of degree q satisfying

$$L_j(x_i) = \begin{cases} 0, & i \neq j, \\ 1, & i = j. \end{cases} \qquad (2.36c)$$

The Lagrange form is useful because it is written in terms of the interpolated values themselves. This is referred to as a **nodal representation**, borrowing from finite element terminology.

For other purposes, though, it may be more convenient to write *the same polynomial* $\phi(x)$ in **Newton's form**

$$\phi(x) = g[x_0] + g[x_0, x_1](x - x_1) + \cdots$$
$$+ g[x_0, x_1, \ldots, x_q](x - x_0)(x - x_1) \cdots (x - x_{q-1}),$$

where the **divided differences** are coefficients defined recursively by

$$g[x_l] = g(x_l),$$
$$g[x_l, \ldots, x_{l+i}] = \frac{g[x_{l+1}, \ldots, x_{l+i}] - g[x_l, \ldots, x_{l+i-1}]}{x_{l+i} - x_l}. \quad (2.37)$$

For instance, quadratic interpolation at the three points x_0, x_1, and x_2 is obtained as follows. We construct the triangular divided difference table

$$g[x_0] = g(x_0), \quad g[x_1] = g(x_1), \quad g[x_2] = g(x_2),$$
$$g[x_0, x_1] = \frac{g(x_1) - g(x_0)}{x_1 - x_0}, \quad g[x_1, x_2] = \frac{g(x_2) - g(x_1)}{x_2 - x_1},$$
$$g[x_0, x_1, x_2] = \frac{g[x_1, x_2] - g[x_0, x_1]}{x_2 - x_0}.$$

Then, for any x we may evaluate the interpolant

$$g(x) = g[x_0] + (x - x_0)g[x_0, x_1] + (x - x_0)(x - x_1)g[x_0, x_1, x_2].$$

Returning to the general case of polynomial interpolation, the interpolation error at any point x is

$$g(x) - \phi(x) = g[x_0, \ldots, x_q, x]\Pi_{i=0}^{q}(x - x_i). \quad (2.38)$$

If the points x_i and x are all in an interval of size $O(h)$ and g has $q + 1$ bounded derivatives, then the interpolation error is $O(h^{q+1})$. If h is small, then

$$g[x_0, \ldots, x_q, x] \approx \frac{g^{(q+1)}(\xi)}{(q + 1)!},$$

where the $(q + 1)$st derivative $g^{(q+1)}$ is evaluated at some nearby point ξ.

The Newton interpolation form is recursive and has two additional important properties:

- The interpolation points are added one at a time. Thus, the interpolant over the $q + 1$ points is an addition of one term to the interpolant over the first q points.

- The points x_l do not have to be in increasing order.

In general, the interpolating function can be written as a linear combination of linearly independent **basis functions** $\phi_0(x), \phi_1(x), \ldots, \phi_q(x)$,

$$\phi(x) = \sum_{j=0}^{q} \alpha_j \phi_j(x).$$

2.12. Reviews

For the Lagrange form (2.36) we have $\phi_j(x) = L_j(x)$ and $\alpha_j = g(x_j)$. For the Newton form $\phi_j(x) = \pi_{i=0}^{j-1}(x - x_i)$, which is a polynomial of degree j, and $\alpha_j = g[x_0, \ldots, x_j]$. Note that the expression (2.38) for the interpolation error is independent of the choice of basis functions, which is just a choice of representation for the same interpolating polynomial.

Any form of polynomial interpolation should be considered only locally: if the points $x_0, \ldots, x_q$ are spread over an interval $[a, b]$ that is not very narrow, then the error (2.38), which is not just an upper bound, is in general uncontrollable and not small. In such a case we can divide the interval into small subintervals by a mesh much like those used for our numerical approximations of differential equations,

$$a = \xi_0 < \xi_1 < \cdots < \xi_J = b,$$

apply polynomial interpolation on each subinterval $[\xi_{j-1}, \xi_j]$, and patch these polynomial pieces together so that the obtained **piecewise polynomial** function $\phi(x)$ has a certain global smoothness. A general rule of thumb is to ensure that $\phi \in C^{m-1}[a, b]$ if the mth derivative of ϕ is to be constructed, say, in the course of approximately solving a differential equation of order m. However, this rule is often broken and lower continuity is sought at breakpoints, e.g., in the contexts of finite elements (Section 3.1.4) and discontinuous collocation [13, 81].

2.12.3 Orthogonal and trigonometric polynomials

Families of orthogonal polynomials are rarely used directly to construct numerial approximants in the sense of Section 2.12.2, but they underlie many important techniques in various applications.

Given a smooth, nonnegative weight function $w(x)$ on $[-1, 1]$, a family of polynomials $\phi_0(x), \phi_1(x), \ldots, \phi_j(x), \ldots$ spanning $\mathcal{L}_2[-1, 1]$ such that $\phi_j(x)$ is a polynomial of degree j is *orthogonal* with respect to the weight function $w(x)$ if

$$\int_{-1}^{1} w(x)\phi_i(x)\phi_j(x)dx = 0 \quad \forall i \neq j.$$

Any polynomial $p(x)$ of degree less than q can be written as a linear combination of the first q orthogonal polynomials,

$$p(x) = \sum_{j=0}^{q-1} \alpha_j \phi_j(x),$$

hence the basis function $\phi_q(x)$ is orthogonal to any polynomial of degree $< q$,

$$\int_{-1}^{1} w(x)p(x)\phi_q(x)dx = 0.$$

Here are two favorite families of orthogonal polynomials.

- For $w \equiv 1$ we obtain the **Legendre polynomials** defined by a three-term recurrence relation

$$\phi_0(x) = 1,$$
$$\phi_1(x) = x,$$
$$\phi_{j+1}(x) = \frac{2j+1}{j+1} x \phi_j(x) - \frac{j}{j+1} \phi_{j-1}(x), \quad j \geq 1.$$

The jth Legendre polynomial has j simple (i.e., distinct) roots $\zeta_i = \zeta_i^{(j)}$, $i = 1, \ldots, j$, that lie inside the interval $(-1, 1)$. They are called **Gaussian points**. Of course we can write a polynomial in terms of its zeros, $\phi_q(x) = c_q \pi_{i=1}^{q}(x - \zeta_i)$, for some nonzero coefficient c_q. Thus, the orthogonality statement above says that for all polynomials of degree $< q$

$$\int_{-1}^{1} p(x)\, \pi_{i=1}^{q}(x - \zeta_i)\, dx = 0$$

with ζ_i the corresponding q Gaussian points. This sets the scene for constructing basic high order methods for numerical integration in Section 2.12.4.

Of course, not all problems are given on the interval $[-1, 1]$. For a general interval $[a, b]$ we can apply a linear transformation to $x \in [-1, 1]$ which *scales* and *translates* it

$$t = \frac{1}{2}[(b-a)x + (a+b)], \text{ or}$$
$$x = \frac{2t - a - b}{b - a}.$$

For example, if $[a, b] = [0, 1]$, then $t = \frac{1}{2}(x+1)$. For $\phi_1(x) = x$ the Gaussian point on $[-1, 1]$ is $\zeta_1 = 0$, and on $[0, 1]$ it is $\zeta_1 = 1/2$, remaining the interval's midpoint. The quadratic Legendre polynomial is $\phi_2(x) = \frac{3}{2}x^2 - \frac{1}{2} = \frac{1}{2}(3x^2 - 1)$, hence the Gaussian points are $\pm\frac{1}{\sqrt{3}}$, and on $[0, 1]$ they are $\zeta_{1,2} = \frac{1}{2}(1 \pm \frac{1}{\sqrt{3}})$.

- For $w(x) = \frac{1}{\sqrt{1-x^2}}$ we have the **Chebyshev polynomials**

$$\phi_0(x) = 1,$$
$$\phi_1(x) = x,$$
$$\phi_{j+1}(x) = 2x\phi_j(x) - \phi_{j-1}(x), \quad j \geq 1.$$

As for the Legendre polynomial, what really matters to us are the zeros of the Chebyshev polynomial, here called **Chebyshev points** and given for the jth polynomial by

$$\zeta_i = \cos\left(\frac{2i-1}{2j}\pi\right), \quad i = 1, \ldots, j.$$

Like the Gaussian points the Chebyshev points are simple and lie in $(-1, 1)$.

2.12. Reviews

The zeros for the $(q+1)$st Chebyshev polynomial have the *min-max* property that

$$(\zeta_1, \zeta_2, \ldots, \zeta_{q+1}) = \operatorname{argmin} \min_{x_0,\ldots,x_q} \max_{-1 \leq x \leq 1} |\pi_{i=0}^{q}(x - x_i)|.$$

Returning to polynomial interpolation and its error expression (2.38) we see that it has a factor that depends on the approximated function g and another factor that depends only on the interpolation points. If we are free to choose these interpolation points, then setting them to be the Chebyshev points minimizes the interpolation error if the g-dependent factor is constant, i.e., if g is a polynomial of degree $q + 1$. There are classical examples that show more generally that polynomial interpolation at Chebyshev points can produce far superior results to interpolating at equidistant points. But the situation has to be somehow special for this to provide a meaningful difference and justify a departure from a simpler uniform mesh treatment.

Of course interpolation and approximation need not be done by (piecewise-) polynomials alone. The class of **trigonometric polynomials** considered in Section 1.3.6 also forms an orthogonal family. A trigonometric polynomial interpolation using $q + 1$ basis functions is particularly effective for *periodic* functions.

2.12.4 Basic quadrature rules

Given the task of approximating an integral

$$\int_a^b f(t)dt$$

for some function $f(t)$ on an interval $[a, b]$, basic quadrature rules are derived by replacing $f(t)$ with an interpolating polynomial $\phi(t)$ and integrating the latter exactly. If there are s distinct canonical interpolation points $0 \leq c_1 < \cdots < c_s \leq 1$, then we can write the interpolating polynomial of degree $< s$ in Lagrange form,

$$\phi(t) = \sum_{j=1}^{s} f(x_j) L_j(t),$$

where L_j are the Lagrange polynomials (2.36b) with $x_j = a + (b-a)c_j$. Note that the notation has slightly shifted with respect to (2.36), with $j \to j+1$ and $q = s - 1$. Then

$$\int_a^b f(t)dt \approx \sum_{j=1}^{s} w_j f(x_j),$$

where the weights w_j are given by

$$w_j = \int_a^b L_j(t)dt.$$

For instance, with $s = 1$, $c_1 = 0$, we have f approximated by its value at a, so the integral is approximated by $(b-a)f(a)$. Likewise, the combination $s = 1$, $c_1 = 1/2$, yields the **midpoint rule** $(b-a) f\left(\frac{a+b}{2}\right)$, and $s = 2$, $c_1 = 0$, $c_2 = 1$, yields the **trapezoidal rule** $(b-a) \frac{f(a)+f(b)}{2}$. The more complex **Simpson's rule** has $s = 3$, $c_1 = 0$, $c_2 = 1/2$, $c_3 = 1$, yielding $\frac{b-a}{6} (f(a) + 4f((a+b)/2) + f(b))$.

The **order** of the quadrature rule is p if the rule is exact for all polynomials of degree $< p$, i.e., if for any polynomial f of degree $< p$,

$$\int_a^b f(t)dt = \sum_{j=1}^s w_j f(c_j).$$

If $b - a = O(h)$, then the error in a quadrature rule of order p is $O(h^{p+1})$. Obviously $p \geq s$, but p may be significantly larger than s if the points c_j are chosen carefully.

The midpoint and trapezoidal rules have order $p = 2$. Simpson's rule has order $p = 4$. Gaussian quadrature at s points has the highest order possible at $p = 2s$.

If the integration interval length $b - a$ is not small, then there is no reason in general to expect an accurate approximation from any of these basic rules. A *composite quadrature* method consists of dividing the interval $[a, b]$ into subintervals, say, of length h each, where the size of h can now be controlled, and applying a basic rule in each of these subintervals. The resulting method is the same as a corresponding collocation–Runge–Kutta method applied to the very special ODE $y' = f(t)$.

2.12.5 Fixed point iteration and Newton's method

For solving an *algebraic nonlinear equation*

$$g(x) = 0,$$

a class of methods is obtained by rewriting the problem as

$$x = \phi(x)$$

and defining a sequence of iterates as follows: x^0 is an initial guess. For a current iterate x^ν, where ν is the iteration counter, we set

$$x^{\nu+1} = \phi(x^\nu), \quad \nu = 0, 1, \ldots.$$

This is called a *fixed point iteration*. The challenge is to choose ϕ that will be easy to evaluate while the whole iteration process converges quickly.

The convergence of a fixed point iteration to a given root x^* at a linear rate is guaranteed if there is a constant ρ, $0 \leq \rho < 1$, such that

$$|\phi'(x)| \leq \rho$$

for all x in a neighborhood of x^* to which x^0 also belongs. The smaller ρ is, the faster the iteration is guaranteed to converge.

Newton's method is a special case of fixed point iteration with

$$\phi(x) = -\frac{g(x)}{g'(x)}.$$

2.12. Reviews

To derive it we write

$$0 = g(x) = g(x^\nu) + g'(x^\nu)(x - x^\nu) + \cdots.$$

Approximating the solution x by neglecting the higher order terms in this Taylor expansion, we define the next iterate $x^{\nu+1}$ as the solution of the linear equation

$$0 = g(x^\nu) + g'(x^\nu)(x^{\nu+1} - x^\nu).$$

This defines an iteration for $x^{\nu+1}$ that converges (locally) quadratically under certain conditions.

We can generalize this directly to a system of m algebraic equations in m unknowns,

$$\mathbf{g}(\mathbf{x}) = \mathbf{0}.$$

Everything remains the same, except that first derivatives are replaced by $m \times m$ Jacobian matrices,

$$\frac{\partial \mathbf{g}}{\partial \mathbf{x}} = \begin{pmatrix} \frac{\partial g_1}{\partial x_1} & \frac{\partial g_1}{\partial x_2} & \cdots & \frac{\partial g_1}{\partial x_m} \\ \frac{\partial g_2}{\partial x_1} & \frac{\partial g_2}{\partial x_2} & \cdots & \frac{\partial g_2}{\partial x_m} \\ \vdots & \vdots & \ddots & \vdots \\ \frac{\partial g_m}{\partial x_1} & \frac{\partial g_m}{\partial x_2} & \cdots & \frac{\partial g_m}{\partial x_m} \end{pmatrix}.$$

For the fixed point iteration we have

$$\mathbf{x}^{\nu+1} = \boldsymbol{\phi}(\mathbf{x}^\nu),$$

and this converges if there is a constant $\rho < 1$ such that in a neighborhood to which both $\mathbf{x}^*$ and $\mathbf{x}^0$ belong,

$$\left\| \frac{\partial \boldsymbol{\phi}}{\partial \mathbf{x}} \right\| \leq \rho.$$

For Newton's method we require the Jacobian matrix $\frac{\partial \mathbf{g}}{\partial \mathbf{x}}$ for the actual iteration, not just for stating a convergence criterion. The method reads

$$\mathbf{x}^{\nu+1} = \mathbf{x}^\nu - \left(\frac{\partial \mathbf{g}}{\partial \mathbf{x}}(\mathbf{x}^\nu) \right)^{-1} \mathbf{g}(\mathbf{x}^\nu), \quad \nu = 0, 1, \ldots.$$

Rather than computing $\mathbf{x}^{\nu+1}$ directly it is better in certain situations (when ill-conditioning is encountered), and never worse in general, to solve the linear system for the difference $\boldsymbol{\delta}$ between $\mathbf{x}^{\nu+1}$ and $\mathbf{x}^\nu$ and then update. Thus, $\boldsymbol{\delta}$ is computed (for each ν) by solving the linear system

$$\left(\frac{\partial \mathbf{g}}{\partial \mathbf{x}} \right) \boldsymbol{\delta} = -\mathbf{g}(\mathbf{x}^\nu),$$

where the Jacobian matrix is evaluated at $\mathbf{x}^\nu$, and the next Newton iterate is obtained by

$$\mathbf{x}^{\nu+1} = \mathbf{x}^\nu + \boldsymbol{\delta}.$$

2.12.6 Discrete and fast Fourier transforms

In this brief description we follow Trefethen [169].

The continuous Fourier transform (1.17), (1.18), so powerful and essential for theory in this book, is not useful for constructing a practical method. For the latter let us consider a uniform *periodic mesh*,

$$0 = x_0 < x_1 < \cdots x_{J-1} < x_J = 2\pi,$$

where the value of the approximated function at x_0 is identified with that at x_J. We have $x_j = jh$, $j = 1, \ldots, J$, with $h = 2\pi/J$.

Let us assume that J is even and that the given values of the function $v(x)$ are $v(x_j) = v_j$, $j = 1, \ldots, J$. The *discrete Fourier transform* (**DFT**) is defined by

$$\hat{v}_l = h \sum_{j=1}^{J} e^{-\iota l j h} v_j, \quad l = -J/2 + 1, \ldots, J/2. \tag{2.39a}$$

Note that the integer variable l has replaced ξ in (1.17). Then the *inverse DFT* reads

$$v_j = \frac{1}{2\pi} \sum_{l=-J/2+1}^{J/2} e^{\iota l j h} \hat{v}_l, \quad j = 1, \ldots, J. \tag{2.39b}$$

We can write the DFT as a matrix transform,

$$\hat{\mathbf{v}} = B\mathbf{v}, \quad \mathbf{v} = B^{-1}\hat{\mathbf{v}},$$

where both the $J \times J$ matrix B and its inverse are explicitly given. The DFT and its inverse transform each cost $O(J^2)$ flops to carry out simplemindedly.

However, the matrices B and B^{-1} are special in that their elements are repetitive, all being just powers of the Jth root of unity, $e^{\iota 2\pi/J}$. Taking advantage of this to economize the computation leads to the *fast Fourier transform* (FFT), one of the most important algorithms ever, that costs only $O(J \log J)$ flops. The details of the algorithm get technical and can be found in many references, including [89, 42].

Let us see how to use the DFT and FFT in order to closely approximate derivatives of a smooth, periodic function $v(x)$. For technical reasons we define $\hat{v}_{-N/2} = \hat{v}_{N/2}$ and replace (2.39b) by

$$v_j = \frac{1}{2\pi} \sum_{l=-J/2}^{J/2} c_l e^{\iota l j h} \hat{v}_l, \quad j = 1, \ldots, J,$$

where $c_l = 1/2$ for $l = \pm J/2$, $c_l = 1$ otherwise. This can be extended into the trigonometric polynomial interpolant

$$p(x) = \frac{1}{2\pi} \sum_{l=-J/2}^{J/2} c_l e^{\iota l x} \hat{v}_l, \quad x \in [0, 2\pi], \tag{2.40}$$

whose derivatives can now be taken and evaluated at $x = jh$. The rth derivative brings out the familiar factor $(\iota l)^r$, so we set $\hat{w}_l = (\iota l)^r \hat{v}_l$. If r is odd, then due to lost symmetry we set $\hat{w}_{J/2} = 0$. An inverse transform of $\hat{w}$ then completes the task. The resulting derivative approximations are called *spectral derivatives*. They can be extremely accurate if v is very smooth in addition to being periodic.

We can form a *differentiation matrix* for the first derivative (and higher) in this way. The $J \times J$ matrix D, when multiplying $\mathbf{v}$, then yields a corresponding vector of J derivative values approximating $\frac{dv}{dx}(jh)$, $j = 1, \ldots, J$. But a direct use of FFT is simpler. In MATLAB, assuming that v is a column array containing the values v_j and that $J2 = 0$ if r is odd, $J2 = J/2$ otherwise, the script

```
v_hat = fft(v);
w_hat = (1i)^r * ([0:N/2-1 J2 -N/2+1:-1].^r)' .* v_hat;
w = real(ifft(w_hat));
```

produces the desired approximations to the rth derivative. The strange row vector that gets raised to the power r simply corresponds to the MATLAB convention of ordering the wave numbers l. If more generally $v = v(s)$, where $s \in [a, b]$, then apply a scaling plus translation transformation to the independent variable. In fact, the only change in the above script (other than in the interpretation of the input and output locations) is that w_hat gets multiplied by $(2\pi/(b-a))^r$.

2.13 Exercises

0. **Review questions**

    (a) Does f have to be linear in y or t in order to apply a linear multistep method for the ODE $y' = f(t, y)$?

    (b) What is a local truncation error? How does it differ from the local error defined in Section 2.4?

    (c) Write down the class of three-stage Runge–Kutta methods.

    (d) Does a method of order p automatically yield a solution error $O(k^p)$ for step size k sufficiently small?

    (e) What is a stiff initial value ODE?

    (f) Define 0-stability, A-stability, and L-stability.

    (g) Why are Adams methods generally considered useful only for nonstiff problems?

    (h) Why is the fixed point iteration considered inappropriate for stiff problems? What is used instead in practice?

    (i) What is a semi-explicit DAE?

    (j) Is implicit trapezoidal a symmetric method? What about the explicit trapezoidal method?

    (k) Why is it difficult in general to find efficient methods for highly oscillatory problems?

(l) What is the cause for the essential added difficulty when solving boundary value ODEs as compared to initial value ODEs?

1. Consider the method

$$y_{n+\theta} = y_n + k\theta f(t_n, y_n),$$
$$y_{n+1} = y_n + kf(t_{n+\theta}, y_{n+\theta}),$$

where $0.5 \leq \theta \leq 1$ is a parameter.

   (a) Show that this is a two-stage Runge–Kutta method. Write it in tableau form.

   (b) Show that the method is first order accurate, unless $\theta = 0.5$ when it becomes second order accurate (and is then called *explicit midpoint*).

   (c) Show that the domain of absolute stability contains a segment of the imaginary axis (i.e., not just the origin) iff $\theta > 0.5$.

2. For a given ODE system

$$y' = f(y),$$

consider the so-called θ-method

$$y_{n+1} = y_n + k[\theta f(y_{n+1}) + (1-\theta)f(y_n)]$$

for some value θ, $0 \leq \theta \leq 1$. This popular family of methods is different from the one considered in Exercise 1.

   (a) Which methods are obtained for the values (i) $\theta = 0$, (ii) $\theta = 1$, and (iii) $\theta = 1/2$?

   (b) Find a range of θ-values, i.e., an interval $[\alpha, \beta]$ such that the method is A-stable for any $\alpha \leq \theta \leq \beta$.

   (c) For what values of θ does the method have stiff decay?

   (d) For a given δ, $0 \leq \delta < 1$, let us call a method δ-damping if

$$|y_{n+1}| \leq \delta |y_n|$$

   for the test equation $y' = \lambda y$ as $\lambda k \to -\infty$. (Thus, if $y_0 = 1$, then for any tolerance TOL > 0, $|y_n| \leq$ TOL after n steps when n exceeds $\frac{\log TOL}{\log \delta}$.)
   Find the range of θ-values such that the θ-method is δ-damping.

   (e) Write the θ-method as a general Runge–Kutta method; i.e., specify A, $\mathbf{b}$, and $\mathbf{c}$ in the tableau

$$\begin{array}{c|c} \mathbf{c} & A \\ \hline & \mathbf{b}^T \end{array}$$

   (f) Using the order conditions given in (2.13) (you don't need to prove them) determine what is the order of the θ-method for each θ.

2.13. Exercises

(g) People often toy with values $\theta = 1/2 + \varepsilon$, where $0 < \varepsilon \ll 1$ is a suitably small parameter. Discuss advantages and disadvantages of choosing such a value.

3. The TR-BDF2 [116] is a one-step method consisting of applying first the trapezoidal scheme over half a step $k/2$ to approximate the midpoint value, and then the BDF2 scheme over one step

$$y_{n+1/2} = y_n + \frac{k}{4}(f(y_n) + f(y_{n+1/2})), \quad (2.41a)$$

$$y_{n+1} = \frac{1}{3}[4y_{n+1/2} - y_n + kf(y_{n+1})]. \quad (2.41b)$$

One advantage is that only two systems of the original size need be solved per time step.

(a) Write the method (2.41) as a Runge–Kutta method in standard tableau form (i.e., find A and **b**). Conclude that it is an instance of a *diagonally implicit Runge–Kutta* (DIRK) method [85, 14].

(b) Show that both the order and the stage order equal 2.

(c) Show that the stability function satisfies $R(-\infty) = 0$: this method is L-stable [85, 113].

(d) Show that this method has stiff decay.

(e) Can you construct an example where this method would fail while the BDF2 method would not?

4. Consider the two-step method

$$y_{n+1} - y_{n-1} = 2k \, f(y_n).$$

This is the generating method for the leap-frog scheme (1.15c).

However, the method does not do well for problems with damping: show that its absolute stability region has an empty interior. Thus, the method is unstable for any initial value ODE of the form $y' = \lambda y$ with $\mathcal{R}e \, \lambda < 0$.

5. Show that the Runge–Kutta step (2.10) can be written as

$$K_i = f\left(t_n + c_i k, \, y_n + k \sum_{j=1}^{s} a_{ij} K_j\right), \quad 1 \leq i \leq s, \quad (2.42a)$$

$$y_{n+1} = y_n + k \sum_{i=1}^{s} b_i K_i. \quad (2.42b)$$

6. Write down explicitly the order conditions (2.13) for $p = 1, 2$, and 3. Elaborate upon methods that satisfy them.

7. A popular belief has it that implicit methods are suitable for stiff problems. In fact, one has to be a bit careful here.

    (a) Construct a consistent implicit linear multistep method that is not suitable for stiff initial value ODEs.

    (b) Construct a consistent implicit Runge–Kutta method that is not suitable for stiff initial value ODEs.

8. Consider the special case of the test equation, $y' = \lambda y$, where λ is real, $\lambda \leq 0$. If $y(0) = 1$, then the exact solution $y(t) = e^{\lambda t}$ remains nonnegative and decays monotonically as t increases. Let us call a discretization method **nonnegative** for a step size k if $y_n \geq 0$ implies $y_{n+1} \geq 0$. Show the following:

    (a) If in addition $z = \lambda k$ is in the absolute stability region, then we have **monotonicity**
    $$y_n \geq y_{n+1} \geq 0.$$
    This guarantees that the qualitatively unpleasant oscillations that the trapezoidal method produces in Figure 2.9 will not arise.

    (b) The forward Euler method is nonnegative only when $z \geq -1$. (NB Always $z \leq 0$.)

    (c) The backward Euler method is unconditionally nonnegative.

    (d) The trapezoidal method is only conditionally nonnegative even though it is A-stable. Find its nonnegativity condition.

    (e) Find the nonnegativity condition for the TR-BDF2 method of Exercise 3. Is it unconditionally nonnegative?

    (f) Can a *symmetric* Runge–Kutta method be unconditionally nonnegative? Justify if not, or give an example if yes.
    (This last item is a bit harder than the rest.)

9. The problem
    $$y' = y^2, \quad y(0) = 1,$$
    has the exact solution
    $$y(t) = \frac{1}{1-t}, \quad 0 \leq t < 1.$$
    This blows up at $t = 1$.

    (a) Considering this problem on the interval $[0, 1-\varepsilon]$ for a parameter $0 < \varepsilon \ll 1$, investigate what Theorem 2.3 guarantees as a function of ε.

    (b) Integrate the problem numerically using (i) forward Euler and (ii) RK4 with a constant step size k. Try this for $k = \varepsilon = .1, .01, .001, .0001$. (The alliance of ε with k is unholy, of course, because the problem is generally assumed defined *before* applying the numerical method. But let us proceed nonetheless.)
    What do you observe? Is the difference between the methods' behavior fundamental?

2.13. Exercises

(c) Some people would say that problems like this simply should not be solved numerically. Try to defend this statement (without necessarily becoming a believer yourself).

10. Consider applying a semi-implicit variant of a BDF method to a stiff, nonlinear ODE system with a smooth solution.

    Show that the resulting method is generally only first order accurate, even if the order of the BDF method is higher.

11. Let us rewrite (2.31) for a second order ODE system

$$x'' = f(x, v), \tag{2.43}$$

where $v = x'$ and we have set $M = 1$ as well. This is equivalent, of course, to the first order ODE form

$$x' = v, \tag{2.44a}$$
$$v' = f(x, v), \tag{2.44b}$$

but we wish to seek more direct approximations for (2.43).

One simple idea is

$$x_{n+1} = 2x_n - x_{n-1} + k^2 f(x_n, v_n), \tag{2.45a}$$
$$v_n = (2k)^{-1}(x_{n+1} - x_{n-1}). \tag{2.45b}$$

However, we wish to define a one-step method that is more flexible when different time steps are required.

The *Newmark* methods comprise a two-parameter family defined by

$$x_{n+1} = x_n + kv_n + \frac{k^2}{2}((1-\beta)f_n + \beta f_{n+1}), \tag{2.46a}$$
$$v_{n+1} = v_n + k((1-\gamma)f_n + \gamma f_{n+1}). \tag{2.46b}$$

(a) Show that (2.45) is second order accurate for x. What is the accuracy order for v?

(b) Show that (2.46) is second order accurate in both x and v for any choice of parameter $0 \le \beta \le 1$ with $\gamma = .5$.

(c) Show that with $\beta = \gamma = .5$ in (2.46) we have the trapezoidal method for (2.44).

(d) Write down the method (2.46) for $\beta = 0$ and $\gamma = .5$. Under what circumstances should this variant be advantageous?

(e) Try out the three specific methods mentioned above. Invent one example where f depends on v and another where it does not. Check convergence orders and absolute error values. Discuss your results.

12. Show that the constrained multibody system formulation (2.30) is a special case of the general DAE form (2.29) for $y = (x, x', z)$. What is the Jacobian matrix of F with respect to y'?

13. Write a program to solve the Black–Scholes model described in Exercise 1.3. You may choose to discretize (1.27)–(1.28) directly, or opt for (1.29) or (1.30) followed by back transformation to u. Justify your choice. The display and assessment of results should in any case be applied to the original $u(t, x)$.

    Use the second order accurate spatial discretization
    $$u_{xx}(t, x_j) \sim \frac{u_{j+1} - 2u_j + u_{j-1}}{h^2},$$
    $$u_x(t, x_j) \sim \frac{u_{j+1} - u_{j-1}}{2h}, \quad \text{if needed}.$$

    This leads to an ODE system in time for $u_1(t), \ldots, u_J(t)$.

    For the time discretization try (i) forward Euler and (ii) the implicit trapezoidal scheme. For each, select a fixed time step k based on accuracy and stability considerations, aiming at an overall accuracy of $O(h^2)$.

    Apply your program for the values $r = 0.1$, $\sigma = 0.2$, $E = 1$, and plot the solution at $0 \leq x \leq 2$ for $T - t = 0, 0.5, 1.0,$ and 1.5. Try $h = .01, .005, .0025,$ and $.00125$ (with the time step decreased accordingly), and assess from the differences whether an $O(h^2)$ accuracy is achieved.

14. In applications often the derivative of the sought function, $y'(t)$, depends not only on the solution y at t but also on y at other locations. For instance, consider
    $$y'(t) = f(t, y(t), y(t - d)),$$
    where d is a known *delay*. This is no longer an ODE but rather a **delay differential equation** (DDE). In general the delay d may be a function, it may be positive or negative, and there may be several delays [86, 22]. But below we concentrate on the instance
    $$y'(t) = -y(t) + y^2(t - 1) + 1, \quad 0 \leq t \leq 10,$$
    $$y(t) = 0, \quad t \leq 0.$$

    (a) Show that y' is discontinuous at $t = 0$ and y'' is discontinuous at $t = 1$.

    (b) Contemplating a fourth order numerical method for this initial value DDE, explain why a Runge–Kutta method is preferred over a multistep one and why it is a good idea to include the integers $l = 0, 1, 2, \ldots, 10$ as mesh points.

    (c) When integrating from t_n to t_{n+1} using (2.10) we need solution values at the points $t_n + c_i k - 1$. Describe a local interpolation scheme to obtain such values to sufficient accuracy.

    (d) Write a code to solve the above problem and plot its solution. You may use software such as MATLAB's dde23 or other programs available on the Internet.

15. Continuing with the DDE of Exercise 14, let
    $$y_j(\tau) = y(\tau + j - 1), \quad 0 \leq \tau \leq 1,$$
    and denote $y_0 \equiv 0$ and $\mathbf{y} = (y_1, \ldots, y_{10})^T$.

2.13. Exercises

(a) Show that the DDE problem can be written as a boundary value ODE (BVODE) without delays,

$$\frac{d\mathbf{y}}{d\tau} = \mathbf{f}(\tau, \mathbf{y}),$$
$$\mathbf{0} = \mathbf{g}(\mathbf{y}(0), \mathbf{y}(1)).$$

(b) The resulting boundary conditions from the BVODE formulation above are not separated. Reformulate the BVODE into one with separated BC by defining $\mathbf{y}(\tau)$ more intelligently.

(c) Solve the resulting BVODE using MATLAB's code bvp4c or any other code. Compare the results to those of Exercise 14.

(d) Discuss pros and cons of the above two approaches, namely, direct discretization vs. reformulation into standard BVODE form, in a more general context including, e.g., more general integration intervals and more general delays.

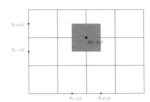

Chapter 3
Finite Difference and Finite Volume Methods

Having given a taste of finite difference methods for PDEs in Section 1.2 and an exposition of methods for ODEs in Chapter 2, we embark in the present chapter upon a wider and more methodical derivation of methods and concepts for PDE discretization.

To recall, we are treating the time variable t as special among the independent variables. To facilitate a gradual development of concepts and methods let us consider discretizing the spatial domain and derivatives first. This gives a *semi-discretization*, which yields an (often very large) set of initial value ODEs. The advantages of the semi-discretization approach are as follows:

1. ODE methods *and software* may be used under certain conditions to solve the problem. In this context the term **method of lines** is commonly used; see Figure 3.1.

2. Different approaches (not just finite differences) for the spatial discretization may be introduced.

3. The analysis for the semi-discrete formulation is sometimes simpler than for the fully discrete formulation.

There are also disadvantages, or at least pitfalls:

1. Not all useful full discretization methods can be naturally obtained in this stepwise way.

2. The approach assumes a certain separation between time and space which may not be present, at least not naturally, especially in tough hyperbolic problems.

3. The analysis of a semi-discrete formulation must take into account the fact that the number of ODEs in (3.2) is "large," i.e., the size of L tends to ∞ as the spatial discretization parameter tends to 0. One cannot simply apply ODE-based knowledge here without care. And then, once this analysis is completed we are still not finished, because the effect of the ensuing time discretization must be accounted for as well.

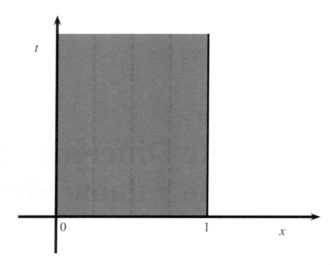

Figure 3.1. *Method of lines. The shaded strip is the domain on which the diffusion PDE is defined. The approximations $v_j(t)$ are defined along the dashed lines.*

We discuss semi-discretizations in Section 3.1, developing a feeling for the derivation of finite difference and finite volume methods in space and the handling of boundary conditions (BC).

In Section 3.2 we apply an ODE discretization to the time variable as well, obtaining a *full discretization*. Combining the techniques acquired for time discretization in Chapter 2 with those developed for semi-discretizations in the present chapter, there is a family of complete discretizations to consider. For the fully discretized problems we then define basic concepts of order, stability, and convergence, and we show in particular how the order of accuracy and the stability of the difference method are expressed in the resulting approximation error for the PDE solution.

> **Note:** Section 3.1 is long, essentially a chapter within a chapter, and we will find ourselves referring back to it throughout the text.

3.1 Semi-discretization

Consider a general, linear PDE of evolution in d space variables,

$$u_t = \mathcal{L}u + q, \qquad \mathbf{x} \in \Omega, \; t > 0, \tag{3.1a}$$

where $\Omega \subset \mathbb{R}^d$, $q = q(t, \mathbf{x})$ is given, $u = u(t, \mathbf{x})$ is sought, and $\mathcal{L}$ is a differential operator of the form

$$\mathcal{L} = \sum_{i_1 + \cdots + i_d \leq m} a_{i_1, i_2, \ldots, i_d} \frac{\partial^{i_1 + \cdots + i_d}}{\partial x_1^{i_1} \partial x_2^{i_2} \cdots \partial x_d^{i_d}}. \tag{3.1b}$$

3.1. Semi-Discretization

The coefficients of $\mathcal{L}$ may depend on t and $\mathbf{x}$. There are initial conditions to be satisfied,

$$u(0, \mathbf{x}) = u_0(\mathbf{x}), \qquad \mathbf{x} \in \Omega, \tag{3.1c}$$

as well as appropriate BC.

Consider next the task of discretizing $\mathcal{L}u + q$ for a fixed t. It may be useful to think of $\mathcal{L}$ as an *elliptic* differential operator when the evolution PDE (3.1a) is parabolic. There are various approaches for performing the discretization, such as finite elements [158, 19, 35, 103, 67], spectral methods [45, 169, 32], and particle methods. We concentrate mostly on finite difference and finite volume methods [137, 116, 110, 76, 164] but also devote a short section in the spirit of Sections 2.8 and 2.11 to finite element methods. Whichever spatial discretization method is used, we end up with an ODE system,

$$\mathbf{v}_t = L\mathbf{v} + \mathbf{q}, \tag{3.2}$$

where $\mathbf{v}$ may be thought of as a vector of unknowns corresponding to the solution u of (3.1) at mesh points, L is a square matrix discretizing $\mathcal{L}$ (L depends on t if $\mathcal{L}$ does), and $\mathbf{q}$ is a corresponding discretization of q. The special form of (3.2) corresponds to the special form of (3.1). Other, more complex and even nonlinear differential operators can be discretized in a similar fashion.

Below we continue with finite difference, and occasionally finite volume, discretizations in space. For the homogeneous problem ($q = 0$) in one space dimension we write such a scheme as

$$\frac{d}{dt} v_j(t) = \sum_{i=-l}^{r} \alpha_i v_{j+i}(t), \tag{3.3}$$

where $\alpha_i = \alpha_i(t, x_j)$ in general.

Example 3.1 (A diffusion problem) For the simple diffusion problem in one space variable x and time t,

$$u_t = u_{xx} + r(x, u),$$
$$u(0, x) = u_0(x), \quad u(t, 0) = g_0(t), \; u(t, 1) = g_1(t),$$

where $\Omega = [0, 1]$ and r is a known (possibly nonlinear) function, we discretize on a uniform spatial mesh with spacing h. Let $v_j(t)$ approximate $u(t, x_j)$, where $x_j = jh$, $j = 0, 1, \ldots, J + 1$. Then replacing $\frac{\partial^2 u}{\partial x^2}$ by a three-point central difference we obtain

$$\frac{dv_j}{dt} = \frac{v_{j+1} - 2v_j + v_{j-1}}{h^2} + r(x_j, v_j), \qquad j = 1, \ldots, J.$$

This ODE is of the form (3.3) with $l = r = 1$. The boundary conditions are used to close the system in an obvious manner, setting $v_0(t) = g_0(t)$ and $v_{J+1}(t) = g_1(t)$.

We have obtained an initial value ODE problem with the initial data $v_j(0) = u_0(x_j)$, $j = 1, \ldots, J$. ∎

If we write the discretization (3.3) in the vector form (3.2), then the obtained matrix L is very large (possibly infinite) and very sparse. For instance, in the case of Example 3.1, L

is tridiagonal. This sparsity is typical for finite difference, finite volume, and finite element methods.

Suppose a finite difference semi-discretization on a uniform mesh with spacing h is employed, yielding (3.2), where **v** is the vector of mesh values: $v_j(t) \approx u(t, x_j)$. Let **u** be the vector of size J composed of the exact values, $u_j(t) = u(t, x_j)$. Denote the **truncation error** for the semi-discrete system (3.2) by

$$\mathbf{d} = \mathbf{u}_t - (L\mathbf{u} + \mathbf{q}). \tag{3.4}$$

The method (3.2) has **order of accuracy** p if $\|\mathbf{d}\|_\infty = O(h^p)$ for all sufficiently smooth solutions. Under certain conditions, namely, stability and sufficient smoothness, this translates to **convergence** of a similar order, i.e.,

$$\|\mathbf{u}(t) - \mathbf{v}(t)\| = O(h^p).$$

In Section 3.1.1 we derive basic discretizations for first and second derivatives. This is followed by the lengthy Section 3.1.2, where **staggered grid** or **staggered mesh**[12] discretizations are developed and demonstrated. These arise from attempts to obtain **compact**, **symmetric** discretizations in space. Several more involved examples in more than one space variable are provided to illustrate more complex issues that arise.

A partial treatment of boundary conditions (BC) is given in Section 3.1.3, first in one dimension and then in several space variables. The latter are treated via a **variational approach**, connecting naturally to Section 3.1.4, and this can be skipped on first reading even though its subject matter is important. In Section 3.1.4 we briefly describe the *finite element method*, reviewing it for the sake of completeness in the vein of Sections 2.8 and 2.11. Nonuniform meshes are considered in Section 3.1.5. A few words regarding stability and convergence end this section, leading to the next.

> **Note:** This section need not be absorbed fully at first pass, as it often proceeds tangentially to much of the rest of the text.

3.1.1 Accuracy and derivation of spatial discretizations

The most basic tool for deriving difference approximations and verifying their order of accuracy is Taylor's expansions; see the review in Section 1.3.1. This approach is intuitive but not always methodical, and there are certain issues that are better highlighted by the alternative approach considered next.

Differentiating polynomial interpolants

Polynomial interpolation provides a general approach for deriving difference formulas; see the review in Section 2.12.2. It is more methodical but less intuitive than manipulating Taylor's expansions. It also works for deriving discretizations on nonuniform mesh stencils (Exercises 2 and 3).

---
[12]The words "grid" and "mesh" are used interchangeably in this book.

3.1. Semi-Discretization

For instance, let $u_i = u(x_i) \forall i$ and consider the straight line interpolation between (x_j, u_j) and (x_{j+1}, u_{j+1}),

$$w(x) = u_j + \frac{u_{j+1} - u_j}{h}(x - x_j).$$

Differentiating this gives

$$w_x(x_j) = \frac{u_{j+1} - u_j}{h},$$

which is the *forward* approximation for $u_x(x_j)$. Polynomial interpolation also yields that the error is first order in h,

$$u_x(x_j) = \frac{u_{j+1} - u_j}{h} - \frac{h}{2} u_{xx}(\eta), \tag{3.5a}$$

for some generic argument η; in this case, $x_j \leq \eta \leq x_{j+1}$.

Likewise, interpolating (x_{j-1}, u_{j-1}), (x_j, u_j), and (x_{j+1}, u_{j+1}) by a quadratic polynomial and differentiating yields the *centered* approximation

$$u_x(x_j) = \frac{u_{j+1} - u_{j-1}}{2h} - \frac{h^2}{6} u_{xxx}(\eta). \tag{3.5b}$$

Here $h = x_j - x_{j-1} = x_{j+1} - x_j$. Differentiating the same quadratic interpolant once more yields a formula used in Example 3.1 for the second derivative,

$$u_{xx}(x_j) = \frac{u_{j+1} - 2u_j + u_{j-1}}{h^2} - \frac{h^2}{12} u_{xxxx}(\eta). \tag{3.5c}$$

This obviously can be generalized to obtain approximations for higher derivatives. Note that for the lth derivative we must use an interpolating polynomial of degree at least l (or risk an approximation that vanishes identically). This in turn implies that *a difference formula for the lth derivative must be based on at least $l + 1$ points.*

A difference formula that is based on precisely $l + 1$ consecutive mesh points (in one dimension) is called **compact**. Such a formula uses a difference stencil of minimal size. The formulas (3.5a) and (3.5c) are both compact, but the centered formula (3.5b) for u_x is not! Noncompact formulas can certainly be used in many situations, as we have already seen in Chapter 1, but occasionally they give rise to spurious modes; see, e.g., Section 7.1.

Notation and "exact" formula

It is useful to introduce the following difference notation:

$$D_+ u_j = u_{j+1} - u_j \quad \textbf{Forward}, \tag{3.6a}$$

$$D_- u_j = u_j - u_{j-1} \quad \textbf{Backward}, \tag{3.6b}$$

$$D_0 u_j = u_{j+1} - u_{j-1} \quad \textbf{Long centered}, \tag{3.6c}$$

$$\delta u_j = u_{j+1/2} - u_{j-1/2} \quad \textbf{Short centered}, \tag{3.6d}$$

$$\mu u_j = \frac{1}{2}(u_{j+1/2} + u_{j-1/2}) \quad \textbf{Short average}, \tag{3.6e}$$

$$E u_j = u_{j+1} \quad \textbf{Translation}. \tag{3.6f}$$

The operators (3.6a)–(3.6e) can be clearly expressed in terms of the translation operator E, e.g.,
$$D_+ = E - I, \; D_- = I - E^{-1},$$
etc. It is possible to combine these operators, e.g.,
$$h^{-2} D_+ D_- u_j = h^{-1} D_+ \frac{u_j - u_{j-1}}{h} = \frac{1}{h}\left[\frac{u_{j+1} - u_j}{h} - \frac{u_j - u_{j-1}}{h}\right]$$
$$= \frac{1}{h^2}\left[u_{j+1} - 2u_j + u_{j-1}\right] = u_{xx} + O(h^2).$$

It is also possible to derive operator identities, such as $\delta^2 = D_+ D_- = D_- D_+$ and $\mu^2 = 1 + \delta^2/4$.[13]

To obtain an exact formula we can write formally
$$Eu(x) = u(x+h) = u + hu_x + \frac{1}{2}h^2 u_{xx} + \cdots$$
$$= \left(I + h\partial_x + \frac{1}{2}h^2 \partial_{xx} + \cdots\right) u(x) = e^{h\partial_x} u(x).$$

Hence $E = e^{h\partial_x}$, implying that
$$\partial_x = h^{-1} \log E. \tag{3.7}$$

Of course, the formula (3.7) has the disadvantage that it requires infinitely many mesh points, but its local approximations may be useful. Below we proceed to derive some such approximations, generalizing the simple formulas (3.5).

For the problematic first derivative, u_x, we can derive *forward* (and likewise *backward*) formulas as follows. Write
$$\log E = \log(I + D_+) = D_+ - \frac{1}{2}D_+^2 + \frac{1}{3}D_+^3 + \cdots.$$

So, the first term alone is the first order (3.5a). A second order one-sided approximation of u_x at x_j is given by
$$u_x = \frac{1}{h}\left(D_+ - \frac{1}{2}D_+^2\right) u + O(h^2) = \frac{1}{h}(u_{j+1} - u_j) - \frac{1}{2h}(u_{j+2} - 2u_{j+1} + u_j) + O(h^2).$$

The disadvantages of this formula are that (i) an extra neighbor is required, which may be especially problematic near boundaries, and (ii) the error constant is not as small as a more centered formula involving the same number of mesh points would yield. But if there is a discontinuity to the left of x_j, then taking one-sided differences into the smooth part of the solution can be a clever move.

To obtain *centered* formulas for u_x we write
$$D_0 = E - E^{-1} = e^{h\partial_x} - e^{-h\partial_x} = 2\sinh(h\partial_x),$$

---
[13] Of course, μ here stands for the short averaging operator, not a ratio of time and space steps. There should not be too much confusion about this in context because we will typically use μ_x or μ_t to denote averaging in x or t, respectively.

3.1. Semi-Discretization

yielding

$$\partial_x = \frac{1}{h}\sinh^{-1}(D_0/2) = \frac{1}{h}\left[\frac{D_0}{2} - \frac{1}{6}\left(\frac{D_0}{2}\right)^3 + \cdots\right].$$

The first term corresponds to the second order (3.5b). The first two terms give a fourth order approximation. However, they require seven neighboring values. This yields problems near the boundary and other potential difficulties as well. Thus, we use short differences and the fact that $\mu^2 = 1 + \delta^2/4$. Since $\mu\delta = D_0/2$ we get

$$\partial_x = \frac{1}{h}\sinh^{-1}(\mu\delta) = \frac{1}{h}\left[\mu\delta - \frac{1}{6}(\mu\delta)^3 + \cdots\right]$$

$$= \frac{1}{h}\left[\mu\delta - \frac{1}{6}\mu(\delta)^3 + \cdots\right]$$

$$= \frac{D_0}{2h}\left(I - \frac{1}{6}D_+D_- + \cdots\right).$$

This gives a fourth order, five-point formula

$$(u_x)_j = \frac{1}{12h}\left[-u_{j+2} + 8u_{j+1} - 8u_{j-1} + u_{j-2}\right] + O(h^4).$$

For the second derivative u_{xx} we won't discuss one-sided schemes because they are rarely used in practice. Since $\partial_x = \frac{2}{h}\sinh^{-1}(\delta/2)$, we square this expression to obtain

$$\partial_{xx} = \frac{4}{h^2}\sinh^{-2}(\delta/2) = \frac{1}{h^2}\left[\delta^2 - \frac{1}{12}\delta^4 + \frac{1}{90}\delta^6 + \cdots\right].$$

Cutting this expansion after s terms gives an approximation of order $2s$ which requires $2s + 1$ mesh points. For $s = 1$ we obtain the standard three-point formula (3.5c) used in Example 3.1. Note that since only even powers of δ appear, we only evaluate u at mesh points.

Implicit schemes

The idea here is more complex, and the following few paragraphs may be skipped without loss of continuity. Suppose we want to derive a high order, centered formula for u_x using as few neighbors as possible, for the reasons indicated above. To obtain a symmetric scheme we must use a point to the left and a point to the right of j, but we don't want to use more. It is possible to derive such a fourth order scheme, but this requires an implicit, nonlocal formula. (See pp. 234–235 of [133].) Note that

$$I - \frac{1}{6}D_+D_- = \left(I + \frac{1}{6}D_+D_-\right)^{-1} + O(h^4),$$

so

$$\partial_x = \frac{1}{2h}\left(I + \frac{1}{6}D_+D_-\right)^{-1}D_0 + \cdots.$$

This is to be interpreted as

$$\left(I + \frac{1}{6}D_+D_-\right) F_j = \frac{1}{2h} D_0 u_j, \tag{3.8}$$

so an entire row of F_j is obtained all at once, with $F_j = (u_x)_j + O(h^4)$, and only three points are used in the formula stencil.

To find the F_j's we add end conditions, obtained by a Taylor series for u_x, and solve a tridiagonal system. This can be done rapidly; see Section 2.12.1. What results is a nonlocal, higher order approximation for u_x. Example 8.7 demonstrates this scheme in action.

For the second derivative u_{xx}, although the explicit derivation is cleaner than for the first derivative, the fourth order approximation still uses five points, rather than three. To recall, in general we cannot expect to use fewer than $i+1$ mesh points for approximating the ith derivative locally, so here we may strive for a compact three-point formula. To obtain a formula accurate to fourth order we may use the same basic idea as for obtaining (3.8). Write

$$h^2 \partial_{xx} = \left(I - \frac{1}{12}\delta^2\right)\delta^2 \approx \left(I + \frac{1}{12}\delta^2\right)^{-1} \delta^2.$$

Thus, solving

$$\left(I + \frac{1}{12}D_+D_-\right) S_j = \frac{1}{h^2} D_+D_- u_j \tag{3.9}$$

for S_j yields a fourth order, implicit compact approximation for $(u_{xx})_j$.

To carry this out we again need to solve a tridiagonal system for the S_j's in terms of the u_j's. For a review on solving such systems see Section 2.12.1.

3.1.2 Staggered meshes

The previous subsection gives some basic tools for deriving difference methods. In practice, it often turns out to be desirable to derive **symmetric**, or **centered**, discretizations, because they automatically yield higher order accuracy (typically second order in the simplest cases), and because they often reproduce certain solution properties; see Chapters 2, 6, and 7. For second derivatives the natural, compact three-point formula (3.5c) is centered, and not much more is needed. But for first derivatives the centered approximation at mesh points is "long," spanning two subintervals instead of one, and this leads at times to undesirable phenomena because spurious high frequency (or large wavenumber) error components may arise.

The "short" difference introduced in (3.6d) then appears to be more suitable, but this implies centering the entire formula in between where the mesh function values are presumed to approximate the exact solution. For example, given the ODE $u_x = q(x)$ we can approximate the integral form

$$u(x_{j+1/2}) - u(x_{j-1/2}) = \int_{x_{j-1/2}}^{x_{j+1/2}} q(x)dx$$

by the midpoint rule

$$\delta v_j = v_{j+1/2} - v_{j-1/2} = hq(x_j). \tag{3.10a}$$

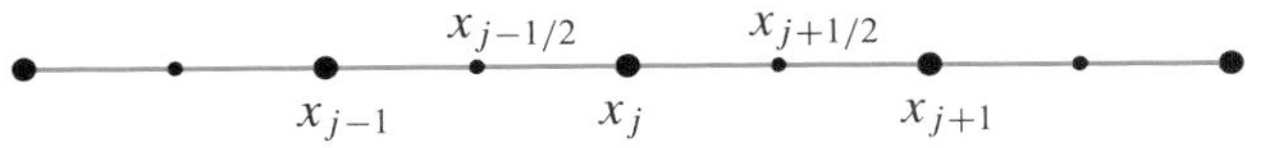

3.1. Semi-Discretization

```
        x_{j-1/2}    x_{j+1/2}
●────●────●────●────●────●
    x_{j-1}   x_j    x_{j+1}
```

Figure 3.2. *A staggered mesh in one dimension. The points x_{j-1}, x_j and x_{j+1} belong to the primal mesh, while the midpoints $x_{j-1/2}$ and $x_{j+1/2}$ belong to the dual mesh.*

We can think of the original mesh points $\{x_j\}$ as defining the **primal mesh** while the midpoints $\{x_{j+1/2}\}$ constitute a **dual mesh**. Note that except for the effect near boundaries, which we assume for the moment to be absent, the discretization (3.10a) is equivalent to the usual midpoint discretization

$$v_{j+1} - v_j = hq(x_{j+1/2}), \qquad (3.10b)$$

because it is just a matter of redefining what the primary mesh is; see Figure 3.2. The approximation is second order accurate.

Example 3.2 (Diffusion equation) Consider the diffusion equation with nonconstant coefficients

$$u_t = (a(x)u_x)_x + q(t,x),$$

where $a(x) > 0$. This type of equation is very popular in applications.

Assuming that a is smooth enough we could rewrite the PDE as $u_t = au_{xx} + a_x u_x + q$ and discretize directly using the formulas of Section 3.1.1,

$$\frac{dv_j}{dt} = \frac{a(x_j)}{h^2}\left[v_{j+1} - 2v_j + v_{j-1}\right] + \frac{a_x(x_j)}{2h}\left[v_{j+1} - v_{j-1}\right] + q(t,x_j).$$

However, this requires specification of a_x and uses a long centered difference unnecessarily.

A better, more elegant way to discretize this PDE uses a staggered mesh. Denoting $w = a(x)u_x$ we have $u_t = w_x + q$, so these are two first order equations in x. We discretize each using a short difference in x. Moreover, to retain a compact overall effect we consider the approximation for w at midpoints and use (3.10b) and (3.10a), respectively, for the two equations

$$a(x_{j+1/2})\frac{v_{j+1} - v_j}{h} = w_{j+1/2},$$

$$\frac{dv_j}{dt} = \frac{w_{j+1/2} - w_{j-1/2}}{h} + q(t,x_j).$$

Eliminating w-values yields the preferred semi-discretization

$$\frac{dv_j}{dt} = h^{-1}\left[a(x_{j+1/2})\frac{v_{j+1} - v_j}{h} - a(x_{j-1/2})\frac{v_j - v_{j-1}}{h}\right] + q(t,x_j).$$

See Exercise 4 for a version of this discretization on a nonuniform mesh. ∎

Example 3.3 (Discontinuous coefficients) The staggered discretization developed in the previous example is particularly useful if $a(x)$ in the diffusion equation is piecewise continuous and bounded, with possible jump discontinuities at isolated points in x, or if it just varies rapidly on the scale of h. We then write down the discretization

$$a_{j+1/2} \frac{v_{j+1} - v_j}{h} = w_{j+1/2}, \qquad (3.11a)$$

$$\frac{dv_j}{dt} = \frac{w_{j+1/2} - w_{j-1/2}}{h} + q(t, x_j), \qquad (3.11b)$$

and thus

$$\frac{dv_j}{dt} = h^{-1} \left[a_{j+1/2} \frac{v_{j+1} - v_j}{h} - a_{j-1/2} \frac{v_j - v_{j-1}}{h} \right] + q(t, x_j), \qquad (3.11c)$$

and ask how to define $a_{j+1/2}$: the exact value of a at the midpoint, even if we knew it, does not necessarily capture the effect of discontinuities or of a significant local variation in the best way.

Recalling that integration is a *smoothing operation*, and given that $w(x)$, unlike $a(x)$, has a bounded derivative, we write

$$u_x = a^{-1}(x) w$$

and integrate this equation over one mesh subinterval in space. Thus, we apply the integration to the rougher (i.e., less smooth) quantity u_x, rather than to the smoother w. Further, we approximate w by its midpoint

$$u(x_{j+1}) - u(x_j) = \int_{x_j}^{x_{j+1}} a^{-1} w \, dx \approx w_{j+1/2} \int_{x_j}^{x_{j+1}} a^{-1} dx,$$

which makes sense because a does not change sign. This yields the above staggered formula (3.11) with

$$a_{j+1/2} = h \left[\int_{x_j}^{x_{j+1}} a^{-1} dx \right]^{-1}. \qquad (3.12a)$$

The obtained formula for $a_{j+1/2}$ employs **harmonic averaging**.

Of course, carrying out precise integration can be excessive in many applications. Assuming, for instance, that $a = a_j$ is constant over each subinterval $[x_{j-1/2}, x_{j+1/2})$ we get the simpler formula

$$a_{j+1/2} = \left[\frac{a_j^{-1} + a_{j+1}^{-1}}{2} \right]^{-1}. \qquad (3.12b)$$

The methodology developed here, often referred to as **finite volume**, can also be used in cases where $a = a(x, u(x))$, i.e., for a nonlinear diffusion equation. Both equations (3.12) generalize directly, where now $a_j = a(x_j, v_j)$.

3.1. Semi-Discretization

What if a is a function of u_x, e.g., $a = \frac{1}{|u_x|}$? Here it is natural to discretize u_x by the centered $(v_{j+1} - v_j)/h$ over the interval $[x_j, x_{j+1}]$. So, the integrand in (3.12a) is constant, the point about harmonic averaging is moot, and

$$a_{j+1/2} = \frac{h}{|v_{j+1} - v_j|}.$$

(Some modification must be made to keep the denominator positive even when u is flat; see Example 3.5.) ∎

If we avoid the elimination step, then the pair (3.11a) and (3.11b) forms a discretization for a couple of first order PDEs where different components of the solution are approximated at different locations on the mesh, hence the name *staggered mesh*. Up to this point, and in the discussion that follows in Section 3.2, we assume one spatial variable for simplicity of exposition and imply a direct extension to more space variables. A staggered mesh is also useful occasionally for one space variable, but it becomes especially important in more dimensions. We demonstrate this next by three more involved examples.

> **Note:** The following three examples are in several space variables.

Example 3.4 (Anisotropic diffusion in two dimensions) The same diffusion equation from Example 3.2 reads in two spatial dimensions,

$$u_t = (au_x)_x + (au_y)_y + q \equiv \nabla \cdot (a\nabla u) + q, \tag{3.13}$$

where $a = a(x, y) > 0$, and $q = q(t, x, y)$ is integrable. Note our notation $\nabla \cdot$ for the divergence operator `div` and ∇ for the gradient operator `grad`. Let us assume for simplicity that this PDE is defined over the unit square spatial domain $\Omega = [0, 1] \times [0, 1]$ for $t \geq 0$, and discretize by a uniform mesh with step $h = 1/(J + 1)$. If a were just a constant, then the semi-discretization is straightforward:

$$\frac{dv_{i,j}}{dt} = \frac{a}{h^2} \left[-4v_{i,j} + v_{i-1,j} + v_{i+1,j} + v_{i,j-1} + v_{i,j+1} \right] + q_{i,j}, \quad 1 \leq i, j \leq J.$$

However, if a varies considerably over the scale of h, particularly if it is discontinuous, then much more care is required in developing an appropriate finite volume discretization.

Thus, as in Example 3.3 we consider a volume (cell) of length h centered at the node (i, j); see Figure 3.3. Defining the **flux** $\mathbf{w} = (w^x, w^y)^T$ we write (3.13) as

$$u_t = w^x_x + w^y_y + q = \nabla \cdot \mathbf{w} + q, \tag{3.14a}$$
$$\mathbf{w} = a\nabla u. \tag{3.14b}$$

Note that the flux is typically an important physical quantity, not just a mathematical invention for the purpose of reduction from (3.13) to (3.14). Often boundary conditions are specified on the normal component of $\mathbf{w}$, rather than on $\frac{\partial u}{\partial n}$. We delay further discussion of boundary conditions to Section 3.1.3.

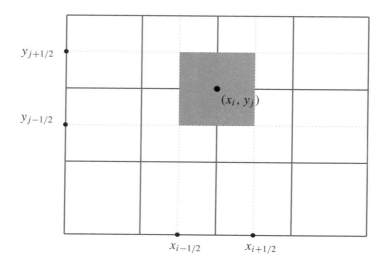

Figure 3.3. *A finite volume in two dimensions.*

The first of these equations looks easy: we integrate (3.14a) over the control volume $V = [x_{i-1/2}, x_{i+1/2}] \times [y_{j-1/2}, y_{j+1/2}]$. The midpoint rule is applied to u_t and q. Moreover, by the *Gauss divergence formula* the integral of $\nabla \cdot \mathbf{w}$ over the control volume is equal to the line integral of $\mathbf{w} \cdot \mathbf{n}$ over the boundary of V, where $\mathbf{n}$ is the outward pointing normal. Writing, e.g.,

$$w^x_{i+1/2,j} = \int_{y_{j-1/2}}^{y_{j+1/2}} w^x(x_{i+1/2}, y)\,dy,$$

and similarly for the other three edges of the control volume, we obtain the staggered mesh expression

$$\frac{dv_{i,j}}{dt} = h^{-1}\left[w^x_{i+1/2,j} - w^x_{i-1/2,j} + w^y_{i,j+1/2} - w^y_{i,j-1/2}\right] + q_{i,j}, \quad 1 \le i, j \le J.$$

For (3.14b) we can divide by a, as in Example 3.3 above, and integrate $u_x = a^{-1}w^x$ in x and $u_y = a^{-1}w^y$ in y. But, where in y should we integrate u_x and where in x should we integrate u_y?! This is where the two-dimensional case can differ painfully from the one-dimensional case described in Example 3.3. A simplistic approach is to interpret $w^x_{i+1/2,j}$ as a point value and thus to integrate u_x in x at $y = y_j$. This yields

$$a_{i+1/2,j}\frac{v_{i+1,j} - v_{ij}}{h} = w^x_{i+1/2,j}.$$

Upon subsequent elimination of the fluxes this yields the formula for the primal mesh unknowns $\{v_{i,j}\}$,

$$\frac{dv_{i,j}}{dt} = h^{-2}\left[a_{i+1/2,j}(v_{i+1,j} - v_{i,j}) - a_{i-1/2,j}(v_{i,j} - v_{i-1,j})\right. \quad (3.15a)$$
$$\left. + a_{i,j+1/2}(v_{i,j+1} - v_{i,j}) - a_{i,j-1/2}(v_{i,j} - v_{i,j-1})\right] + q_{i,j}, \quad 1 \le i, j \le J,$$

3.1. Semi-Discretization

where, e.g.,

$$a_{i+1/2,j} = h \left[\int_{x_i}^{x_{i+1}} a^{-1}(x, y_j) dx \right]^{-1}. \tag{3.15b}$$

However, (3.15) is only what we can best do as an extension of the one-dimensional case. If a really varies wildly over the entire control volume in a way that is not miraculously aligned with the axes x or y, then the above line integrals may not capture some essential part of such a variation. Techniques leading to 2×2 tensor coefficients for a have therefore been proposed, and concepts such as **upscaling** and **homogenization** become relevant [109, 123]. The plot thickens quickly, and we leave it at that. ∎

Example 3.5 (Denoising, diffusion, and total variation) Consider the denoising problem depicted in the two upper subfigures of Figure 3.4. Given the noisy image on the upper right, u_0, we would like to recover the clean image u to the extent possible.

One way to do this is to smooth the noise by applying a diffusion filter. Recall from Chapter 1 that the heat operator smooths rough initial data (see in particular Figure 1.4).

Figure 3.4. *Denoising: The "true image" (upper left) is corrupted by adding 20% white noise (upper right) and the latter is fed as data u_0 to a denoising program using isotropic diffusion (lower left) and anisotropic diffusion (lower right). Anisotropic diffusion produces a more pleasing result.*

Thus we set out to solve

$$u_t = u_{xx} + u_{yy} \, [= \nabla \cdot \nabla u],$$
$$u(0, x, y) = u_0(x, y).$$

At the boundary the so-called natural boundary conditions are specified; see Section 3.1.3. The space discretization of the Laplacian operator is as in Example 3.4, specifically (3.15), for $a \equiv 1$. The mesh is determined by the resolution of the original picture, here being 256×256 on a square spatial domain. The natural boundary conditions are discretized as described in Section 3.1.3 below: this is not central to the present example.

Note that the PDE above does not directly describe a physical process to be simulated at all cost. Rather, it describes a diffusion analogy to denoising. The PDE is part of the *computational model* here. In particular, we have an instance of **artificial time** [9]. Alternative computational models can be formulated for the same denoising purpose.

A forward Euler discretization of the diffusion equation reads

$$u^{n+1} = u^n + k\nabla \cdot \nabla u^n, \quad n = 0, 1, 2, \ldots,$$

and this can be interpreted as a *steepest descent* step for the *minimization* problem

$$\min_u \frac{\lambda}{2}\|u - u^0\|_2^2 + \int_\Omega |\nabla u|^2 dx dy$$

with $|\nabla u| = \sqrt{u_x^2 + u_y^2}$, where we let the nonnegative parameter λ vanish and start the iteration from the data, $u^0 = u_0$. The step size k can now be determined by a *line search*. If you are not familiar enough with the Euler–Lagrange equations, please see (3.18)–(3.19). But if you are also not familiar with optimization techniques, then we suggest to just trust us for now about these statements, as a foray into numerical optimization [139] could take a while.

Clearly if we let "time" (i.e., n in the iteration) grow a lot, we'll reach a very flat, undesirable solution, so this integration must be stopped before it is too late. Unfortunately, the result at the bottom left of Figure 3.4 indicates that even after a few (six) steps there is significant smearing of edges in the reconstructed image. Indeed, what we want to smooth out is the random noise, not the edges!

Thus, we define an **anisotropic diffusion** filter that would smooth only in directions where u does not vary too abruptly. The operator $\nabla \cdot \nabla u$ is replaced by $\nabla \cdot (a\nabla u)$ as in (3.13), with $a(x, y)$ small (but still strictly positive) precisely where $|\nabla u|$ is large. There are various ways for doing that [142, 145, 151], and we show here one called **total variation** (TV), where we set $a = 1 / |\nabla u|$. This leads to the PDE

$$u_t = \nabla \cdot \left(\frac{\nabla u}{|\nabla u|} \right).$$

We mention in passing that the operator on the right-hand side of the latter PDE corresponds to *mean curvature*; see, e.g., [140] for details.

A further insight into TV can be obtained by considering the minimization problem that corresponds to the diffusion model above. Across image edges u is discontinuous, hence

$|\nabla u|$ is unbounded, containing a δ-function. But such a function, although integrable, is *not* square integrable! The smearing across edges apparent in the lower left subfigure of Figure 3.4 is the result of the compensation by the discrete algorithm for this lack of integrability in the limit $h \to 0$. Replacing $\int_\Omega |\nabla u|^2 dxdy$ by the *total variation* integral

$$\int_\Omega |\nabla u| dxdy$$

(see also Section 10.3) therefore yields better, sharper results. Modifying the minimization problem accordingly and applying a steepest descent step then corresponds to the TV PDE presented above.

In practice the TV expression is modified in regions where u is very flat, to keep $a > 0$ bounded, either by adding a small positive constant under the square root sign or by switching to least squares. This results in $\nabla \cdot \left(\frac{\nabla u}{\sqrt{u_x^2+u_y^2+\epsilon^2}}\right)$ or $\nabla \cdot \left(\frac{\nabla u}{\max(|\nabla u|,\gamma)}\right)$, respectively, appearing in the TV PDE, where ϵ and γ are suitable positive constants; see [8].

Now the choice of expression for $a_{i+1/2,j}$ becomes uncomfortable, because the expression couples derivatives in both x and y directions. The most common choice for the unmodified TV part is

$$a_{i+1/2,j} = a_{i,j+1/2} = h\left[(u_{i+1,j} - u_{i,j})^2 + (u_{i,j+1} - u_{i,j})^2\right]^{-1/2},$$

and this leads to the reconstruction depicted at the lower right subfigure of Figure 3.4. The result is clearly better than what is obtained using isotropic diffusion.

Of the various possibilities available in the literature for anisotropic diffusion, TV enjoys the most solid theoretical backing. However, in practice it does not always perform best. A 2×2 tensor of coefficients for a has also been proposed [181] in which one tries to smooth only along, and not across, edges. ∎

Example 3.6 (Maxwell's equations in three dimensions) Maxwell's equations for electromagnetic fields read

$$\mu \mathbf{H}_t = -\nabla \times \mathbf{E},$$
$$\epsilon \mathbf{E}_t = \nabla \times \mathbf{H} - \sigma \mathbf{E},$$

where $\mathbf{E}(t, x, y, z) = (E1, E2, E3)^T$ is the electric field in three space dimensions, $\mathbf{H}(t, x, y, z) = (H1, H2, H3)^T$ is the magnetic field, ϵ and μ are known constants, and $\sigma = \sigma(x, y, z)$ is a known function [163, 30]. These six differential equations form in general a hyperbolic system. However, if we are allowed to set $\epsilon \mathbf{E}_t = 0$ (this is called the *quasi-static approximation*), then the system becomes parabolic.

The `curl` operator $\nabla \times$ can be viewed as a skew-symmetric matrix of first order derivatives,

$$\nabla \times = \begin{pmatrix} 0 & -\partial_z & \partial_y \\ \partial_z & 0 & -\partial_x \\ -\partial_y & \partial_x & 0 \end{pmatrix}.$$

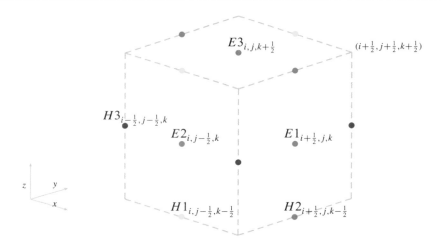

Figure 3.5. *A staggered mesh cell for Maxwell's equations.*

We wish to design a semi-discretization which is centered and uses only short differences. Thus, for instance, the equation

$$\mu H1_t = E2_z - E3_y$$

indicates that $E2$ and $E3$ should be placed on a staggered mesh in such a way that their short differences in the z and y directions, respectively, are centered at the same point where $H1$ is defined. For a uniform mesh consisting of cubic cells, the arrangement of the unknowns is depicted in Figure 3.5. The resulting semi-discrete equations read

$$\mu \frac{dH1_{i,j+\frac{1}{2},k+\frac{1}{2}}}{dt} = \frac{E2_{i,j+\frac{1}{2},k+1} - E2_{i,j+\frac{1}{2},k}}{h} - \frac{E3_{i,j+1,k+\frac{1}{2}} - E3_{i,j,k+\frac{1}{2}}}{h},$$

$$\epsilon \frac{dE1_{i+\frac{1}{2},j,k}}{dt} = -\frac{H2_{i+\frac{1}{2},j,k+1/2} - H2_{i+\frac{1}{2},j,k-\frac{1}{2}}}{h} + \frac{H3_{i+\frac{1}{2},j+\frac{1}{2},k} - H3_{i+\frac{1}{2},j-\frac{1}{2},k}}{h}$$
$$-(\sigma E1)_{i+\frac{1}{2},j,k},$$

etc. Note that each of these six difference equations is centered at a different (though nearby) location on the mesh; see Exercise 12.

A resulting bonus of this staggered mesh discretization is that the discretization of the operator $\nabla \cdot \nabla \times$ (where $\nabla \cdot$ is the divergence operator, $\nabla \cdot \mathbf{H} = H1_x + H2_y + H3_x$) vanishes, like the operator itself. This is a property of invariance reproduced by the discretization scheme. ■

3.1.3 Boundary conditions

In Example 1.2 we demonstrated the treatment of periodic boundary conditions (BC). In this section we discuss basic questions regarding the handling of other simple BCs. More complex situations are delayed until Chapter 8.

3.1. Semi-Discretization

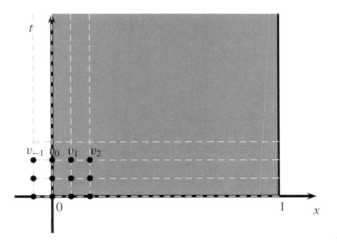

Figure 3.6. *Location of unknowns for an initial-boundary value problem. Beyond the boundary at $x = 0$ are ghost unknowns v^n_{-1}, $n = 0, 1, \ldots$.*

Let us first consider the simple heat equation in a strip,

$$u_t = u_{xx}, \qquad 0 \le x \le 1, \, t > 0,$$
$$u(0, x) = u_0(x);$$

see Figure 3.6. We need BCs both at $x = 0$ and at $x = 1$ for $t > 0$. It can be shown that the problem is well-posed under the boundary conditions

$$\alpha u(t, 0) + \gamma u_x(t, 0) = g_0(t),$$
$$\beta u(t, 1) + \delta u_x(t, 1) = g_1(t), \qquad (3.16)$$

provided that $\alpha^2 + \beta^2 > 0$. This is good news numerically, because we generally need boundary conditions at both ends of the interval in x.

In Example 3.1 we examined the case of Dirichlet BC with the standard three-point formula for u_{xx}. For instance, set $r = 0$ there to obtain

$$\frac{dv_j}{dt} = \frac{v_{j+1} - 2v_j + v_{j-1}}{h^2}.$$

This formula is second order accurate, and in case of Dirichlet BC ($\alpha = \beta = 1$, $\gamma = \delta = 0$ in (3.16)) it is applied for $j = 1, \ldots, J$, where $h(J + 1) = 1$, with $v_0(t) = g_0(t)$ and $v_{J+1}(t) = g_1(t)$.

The situation is a bit more complex for Neumann or mixed BC. Consider the case of Neumann BC at $x = 0$, i.e., $\alpha = 0$, $\gamma = 1$, say, with Dirichlet BC at the other end $x = 1$. Then we must approximate the BC at $x = 0$ appropriately. The approximation order of the BC formula must at least match the order of accuracy of the interior formula, or else a deterioration in the global order of convergence may result, at least locally (see the error bound (3.36)). In the present case the order is 2 and so we can use the scheme

$$\frac{v_1 - v_{-1}}{2h} = g_0(t), \qquad (3.17)$$

$$\frac{dv_j}{dt} = \frac{v_{j+1} - 2v_j + v_{j-1}}{h^2}, \qquad j = 0, 1, \ldots, J.$$

In other words, we define a **ghost unknown** v_{-1} (see Figure 3.6) and apply at the boundary $x = 0$ both discretizations, for the PDE and for the BC.

This allows a local elimination of the ghost unknown v_{-1} for each t. The obtained ODE system is subsequently discretized in time. Alternatively, we can consider the equations of (3.17) without prior elimination. The system resulting from the method of lines is then seen as a system of **differential-algebraic equations** (DAEs) in time. There are packages based on BDF discretizations and on collocation at Radau points which are designed to solve initial value DAE problems—see Section 2.8 and [14]. While for the simple prototype problem considered here the elimination of the ghost unknown is straightforward, in more complex problems and spatial discretizations the more general DAE formulation is attractive because it provides for a higher level mode of programming.

The elimination of v_{-1} from (3.17) yields

$$v_{-1} = v_1 - 2hg_0(t);$$

hence

$$\frac{dv_0}{dt} = \frac{2v_1 - 2v_0 - 2hg_0(t)}{h^2}.$$

It may appear as if this formula is only first order accurate, whereas (3.17) is obviously second order. However, the elimination of v_{-1} in itself of course does not yield an order reduction. It is important when assessing order of accuracy to consider the local truncation error in the PDE and the BC discretizations separately, with the formulas written in divided difference form, as in (3.17). The elimination of v_{-1} involves multiplication by h, hence the misleading formal first order resulting from it.

On the other hand, suppose we were to replace the centered difference formula by a forward formula,

$$\frac{v_1 - v_0}{h} = g_0(t).$$

This carries with it the benefit of no ghost unknowns (the PDE discretization need not be applied at the boundary $j = 0$) and allows local elimination of v_0 as for the Dirichlet BC. However, the resulting scheme is unfortunately only first order accurate indeed. We can consider a second order one-sided formula for $u_x(t, 0)$ based on v_0, v_1, and v_2 (see Section 3.1.1) and eliminate v_0 from that. But the resulting matrix L in (3.2) is no longer tridiagonal. Using the ghost unknown is therefore popular in practice, although it is not always advisable either; see Chapter 8.

The above ideas extend to more space variables, barring tricky geometry issues. Consider first the diffusion equation on a square (or cubic) spatial domain. Handling Dirichlet BC is straightforward: we use the knowledge of the values of u on the boundary to eliminate the corresponding v-values. For a Neumann BC it is useful to imagine the same control

3.1. Semi-Discretization

volume as in Figure 3.3 applied around a boundary node, i.e., with half the volume extending outside the actual domain. The reasoning of Example 3.4 can be directly applied to eliminate ghost unknowns, much like the one-dimensional case (Exercise 6).

The situation becomes more complex with domains whose boundary does not align well with coordinate axes. This is where finite elements shine. But even without finite elements it is possible to cook up local interpolation schemes involving mesh points (nodes) near the boundary where solution values are given in the Dirichlet case (Exercise 7). For Neumann conditions a similar ghost unknown trick can be devised in principle along the normal to the boundary.

Natural and essential BCs

Let us ignore time dependence for the moment and consider a steady-state, elliptic PDE, say, in two spatial dimensions. Such PDEs often arise from a **variational principle**, i.e., as a necessary condition for a functional minimization problem. For instance, the problem

$$\min_u \int_\Omega \left[a|\nabla u|^2 + bu^2 - 2uq\right] dx dy, \tag{3.18}$$

where $a(x, y) > 0$, $b(x, y) \geq 0$, and q are square integrable (i.e., they are functions in $\mathcal{L}_2(\Omega)$), and $u(x, y)$ is required to have bounded first derivatives (i.e., $u \in \mathcal{H}_1(\Omega)$), has the necessary condition called the **Euler–Lagrange equation**,

$$-\nabla \cdot (a\nabla u) + bu = q \quad \text{in } \Omega, \tag{3.19a}$$

$$\frac{\partial u}{\partial n}|_{\partial\Omega} = 0. \tag{3.19b}$$

The homogeneous Neumann BCs (3.19b) are therefore *natural*. If the minimization problem (3.18) is subject to some Dirichlet BC, say, that u vanish on part of the boundary $\partial\Omega_1 \subset \partial\Omega$, then the PDE problem (3.19) would be subject to the same, *essential* BC, and the boundary condition (3.19b) becomes

$$u|_{\partial\Omega_1} = 0, \quad \frac{\partial u}{\partial n}|_{\partial\Omega_2} = 0, \tag{3.19c}$$

where $\partial\Omega_2 = \partial\Omega \setminus \partial\Omega_1$.

Next, to solve a PDE problem of the form (3.19) we can discretize the minimization formulation (3.18). Assuming for notational simplicity that Ω is the unit square and $\partial\Omega_2 = \partial\Omega$, i.e., no essential BC, and using a simple centered discretization for the integrals on each mesh element, we obtain

$$\begin{aligned}\min_v \quad & \frac{1}{2}\sum_{i,j=0}^{J}[a_{i+1/2,j}(v_{i+1,j} - v_{i,j})^2 + a_{i+1/2,j+1}(v_{i+1,j+1} - v_{i,j+1})^2 \\ & + a_{i,j+1/2}(v_{i,j+1} - v_{i,j})^2 + a_{i+1,j+1/2}(v_{i+1,j+1} - v_{i+1,j})^2] \\ & + \frac{h^2}{4}\sum_{i,j=0}^{J}[b_{i,j}v_{i,j}^2 - 2q_{i,j}v_{i,j} + b_{i+1,j}v_{i+1,j}^2 - 2q_{i+1,j}v_{i+1,j} \\ & + b_{i,j+1}v_{i,j+1}^2 - 2q_{i,j+1}v_{i,j+1} + b_{i+1,j+1}v_{i+1,j+1}^2 - 2q_{i+1,j+1}v_{i+1,j+1}],\end{aligned} \tag{3.20}$$

i.e., an algebraic minimization problem for $(J+2)^2$ unknowns $\{v_{i,j}\}$. The necessary conditions for the latter minimization problem yield a discretization for the PDE problem (3.19), *including the natural BC*. In short, we discretize and then minimize, rather than the other way around; see Section 11.1.

The resulting set of necessary conditions corresponds to that derived in Example 3.4. At interior mesh points we have

$$-\left[a_{i+1/2,j}(v_{i+1,j} - v_{i,j}) - a_{i-1/2,j}(v_{i,j} - v_{i-1,j})\right.$$
$$+ a_{i,j+1/2}(v_{i,j+1} - v_{i,j}) - a_{i,j-1/2}(v_{i,j} - v_{i,j-1})\Big]$$
$$+ h^2 b_{i,j} v_{i,j} = h^2 q_{i,j}, \quad 1 \leq i, j \leq J.$$

The natural BC discretization can be obtained alternatively as described earlier in this subsection using a control volume at the boundary. Indeed, (3.20) is viewed as writing the original integral over the unit square as the sum of integrals over mesh control volumes which are subsequently discretized in an obvious way.

Writing the necessary conditions as

$$A\mathbf{v} = \mathbf{q},$$

where $\mathbf{v}$ is the mesh function $\mathbf{v}$ reshaped as a vector, it is clear that A is symmetric, although the discretization (3.20) involves only forward differences. If $b > 0$ or $\partial \Omega_1$ is not empty, then A is also positive definite. Otherwise A is only positive semi-definite with a constant null-space, i.e., $\mathbf{v}$ is determined only up to a constant as is the exact solution u of (3.19).

3.1.4 The finite element method

The finite element method (FEM) is in fact an approach leading to a class of discretization methods that provide an alternative to the finite difference and finite volume families. A structured approach, it usually gives rise to more solid mathematical theory, which makes it the darling of many a mathematician. Its practical utility lies mainly in the general and elegant treatment of domains of general shapes whose boundaries do not align with coordinate axes, as is often the case with applications in several space variables.

The derivation in Section 3.1.3 through functional minimization is fundamental also to basic, so-called *conforming* finite element methods. For instance, the domain Ω can be triangulated, which adds great flexibility when domains with complex geometry are encountered; see Figure 3.7. An approximate solution is sought in the form of a piecewise polynomial function over the elements. On each triangle this solution reduces to a polynomial of a certain degree, and these polynomials are patched together along interelement boundaries. The requirement that the resulting piecewise polynomial only be continuous with bounded first derivatives (and not continuously differentiable), which is enabled by the minimization formulation, is of great practical importance. Finite volume methods can mimic the finite element construction as well, although finite element theory is more elegant.

However, the FEM approach can also be more cumbersome to derive and more expensive to carry out than finite difference and finite volume methods, hence it is rarely used for time or one space variable discretizations. Its main application has traditionally been for elliptic-type problems that may of course arise from semi-discretizations of time-dependent problems in more than one space variable.

3.1. Semi-Discretization

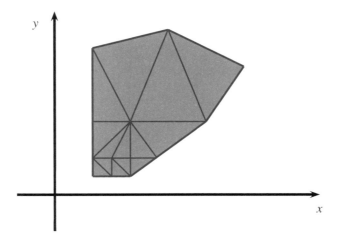

Figure 3.7. *A triangulation of a nonrectangular domain in two dimensions.*

Therefore we give here a brief introduction to the FEM. Many textbooks have been written on this subject, including [158, 104, 67, 73, 35, 59]. Needless to say, our few pages are not meant to replace any FEM textbook. Nonetheless, it is important to note also that most issues with which this book is concerned arise in the context of FEM methods for time-dependent problems as well.

To focus attention let us consider the Poisson equation on a domain Ω in two dimensions,

$$-(u_{xx} + u_{yy}) \equiv -\Delta u = q, \quad (x, y) \in \Omega. \tag{3.21}$$

The domain boundary $\partial\Omega$ is assumed smooth and a combination of homogeneous Dirichlet and Neumann BCs as in (3.19) is imposed on it. To be specific, assume BCs in the form (3.19c). The given right-hand-side function $q(x, y)$ is assumed to be square integrable, $q \in \mathcal{L}_2$. Then $u \in \mathcal{H}_2$; see Section 1.3.3.

In the FEM the domain Ω is replaced by a union of subdomains of simple geometric shapes—triangles or rectangles in two dimensions, as in Figure 3.7. These are called **elements**. Let h represent the spatial mesh width, say, the diameter of the largest element. In general there is a "skin," not shown in the figure, that is the difference between Ω and the union of elements Ω_h approximating it, which is left out by this process, possibly adding an $O(h)$ error component to the approximation. A globally continuous piecewise polynomial approximate solution is then sought. It is important to ensure that no very small angles arise in such a triangulation, since these would introduce a large approximation error in the computed solution [158].

We are looking for a solution u_h that lies in the space V_h of all continuous piecewise polynomials that satisfy the Dirichlet (essential) part of the homogeneous BC (3.19c).

The functions in V_h have bounded first derivatives, but second derivatives may be generally unbounded, unlike those of the exact solution u. However, the space in which u is sought can be correspondingly enlarged by considering the **variational form**. If V is the space of all functions that satisfy the essential BC such that v, v_x, and v_y are square integrable (see the review of function spaces in Section 1.3.3), then for each $v \in V$ we can integrate by parts to obtain

$$\iint_\Omega -(u_{xx} + u_{yy})v \, dx dy = \iint_\Omega (u_x v_x + u_y v_y) dx dy.$$

Denote[14]

$$b(u, v) = \iint_\Omega (u_x v_x + u_y v_y) dx dy, \quad (u, v) = \iint_\Omega uv \, dx dy. \qquad (3.22a)$$

Then (3.21) has the variational formulation: find $u \in V$ such that for all $v \in V$

$$b(u, v) = (q, v). \qquad (3.22b)$$

Obviously the solution of our problem (3.21) satisfies the weaker form (3.22). More interestingly, the converse can be shown to hold as well.

Now, the **Galerkin** FEM discretizes the variational form (3.22). Thus we seek $u_h \in V_h$ such that for any $v_h \in V_h$

$$b(u_h, v_h) = (q, v_h). \qquad (3.23)$$

This FEM is called **conforming** since $V_h \subset V$.

How is the computation mandated by (3.23) carried out in practice? We describe the piecewise polynomial functions using a **basis**. Let the functions $\phi_i(x, y) \in V_h$, $i = 1, \ldots, J$, be such that any $v_h \in V_h$ can be uniquely written as

$$v_h(x, y) = \sum_{i=1}^{J} \alpha_i \phi_i(x, y)$$

for a set of coefficients α_i. Then we are seeking the coefficients α_i^* that define u_h by requiring

$$\sum_{i=1}^{J} b(\phi_i, \phi_j) \alpha_i^* = b\left(\sum_{i=1}^{J} \alpha_i^* \phi_i, \phi_j\right) = (q, \phi_j), \quad j = 1, \ldots, J.$$

This yields a set of J linear equations for the J coefficients α_i^*.

Assembling this linear system of equations, the resulting matrix A is called the **stiffness matrix** for historical reasons dating back to the days when civil engineers invented this method. In the (i, j)th position the stiffness matrix has $b(\phi_i, \phi_j)$. Of course, we seek a basis such that assembling A and solving the resulting linear algebraic equations are as efficient as possible. For a piecewise linear approximate solution on a triangulation as in Figure 3.7 this is achieved by associating with each vertex, or *node* (x_i, y_i), a **roof function** (or **tent function**, depending perhaps on what comfort level one is used to) $\phi_i \in V_h$ that vanishes at all nodes other than the ith one and satisfies $\phi_i(x_i, y_i) = 1$. This gives three

---

[14]We are using a different notation here than elsewhere, in that v is not the approximation for u. Rather, u_h is. This is standard notation in the FEM literature.

3.1. Semi-Discretization

interpolation conditions at the vertices of each triangle, determining a linear polynomial, and the global solution pieced together from these triangle contributions is thus once differentiable. Then, unless vertex j is an immediate neighbor of vertex i we have $b(\phi_i, \phi_j) = 0$ because the regions on which these two functions do not vanish do not intersect. Moreover, the construction of nonzero entries of A becomes relatively simple, although it still involves quadrature approximations for the integrals that arise.

Note that
$$\alpha_i^* = u_h(x_i, y_i).$$

This is called a **nodal** method. The obtained scheme has many characteristics of a finite difference method, including sparsity of A, with the advantages that the mesh has a more flexible shape. Note also that A is symmetric positive definite, as in the finite volume approach through minimization described earlier in Section 3.1.3.

The assembly of the FEM equations, especially in more complex situations than for our model problem (3.21), can be significantly more costly. Moreover, for different PDEs such as in Examples 3.4 to 3.6, different elements (e.g., **mixed** finite elements [35]) are required. Nevertheless, for problems with complex geometries the FEM is the approach of choice, and most general packages for solving elliptic boundary value problems are based on it.

In recent years a family of nonconforming FEM called **discontinuous Galerkin** has become a red-hot research area. As the name implies, these methods do allow solution discontinuities in between elements, with penalties on the jumps to ensure they remain minuscule where the exact solution is smooth. This gives an opportunity, in addition to providing flexibility in other aspects, for entertaining actual discontinuities in the solution, the like of which often occurs in nonlinear hyperbolic-type PDEs. So, here is a platform for designing methods that both capture discontinuities and work on complex spatial geometries. See, e.g., [51, 50].

3.1.5 Nonuniform meshes

The finite difference and finite volume meshes considered in this section so far are all uniform in space: the width of all spatial steps is the same h. There are circumstances, however, where a nonuniform mesh may be more effective, requiring significantly fewer points than a uniform mesh to represent the sought solution adequately. Example 2.6 provides one such instance; see Figures 2.8 and 2.9. (The fact that the independent variable there is "time" is immaterial here.) Essentially, we want more mesh points in regions of the domain Ω where the solution varies rapidly and fewer points where u varies slowly. Other situations where we would want a denser mesh in particular subdomains are singularities in the PDE, noncartesian coordinate systems, and local boundary articulation in more than one space variable.

Questions arise as to how to estimate the local solution error, or at least obtain a local error indicator based on which a nonuniform mesh may be specified. However, available algorithms in the much more complex PDE context are typically significantly cruder than those treated in Section 2.4 for the initial value ODE case. We delay further discussion of these issues until Section 11.2. Here we concentrate on maintaining nonuniform spatial meshes and discretizing function derivatives on them.

In one space variable things are relatively straightforward. The spatial domain $\Omega = [0, 1]$, say, can be subdivided by a nonuniform mesh,

$$0 = x_0 < x_1 < \cdots < x_{J+1} = 1, \tag{3.24}$$

with the widths $h_j = x_{j+1} - x_j$, $j = 0, 1, \ldots, J$, not necessarily all equal. A one-step discretization is defined with no serious change from the uniform mesh case. For discretizations spanning more than one subinterval see Exercises 2, 3, and 4.

The nonuniform mesh may be viewed as having been obtained by a **coordinate transformation** from a uniform mesh. Thus, denote $h = 1/(J + 1)$ and let $\Psi : [0, 1] \longrightarrow [0, 1]$ be a sufficiently smooth, monotonically increasing function satisfying

$$\Psi(jh) = x_j, \quad j = 0, 1, \ldots, J + 1. \tag{3.25a}$$

Often the nonuniform mesh is actually obtained by selecting an appropriate map Ψ. Otherwise we may envision a sufficiently smooth spline interpolant, say. Since $\Psi(\xi)$ is monotonic[15] it has an inverse

$$\Phi = \Psi^{-1}. \tag{3.25b}$$

Figure 3.8 depicts such a mesh transformation.

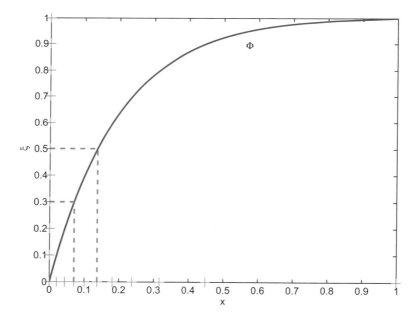

Figure 3.8. *A nonuniform mesh transformation. Here* $\xi = \Phi(x) = c(1 - e^{-\alpha x})$, *where* $\alpha = 5$ *and c is a constant such that* $\Phi(1) = 1$.

---

[15]Note that ξ is just the argument of the mesh function. There is no wave number in sight so no notational confusion should result.

3.1. Semi-Discretization

Writing $\xi = \Phi(x)$ we have in the new independent variable ξ

$$\frac{\partial u}{\partial x} = \left(\frac{\partial \xi}{\partial x}\right)\frac{\partial u}{\partial \xi} = \Phi'(\Psi(\xi))\frac{\partial u}{\partial \xi},$$

$$\frac{\partial^2 u}{\partial x^2} = \Phi'(\Psi(\xi))\frac{\partial}{\partial \xi}\left(\Phi'(\Psi(\xi))\frac{\partial u}{\partial \xi}\right).$$

Thus, a discretization on a nonuniform mesh of u_x and u_{xx} is equivalent to the discretization on a uniform mesh of the modified expressions in ξ given by the right-hand sides of the above identities.

Example 3.7 The ODE

$$\frac{d^2 u}{dx^2} = q(x)$$

for some given function q, when discretized on a nonuniform mesh (3.24), becomes the ODE

$$\frac{d}{d\xi}\left(a(\xi)\frac{du}{d\xi}\right) = \frac{q(\Psi(\xi))}{a(\xi)}, \quad a(\xi) = \Phi'(\Psi(\xi)),$$

discretized on the corresponding uniform mesh. For the coordinate transformation function depicted in Figure 3.8,

$$x = \Psi(\xi) = -\frac{1}{\alpha}\ln(1 - \xi/c), \quad c = \frac{1}{1 - e^{-\alpha}},$$

where $\alpha > 0$ is a parameter, we have the inverse transformation

$$\xi = \Phi(x) = c\left(1 - e^{-\alpha x}\right),$$

hence $\Phi'(x) = \alpha c e^{-\alpha x} \geq 0$, and

$$a(\xi) \equiv \Phi'(\Psi(\xi)) = \alpha(c - \xi).$$

Note that a is positive, bounded away from 0, and grows in magnitude linearly with α.

The price to pay for moving to a uniform mesh is a problem with variable coefficients, and the variable coefficient becomes steeper the further the mesh is from being uniform. ∎

More complications arise in several space variables. The simplest extension of the uniform case is where a cartesian mesh has different uniform step sizes in different directions, say, h^x in the x-direction, h^y in the y-direction, and h^z in the z-direction. For instance, in geophysical applications often the depth variable z is different in scale from the horizontal variables x and y, so a different mesh is naturally required for its discretization. Such an extension of the uniform case is straightforward, because our derivatives are usually in one direction only: in $u_{xx} + u_{yy}$ the term u_{xx} is discretized on a uniform mesh in x with step h^x while u_{yy} is discretized on a uniform mesh in y with step h^y.

The situation remains essentially similar so long as the mesh is in **tensor product** form. This means that there are one-dimensional meshes like (3.24) for each spatial coordinate.

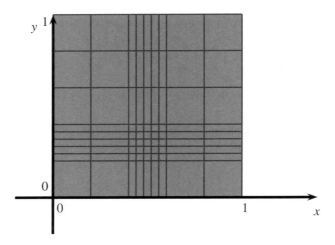

Figure 3.9. *This tensor product mesh should be fine when both* $.4 \leq x \leq .6$ *and* $.2 \leq y \leq .4$, *but instead it is fine when either* $.4 \leq x \leq .6$ *or* $.2 \leq y \leq .4$.

Each mesh point can be written, say, in two dimensions, as (x_i, y_j), with x_i from the x-mesh and y_j from the y-mesh. Bookkeeping is still easy for such a mesh and derivative discretizations are as discussed above and in the exercises for the one-dimensional case.

However, the tensor product mesh may not always be adequate. For one thing the domain Ω need not be rectangular; see, e.g., Figure 3.7. Even if it is, the mesh may be fine not only where it should be; see Figure 3.9. In contrast, finite element meshes are naturally local in several space variables. In Figure 3.7 the smaller triangles do not affect the size or shape of the larger triangles. This ability to refine the mesh in a relatively local manner is even more important in three dimensions.

The finite element mesh of Figure 3.7 is not rigidly ordered; indeed it is **unstructured**. Care must be taken when refining unstructured meshes to avoid small and large obtuse angles in triangular (or tetrahedral in three-dimensional) elements. This flexibility unfortunately also means that the data structure required to store and manipulate the mesh (e.g., when refining it) is more involved, and that the sparsity of the stiffness matrix may be complicated; see Section 3.1.4. The latter, in turn, may adversely affect the efficiency of an iterative algorithm for solving linear equations that may arise for such a matrix; see Section 9.2.1.

In the finite volume context it is better then to allow only a controlled refinement. In a **Quadtree** data structure the mesh is defined by means of successive subdivisions that may be carried out locally. Starting from a square mesh, say, a rectangular finite volume (or cell, or "element") in two dimensions may be divided if need be into four equal subrectangles. This subdivision operation may be repeated recursively if needed; see Figure 3.10. Each node of the corresponding tree points to a mesh cell, and the subdivision operation of a cell is recorded as the tree node having four children. The tree leaves correspond to the actual resulting mesh [93]. If h is the maximum width of all rectangular elements and h_{min} is the width of the finest one, then
$$h = 2^l h_{min},$$
where l is the depth of the tree.

3.1. Semi-Discretization

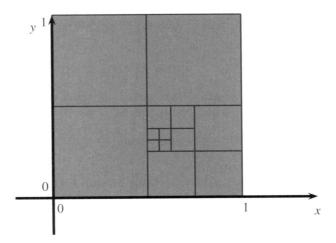

Figure 3.10. *A Quadtree. The root of the tree is associated with the original domain, here the unit square. It has four children nodes, three of which are leaves. The remaining, southeast child has four children, of which the southwest one further has four children. The depth of the tree is therefore $l = 3$.*

The ratio of largest to smallest neighboring cell widths is called the **local mesh ratio**, to be distinguished from the **global mesh ratio**, which is the ratio of largest to smallest over all cell widths. In Figure 3.10 both ratios equal 8. Often in practice the refinement is restricted so that the local mesh ratio is at most 2, for reasons related both to the handling of the data structure and to the accuracy of the computed results. The global ratio can be much larger.

In three dimensions each cell is a box, and a simple subdivision as above replaces such a cell by $2^3 = 8$ subboxes. The corresponding data structure is then an **Octree**. The details get hairier but in principle the situation is analogous to that in two dimensions with the Quadtree.

To discretize on a Quadtree or an Octree we may generally regard the values of a function $u(x, y)$ or $u(x, y, z)$ as given at the nodes or at the cell centers. Supposing that there are u-values at the nodes of the mesh, the discretization of u_x or u_{yy} is again along x or y directions separately, and the one-dimensional methodology applies. The situation becomes more complex, however, when a *staggered mesh* as in Examples 3.4–3.6 is contemplated. Here the flux (for instance) is defined at cell edges, and there is the question of how to center it and the resulting approximations properly. We leave this to Exercise 9; see [122, 79].

Example 3.8 Consider the problem

$$\min_u I(u) = \int_\Omega \left[|\nabla u|^2 - 2uq\right] dxdy$$

on the unit square with homogeneous Dirichlet BC

$$u|_{\partial\Omega} = 0.$$

The necessary and sufficient condition for the minimum is the Euler–Lagrange equation

$$-\Delta u = q,$$
$$u|_{\partial\Omega} = 0.$$

This is obviously a special case of (3.19). Assume that the solution u is smooth and has as many bounded derivatives as we wish; for instance, let

$$u(x, y) = \sin(\pi x)\sin(\pi y),$$

obtained if $q(x, y) = 2\pi^2 \sin(\pi x)\sin(\pi y)$.

Now, consider the approach of discretizing the integral on a Quadtree mesh followed by minimizing the discrete equations. The unit square is subdivided into a union of square cells c_l, $l = 1, \ldots, L$, so

$$I(u) = \sum_{l=1}^{L} \int_{c_l} \left[|\nabla u|^2 - 2uq\right] dxdy,$$

and we can approximate the integral over each cell to second order by applying the trapezoidal rule. Denoting the solution values at the four corners of the lth cell by u_l^{NE}, u_l^{NW}, u_l^{SE}, and u_l^{SW}, and the cell's width by h_l, we have

$$\int_{c_l} \left[|\nabla u|^2 - 2uq\right] dxdy$$
$$= \frac{1}{2}\left[(u_l^{NE} - u_l^{NW})^2 + (u_l^{SE} - u_l^{SW})^2\right] + \frac{1}{2}\left[(u_l^{NE} - u_l^{SE})^2 + (u_l^{NW} - u_l^{SW})^2\right]$$
$$- \frac{h_l^2}{2}\left[u_l^{NE} q_l^{NE} + u_l^{NW} q_l^{NW} + u_l^{SE} u_l^{SE} + u_l^{SW} q_l^{SW}\right] + O(h_l^4).$$

Adding these approximations, we obtain a discrete approximation $I_h(u)$ which satisfies $I_h(u) = I(u) + O(h^2)$, where h is the maximum cell size.

Next we then look for a mesh function v that minimizes the approximate integral, i.e., $v = \operatorname{argmin}_w I_h(w)$. Naturally, we expect

$$\|v - Pu\| = O(h^2),$$

where P projects the exact solution u on the Quadtree mesh by simple injection.

This is precisely what happens when the mesh is uniform. The equations resulting from equating the gradient of $I_h(v)$ to 0 are those obtained by the usual five-point difference formula, i.e., (3.15a) in the steady state with $a \equiv 1$ there. The calculated maximum error behaves like $O(h^2)$.

However, Haber has pointed out that when the Quadtree mesh is not uniform an unwelcome surprise arises: the observed errors are significantly larger and the maximum error is now only $O(h)$. How can that be?!

Let us take a closer look at the part of the unit square depicted in Figure 3.11 with $h_1 = h$. The red point is a **dangling node**, a phenomenon that occurs neither for a uniform

3.1. Semi-Discretization

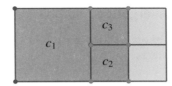

Figure 3.11. *Discretizing Δu on a Quadtree. At the red point the obtained stencil uses only solution values at the green points in a way that would yield second order accuracy only if the solution value at the cell center of c_1 equals $u_2^{NE} = u_3^{SE}$ up to $O(h^2)$.*

mesh nor for the triangle mesh of Figure 3.7. This node is a legitimate vertex of the cells c_2 and c_3, but it is not participating in defining the approximation on c_1 even though it is on its edge. Now, looking at the equation formed by $\frac{\partial I_h(v)}{\partial v_*} = 0$, where $v_* = v_2^{NW} = v_3^{SW}$ is at the red point, we obtain

$$4v_* - v_2^{SW} - v_3^{NW} - (v_2^{NE} + v_3^{SE}) = h^2 q_*.$$

Noting further that $v_2^{NE} = v_3^{SE}$, we obtain the same discretization stencil as for the *homogeneous Neumann BC* on a uniform mesh, depicted in Figure 3.6 and (3.17) with $g_0 = 0$ there. The dangling node has created an artificial boundary across the wall on which it is placed! Now, in general of course the value of u at the center of c_1, which plays "ghost unknown" here, is only equal to the value of u_2^{NE} up to $O(h)$, not $O(h^2)$. Hence we have only an $O(h)$ truncation error at the red point.

How do we fix this?

1. We can fix the value of v_* as an interpolation of its closest green and/or blue neighbors, or perhaps only those which share the problematic edge with it. However, this generates a nonsymmetric matrix to invert.

2. We can move to FEM as follows. Replace c_1 by three triangles, adding edges from the red point to the blue points in Figure 3.11. Replace also the smaller squares by two triangles each. This creates a triangle mesh on which a piecewise linear Galerkin FEM can be defined, for instance. The resulting stiffness matrix is symmetric positive definite. The Quadtree is still useful in keeping tracks of activities on this mesh.

3. Do nothing and live with the reduced order. The rationale here is that the example we have used is artificially bad in that there has to be a reason for refinement in the first place. In the present example, in fact, we have none (except of course for the demonstrative aspect). Now, if u has large gradients across the wall on which the dangling node is placed, suggesting mesh refinement there, then the difference between the first order and second order truncation errors is less noticeable. Also, there are applications such as certain inverse problems in which other error components are not smaller than this $O(h)$ term anyway. ∎

3.1.6 Stability and convergence

Earlier in this section we defined order of accuracy p of a semi-discretization (3.2). If we let the error vector at mesh points be

$$\mathbf{e}(t) = \mathbf{u} - \mathbf{v},$$

then from (3.4) the error satisfies

$$\mathbf{e}_t = L\mathbf{e} + \mathbf{d},$$

where $\mathbf{d}$ is the vector of truncation errors. The accuracy of the discretization scheme implies that the inhomogeneity $\mathbf{d}$ in the differential equation which $\mathbf{e}$ satisfies shrinks like h^p as $h \to 0$.

The semi-discretization (3.2) starting from exact initial values is *convergent* if $\mathbf{e} \to \mathbf{0}$ as $h \to 0$. In fact, as indicated earlier, for a method of order p we would like to obtain a result of the form $\|\mathbf{e}\| = O(h^p)$.

Analogously to the well-posedness requirement (1.7) for the homogeneous initial value PDE (i.e., with $q = 0$), we say that the semi-discretization is *stable* if there are constants $\hat{K}$, $\hat{\alpha}$, and h_0 such that for all $0 < h \leq h_0$

$$\|\mathbf{v}(t)\| \leq \hat{K} e^{\hat{\alpha} t} \|\mathbf{v}(0)\| \quad \forall \, \mathbf{v}(0) \in l_2. \tag{3.26}$$

If the large ODE system (3.2) is *stable*, then indeed we have convergence of order p. The proof, and indeed the entire derivation, are sufficiently similar to the full discretization case that we refer to the treatment in Section 3.2.2 rather than repeating its variation here.

3.2 Full discretization

An obvious way to obtain a full discretization scheme for a time-dependent PDE such as (3.1) is to discretize in time the corresponding semi-discretization (3.2). There is a lot of knowledge from the ODE literature about methods that may prove suitable here, and the information provided in Chapter 2 should come in handy. In recent years, BDF discretizations for parabolic problems and certain explicit Runge–Kutta methods for hyperbolic problems have become popular.

Corresponding to (3.3) we write a general explicit difference scheme at time $t = t_n$ as

$$v_j^{n+1} = \sum_{i=-l}^{r} \beta_i v_{j+i}^n. \tag{3.27a}$$

Forming the mesh function $v^n = \{v_j^n\}$ as in Section 1.2 this can be expressed as

$$v^{n+1} = Q v^n, \tag{3.27b}$$

where Q is a linear operator on l_2. The index j can have a finite or an infinite range, and correspondingly Q can have a representation as a finite (but very large) or an infinite sparse matrix. For the pure initial value problem (Cauchy problem) we write

3.2. Full Discretization

$$Q = \begin{pmatrix} \ddots & \ddots & \ddots & \ddots & & & & \\ & \ddots & \ddots & \ddots & \ddots & & & \\ & & \ddots & \ddots & \ddots & \ddots & & \\ & & & \ddots & \ddots & \ddots & \ddots & \\ & \beta_{-l} & \ddots & \beta_{-1} & \beta_0 & \beta_1 & \ddots & \beta_r \\ & & \ddots & \ddots & \ddots & \ddots & \ddots & \\ & & & \ddots & \ddots & \ddots & \ddots & \ddots \\ & & & & \ddots & \ddots & \ddots & \ddots \end{pmatrix}.$$

For an implicit difference scheme we have similarly

$$\sum_{i=-l}^{r} \gamma_i v_{j+i}^{n+1} = \sum_{i=-l}^{r} \beta_i v_{j+i}^{n}, \tag{3.27c}$$

or

$$Q_1 v^{n+1} = Q_0 v^n, \tag{3.27d}$$

where we assume that Q_1 is an invertible operator. Thus, with $Q = Q_1^{-1} Q_0$ we get the representation (3.27b) for both explicit and implicit schemes, although in the implicit case only Q_0 and Q_1, but not Q, are sparse.

For *multistep schemes*, e.g.,

$$v_j^{n+1} = \sum_{i=-l}^{r} \beta_i v_{j+i}^n + \sum_{i=-l}^{r} \gamma_i v_{j+i}^{n-1}, \tag{3.28a}$$

we can obtain the notation (3.27b) by converting to a system over one step (just for formulation and analysis purposes!),

$$v_j^{n+1} = \sum_{i=-l}^{r} \beta_i v_{j+i}^n + \sum_{i=-l}^{r} \gamma_i w_{j+i}^n,$$
$$w_j^{n+1} = v_j^n. \tag{3.28b}$$

Example 3.9 For the hyperbolic system of s PDEs,

$$\mathbf{u}_t = A(t, x) \mathbf{u}_x,$$

the leapfrog scheme reads

$$\mathbf{v}_j^{n+1} = \mathbf{v}_j^{n-1} + \mu A_j^n (\mathbf{v}_{j+1}^n - \mathbf{v}_{j-1}^n),$$

$$\mathbf{v}_j^{n+1} = \mathbf{v}_j^{n-1} + \mu A_j^n (\mathbf{v}_{j+1}^n - \mathbf{v}_{j-1}^n)$$

$$(1 + \tfrac{4}{3}\mu) v_j^{n+1} - \tfrac{2}{3}\mu (v_{j+1}^{n+1} + v_{j-1}^{n+1})$$
$$= \tfrac{4}{3} v_j^n - \tfrac{1}{3} v_j^{n-1}$$

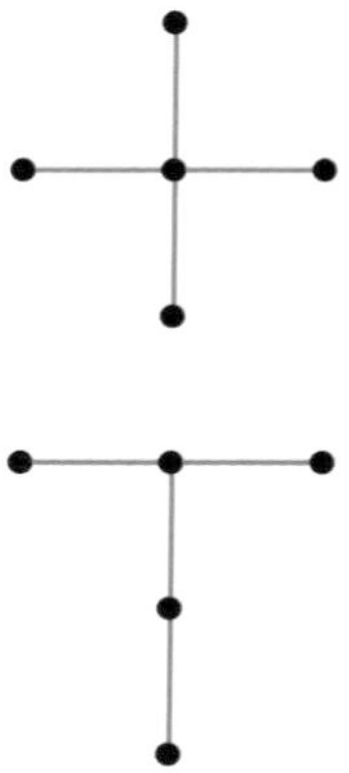

Figure 3.12. *Computational molecules: leapfrog for a hyperbolic system and two-step BDF for the heat equation.*

where $\mu = \frac{k}{h}$; see Figure 3.12. Here $l = r = 1$. Using (3.28b) this becomes a system of size $2s$ at each time step,

$$\begin{pmatrix} \mathbf{v} \\ \mathbf{w} \end{pmatrix}_j^{n+1} = \begin{pmatrix} 0 & I \\ I & 0 \end{pmatrix} \begin{pmatrix} \mathbf{v} \\ \mathbf{w} \end{pmatrix}_j^n + \begin{pmatrix} \mu A_j^n & 0 \\ 0 & 0 \end{pmatrix} \begin{pmatrix} \mathbf{v} \\ \mathbf{w} \end{pmatrix}_{j+1}^n - \begin{pmatrix} \mu A_j^n & 0 \\ 0 & 0 \end{pmatrix} \begin{pmatrix} \mathbf{v} \\ \mathbf{w} \end{pmatrix}_{j-1}^n.$$

The resulting matrix Q is *block-tridiagonal* with $2s \times 2s$ blocks.

Next, for the heat equation

$$u_t = u_{xx}$$

we can apply the two-step implicit BDF method (see Figure 3.12),

$$\left(1 + \frac{4}{3}\mu\right) v_j^{n+1} - \frac{2}{3}\mu \left(v_{j+1}^{n+1} + v_{j-1}^{n+1}\right) = \frac{4}{3} v_j^n - \frac{1}{3} v_j^{n-1},$$

where $\mu = \frac{k}{h^2}$. This is formulated as (3.27c) and as (3.27b) in Exercise 10. ∎

3.2.1 Order, stability, and convergence

Let us write the general difference scheme (3.27c) in a divided difference form,

$$k^{-1} \left(\sum_{i=-l}^{r} \gamma_i v_{j+i}^{n+1} - \sum_{i=-l}^{r} \beta_i v_{j+i}^n \right) = 0.$$

The **local truncation error** is the amount by which the exact solution $u(t, x)$ fails to satisfy this equality:

$$\tau(t, x) = k^{-1} \left[\sum_{i=-l}^{r} \gamma_i u(t+k, x+ih) - \sum_{i=-l}^{r} \beta_i u(t, x+ih) \right]. \quad (3.29)$$

3.2. Full Discretization

As in Section 1.2 we relate to the mesh function v as if it were defined at every point, $v = v(t, x)$, and define

- The difference method is **accurate of order** (p_1, p_2) if for any sufficiently smooth function $u(t, x)$

$$\|\tau(t)\| = \|\tau(t, \cdot)\| \le c(t) (k^{p_1} + h^{p_2}), \tag{3.30}$$

where $c(t)$ is bounded on any fixed finite interval in time, $0 \le t \le t_f$, independently of k and h.

- The difference method is **consistent** if $\|\tau(t)\| \to 0$ as $k, h \to 0$. Thus, any method accurate to first order or higher in both t and x is consistent.

Example 3.10 Continuing with Example 3.1, where we set $r \equiv 0$ and $g_0 \equiv g_1 \equiv 0$, we have the ODE system

$$\frac{dv_j}{dt} = \frac{v_{j+1} - 2v_j + v_{j-1}}{h^2},$$

$$v_j(0) = u_0(x_j)$$

for $j = 1, \ldots, J$. Note that for each t a straightforward manipulation of Taylor expansions yields

$$\frac{u_{j+1} - 2u_j + u_{j-1}}{h^2} = u_{xx}(t, x_j) + \frac{h^2}{12} u_{xxxx}(t, x_j) + O(h^4).$$

- Discretizing the time variable using an explicit Euler method (2.5) gives the scheme

$$\frac{1}{k}(v_j^{n+1} - v_j^n) = \frac{1}{h^2}(v_{j+1}^n - 2v_j^n + v_{j-1}^n),$$

or

$$v_j^{n+1} = v_j^n + \mu(v_{j+1}^n - 2v_j^n + v_{j-1}^n), \tag{3.31a}$$

where $\mu = k/h^2$; see Figure 3.13. This scheme is explicit and compact. By Taylor's expansion about (t_n, x_j),

$$\tau(t_n, x_j) = \frac{1}{k}(u_j^{n+1} - u_j^n) - \frac{1}{h^2}(u_{j+1}^n - 2u_j^n + u_{j-1}^n)$$

$$= u_t + \frac{k}{2} u_{tt} + O(k^2) - u_{xx} - \frac{h^2}{12} u_{xxxx} + O(h^4).$$

Since $u_t = u_{xx}$, we see that $\tau = O(k) + O(h^2)$ provided u_{tt} and u_{xxxx} are bounded, i.e., the method has order $(1, 2)$.

If we freeze h for a moment and write the ODE system as in (3.2) in matrix form

$$\frac{d\mathbf{v}}{dt} = L\mathbf{v},$$

$$v_j^{n+1} = v_j^n + \mu(v_{j+1}^n - 2v_j^n + v_{j-1}^n)$$

$$(1+\mu)v_j^{n+1} - \tfrac{1}{2}\mu(v_{j+1}^{n+1} + v_{j-1}^{n+1}) =$$
$$(1-\mu)v_j^n + \tfrac{1}{2}\mu(v_{j+1}^n + v_{j-1}^n)$$

$$(1+2\mu)v_j^{n+1} - \mu(v_{j+1}^{n+1} + v_{j-1}^{n+1}) = v_j^n$$

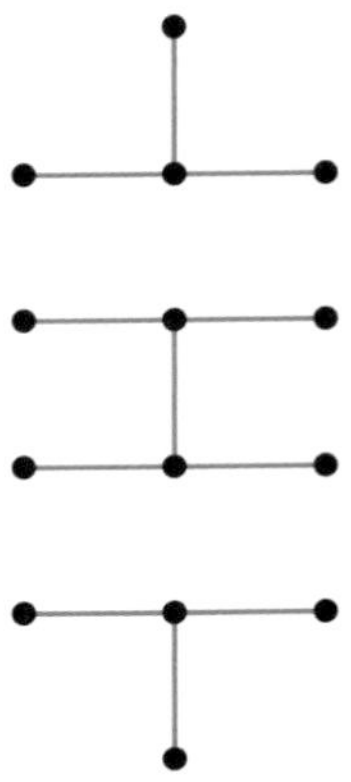

Figure 3.13. *Simple explicit (Euler) and implicit (Crank–Nicolson and backward Euler) discretizations for the heat equation.*

then the eigenvalues of the tridiagonal $J \times J$ matrix L can be calculated exactly. As it turns out, they are all negative and the largest in magnitude is $\approx -4/h^2$. The absolute stability requirement for the forward Euler method then yields $k \leq h^2/2$, i.e., we need $\mu \leq 1/2$ for ODE stability. So it is tempting to say that the method is simply second order in h, because we must maintain $k = O(h^2)$. However, such language may cause a confusion of the concepts of accuracy and stability. Moreover, if the solution becomes smoother as t increases, then we may want to take large time steps because the constant in the $O(k)$ term may be very small. Then the restriction on μ becomes a serious burden.

- Even before properly defining stability in the present context, one intuits that an A-stable method should not have the forward Euler restriction on the time step. Such is the trapezoidal method (2.8) which yields the *Crank–Nicolson scheme*

$$\frac{1}{k}(v_j^{n+1} - v_j^n) = \frac{1}{2h^2}(v_{j+1}^{n+1} - 2v_j^{n+1} + v_{j-1}^{n+1} + v_{j+1}^n - 2v_j^n + v_{j-1}^n),$$

or

$$(1+\mu)v_j^{n+1} - \frac{\mu}{2}(v_{j+1}^{n+1} + v_{j-1}^{n+1}) = (1-\mu)v_j^n + \frac{\mu}{2}(v_{j+1}^n + v_{j-1}^n); \quad (3.31b)$$

see Figure 3.13. To find the local truncation error we expand in Taylor series about the center of the molecule, $(t_{n+1/2}, x_j)$. This yields

$$\tau = \frac{k^2}{24}u_{ttt} - \frac{k^2}{8}u_{xxtt} + O(k^4) + \frac{h^2}{12}u_{xxxx} + O(h^2 k^2).$$

The local truncation error is $O(k^2) + O(h^2)$ for u sufficiently smooth, so the order is (2, 2). The method turns out to be *unconditionally stable* (see Example 4.1).

However, the method is also *implicit*: at each time step a system of algebraic equations must be solved for the unknown values at level $n+1$. The algorithm for tridiagonal

3.2. Full Discretization

systems described in Section 2.12.1 is useful. For more space variables this difficulty is more serious; see Section 9.2, especially Example 9.1.

- Instead of forward Euler we could use the *backward Euler* discretization (2.7) for time. This method has *stiff decay* (see Section 2.6), so we may expect not only unconditional stability but also a decay in high frequency error components, corresponding to a similar decay in the exact solution modes. We will revisit this important idea later, in Chapter 5. The formula (see Figure 3.13) is implicit and reads

$$(1 + 2\mu)v_j^{n+1} - \mu(v_{j+1}^{n+1} + v_{j-1}^{n+1}) = v_j^n. \qquad (3.31c)$$

The local truncation error is derived analogously to that for the forward Euler method, except that the Taylor expansions are carried out about (t_{n+1}, x_j). The obtained order is $(1, 2)$.

- The two-step BDF method discussed in Example 3.9 has order $(2, 2)$ and otherwise similar properties to those of the backward Euler discretization. ■

> **Note:** Continuing with a general full discretization scheme we next define stability and convergence and examine their relationship.

In Example 1.2 we analyzed stability of methods for the advection equation on a somewhat intuitive, introductory level. In Example 3.10 we used the concept of ODE absolute stability before defining PDE stability to motivate different time discretizations for the heat equation. We next generalize and define stability properly. Note that by these examples we may expect some methods to be stable for any positive h and k, some to never be stable, and some to be stable only when

$$k \leq S(h) \quad \forall\, 0 \leq h \leq h_0, \qquad (3.32)$$

where S is a continuous function that satisfies $S(0) = 0$ and $S(z) > 0$ for $z > 0$, and h_0 is some suitable positive constant. We call S a **stability function** and allow it to be unbounded in general. For the scheme (3.31a), $S(z) = z^2/2$; see Figure 3.14. Note that S depends both on the discretization method and on the PDE being discretized. We next define the following:

- The difference method is **stable** if there are constants $\tilde{K}, \tilde{\alpha}$, and $h_0 > 0$ independent of h and k and a stability function S such that for all $0 \leq h \leq h_0$ and $0 \leq k \leq S(h)$

$$\|v(t)\| \leq \tilde{K} e^{\tilde{\alpha} t} \|v(0)\|. \qquad (3.33a)$$

Compare this to the well-posedness requirement (1.7).[16] If $S(h)$ is bounded for $0 \leq h \leq h_0$, then the method is **conditionally stable**. Otherwise it is **unconditionally stable**.

---

[16] Strictly speaking, as for well-posedness we must require uniformity with respect to the starting time, i.e.,

$$\|v(t)\| \leq \tilde{K} e^{\tilde{\alpha}(t-t_0)} \|v(t_0)\|$$

for any $0 \leq t_0 \leq t$. But we set $t_0 = 0$ in (3.33a) for notational simplicity.

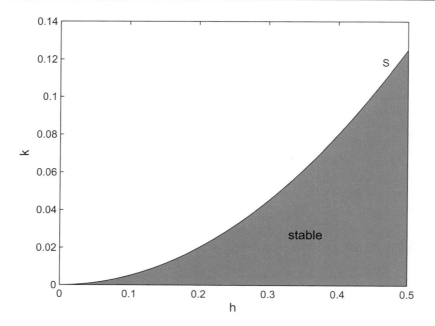

Figure 3.14. *Stability function for forward Euler applied to the heat equation (Example* 3.10*).*

The method is further said to be **strictly stable** if $\tilde{\alpha} \leq \alpha$.

- The method is **convergent** if

$$u(t,x) - v(t,x) \to 0, \qquad \text{as } k, h \to 0. \tag{3.33b}$$

A fundamental theorem in the numerical analysis of differential equations is the **Lax equivalence theorem**:[17]

Theorem 3.11. *If the linear evolutionary PDE* (3.1) *is well-posed and the difference method is consistent, then*

$$\text{convergence} \iff \text{stability}.$$

The complete proof of this theorem can be found in Richtmyer and Morton [144]. See also [159].

[17]The name *Lax–Richtmyer theorem* is also commonly used for Theorem 3.11.

3.2. Full Discretization

Let us concentrate on the more useful aspect of the equivalence theorem, Theorem 3.11: we want to bound the convergence error for a given, stable scheme. For the linear initial value problem (1.5) we write

$$u(t, x) = \mathcal{S}(t, t_0)u(t_0, x),$$

where $\mathcal{S}(t, t_0)$ is the linear solution operator. The well-posedness property is written as

$$\|\mathcal{S}(t, t_0)\| \leq K e^{\alpha(t-t_0)}.$$

Now, for inhomogeneous problems

$$u_t = P(\partial_x, t, x)u + q(t, x), \qquad -\infty < x < \infty, \ t > 0,$$

Duhamel's principle (see [76]) gives

$$u(t, x) = \mathcal{S}(t, 0)u_0(x) + \int_0^t \mathcal{S}(t, \xi)q(\xi, x)d\xi.$$

Hence

$$\|u(t)\| \leq K \left[e^{\alpha t} \|u_0\| + \phi(\alpha, t) \max_{0 \leq \xi \leq t} \|q(\xi, \cdot)\| \right],$$

where

$$\phi(\alpha, t) = \begin{cases} (e^{\alpha t} - 1)/\alpha, & \alpha \neq 0, \\ t, & \alpha = 0. \end{cases} \tag{3.34}$$

This bounds the solution u in terms of the data u_0 and q.

Consider next the approximate solution, where we have analogous results. For the homogeneous problem,

$$v(t, x) = \mathcal{S}_{k,h}(t, t_0)v(t_0, x),$$

where $\mathcal{S}_{k,h}(t, t_0)$ is the discretized solution operator. The stability property (3.33a) now gives

$$\|\mathcal{S}_{k,h}(t, t_0)\| \leq \tilde{K} e^{\tilde{\alpha}(t-t_0)}$$

for all $0 \leq h \leq h_0$, $k \leq S(h)$. For the inhomogeneous problem, writing in operator form

$$Q_1 v(t+k, x) = Q_0 v(t, x) + k q(t, x), \tag{3.35}$$

we have, similarly to Duhamel's principle, that after N time steps with $t = t_N$

$$v(t, x) = \mathcal{S}_{k,h}(t, 0)v(0, x) + k \sum_{n=0}^{N} \mathcal{S}_{k,h}(t, nk) Q_1^{-1} q(nk, x).$$

Thus

$$\|v(t)\| \leq \tilde{K} \left[e^{\tilde{\alpha} t} \|v^0\| + M\phi(\tilde{\alpha}, t) \max_{0 \leq \xi \leq t} \|q(\xi, \cdot)\| \right],$$

where $\quad M = \max_{0 \leq \xi \leq t} \|Q_1^{-1}\|.$

Having successfully expressed the dependence of the numerical solution on the data, we now turn to the error $e(t, x) = u(t, x) - v(t, x)$. The error function $e(t, x)$ satisfies the inhomogeneous difference scheme (3.35) with the local truncation error as the inhomogeneity, $q = \tau$. Thus, stability and accuracy yield convergence with the bound

$$\|u(t) - v(t)\| \leq \hat{K} M \max_{0 \leq \xi \leq t} \|c(\xi)\| (k^{p_1} + h^{p_2}) \phi(\tilde{\alpha}, t). \tag{3.36}$$

This shows that the order of accuracy also features in the bound on the solution error itself, and it highlights the importance of keeping $\tilde{\alpha}$ small if at all possible!

It is important to realize that whereas the local truncation error is *local* in space and time, the global error bound in (3.36) is *global*. The fact that this bound holds everywhere in the PDE's domain of definition also implies that it is hard to use directly in order to refine the mesh locally, for instance, because it is not sufficiently sensitive to local changes. The local truncation error or other local error indicators are more suitable in this regard. Indeed, the arguments leading to (3.36) suggest that it is a good idea to try to make the local truncation error roughly equal everywhere. A rigid implementation of such a criterion may perhaps be too fancy in practice, but it gives an idea of the *goal* for local mesh refinement; see Sections 3.1.5 and 11.2.

3.2.2 General linear stability

We have seen that the two elements controlling convergence in (3.36) are accuracy and stability. It is relatively simple to estimate the local truncation error and to derive schemes with a desirable accuracy order, at least in the simple cases considered hitherto. To get a handle on stability is harder. We can write the general form (3.27b), where $Q = Q(nk, h)$, as

$$v^{n+1} = Q(nk, h) v^n = \cdots = \Pi_{l=0}^{n} Q(lk, h) v^0 = \mathcal{S}_{k,h}(t_{n+1}, 0) v^0.$$

The stability condition is

$$\|\Pi_{l=0}^{n-1} Q(lk, h)\| \leq \tilde{K} e^{\tilde{\alpha} n k} \qquad \forall\, h, n, k \leq S(h),\ nk \leq t_f,$$

and if Q does not depend on t (although it still depends on k), then the stability condition becomes

$$\|Q(h)^n\| \leq \tilde{K} e^{\tilde{\alpha} n k}.$$

This is satisfied if

$$\|Q\| \leq e^{\tilde{\alpha} k} = 1 + O(k), \tag{3.37}$$

provided k and h are small enough with $k \leq S(h)$. The bound (3.37) is a bit more revealing. In fact, we can get rid of $O(k)$ terms, so in the case that h/k is bounded we can even eliminate the dependence on h. Indeed, we have the following theorem.

3.2. Full Discretization

> **Theorem 3.12.** *If the scheme*
> $$v^{n+1} = Qv^n$$
> *is stable and $\tilde{Q}$ is a bounded operator, then the scheme*
> $$v^{n+1} = (Q + k\tilde{Q})v^n$$
> *is also stable.*

The proof is obvious by (3.37). This theorem was first given by Strang; see [144].

Example 3.13 Theorem 3.12 is useful to establish stability for some PDEs with lower order terms. For instance, applying a scheme similar to (1.15a) for the PDE

$$u_t = u_x + b(x)u,$$

we have

$$v_j^{n+1} = v_j^n + \mu(v_{j+1}^n - v_j^n) + kb(x_j)v_j^n,$$

where $\mu = k/h$. Assuming that $k \leq h$ we have shown stability for the scheme with $b = 0$ in Chapter 1. But the latter addition is just a low order term which cannot affect stability. Thus, the extended scheme is also stable under the restriction that $\mu \leq 1$ is fixed.

On the other hand, for a parabolic equation such as

$$u_t = u_{xx} + au_x + bu,$$

using, say, an extension of the Euler method with a centered discretization is space, we have

$$v_j^{n+1} = v_j^n + \mu(v_{j+1}^n - 2v_j^n + v_{j-1}^n) + \frac{ak}{2h}(v_{j+1}^n - v_{j-1}^n) + kbv_j^n.$$

Here, keeping $\mu = k/h^2$ fixed, the term bu gives an $O(k)$ perturbation and has no effect on stability, but the term au_x gives an $O(h)$ perturbation which cannot simply be waved away in the same fashion. ∎

In summary, to show stability we essentially need to show that Q is **power bounded**,

$$\|Q^n\| \leq \bar{K} \qquad \forall n.$$

This task is not simple at all, however: recall that the matrix Q is a very large, if not downright infinite in size. In fact, a general boundedness criterion is not known. It is useful to review Section 1.3.5 at this point.

There are two practical ways to proceed.

- Consider constant coefficient problems. This allows eigenvalue-based and Fourier-based analyses. We consider such approaches in the next chapter. The results are simple and general, but they usually yield only necessary conditions for stability.

- Freeze coefficients and explore dissipativity, or apply some "energy method." This is discussed in Chapter 5. Energy estimates provide sufficient conditions for stability, even for some nonlinear problems, but finding such estimates is an art, applicable only to special cases and depending on the ingenuity of the analyst.

3.3 Exercises

0. **Review questions**

    (a) What is a semi-discretization and why is it useful?

    (b) Review the difference notation (3.6) and show that $\delta^2 = D_+ D_- = D_- D_+$ and $\mu^2 = 1 + \delta^2/4$.

    (c) What is an implicit scheme for approximating a derivative (in contrast to a time-dependent differential equation)?

    (d) What is a staggered mesh and why is it useful?

    (e) Define a ghost unknown and explain its use for discretizing Neumann boundary conditions.

    (f) What are natural and essential BCs? In what way are these names suggestive or intuitive?

    (g) Describe a tensor product mesh, an unstructured mesh, a triangle mesh, and an Octree mesh.

    (h) For a full discretization define order, consistency, stability, and convergence.

    (i) What is a stability function? What is the stability function for the Crank–Nicolson scheme?

    (j) Write down the Lax equivalence theorem and explain in what sense the obtained convergence error bound is global while the local truncation error is local.

1. (a) For the heat equation, let

    $$a(z) = h^2 \sum_{i=-l}^{r} \alpha_i z^i.$$

    Show that the semi-discretization method (3.3) is of order p if and only if (iff) there exists a constant $c \neq 0$ such that

    $$a(z) = (\log z)^2 + c(z-1)^{p+2} + O(|z-1|^{p+3}),$$

    as $z \to 1$.

    (b) For the advection equation, let

    $$a(z) = h \sum_{i=-l}^{r} \alpha_i z^i.$$

3.3. Exercises

Show that the semi-discretization method (3.3) is of order p iff there exists a constant $c \neq 0$ such that

$$a(z) = \log z + c(z-1)^{p+1} + O(|z-1|^{p+2}),$$

as $z \to 1$.

2. Suppose that the points x_{j-1}, $x_j = x_{j-1} + h_{j-1}$, and $x_{j+1} = x_j + h_j$ are used to derive a second order formula for $u_x(x_j)$ which holds even when $h_{j-1} \neq h_j$. A careless approach, which has actually seen use in the research literature, would suggest the formula

$$u_x(x_j) \approx \frac{u_{j+1} - u_{j-1}}{h_{j-1} + h_j}, \qquad (3.38a)$$

which quickly generalizes the familiar centered three-point formula (3.5b).

Instead, consider the formula

$$u_x(x_j) \approx \frac{h_j - h_{j-1}}{h_{j-1} h_j} u_j + \frac{1}{h_{j-1} + h_j} \left(\frac{h_{j-1}}{h_j} u_{j+1} - \frac{h_j}{h_{j-1}} u_{j-1} \right). \qquad (3.38b)$$

(a) Apply the two formulas (3.38) for the function $u(x) = e^x$ at $x_j = 0$ with $h_{j-1} = .5h$, $h_j = h$. Record the errors using both methods for $h = 10^{-l}$, $l = 1, \ldots, 5$.

(b) Derive an error expression for the formula (3.38b), assuming that u is smooth (i.e., that it has all the derivatives you encounter, bounded).

(c) It should be apparent from your numerical results that the formula (3.38a) is only first order accurate. But why is this *necessarily* so?

Formulating the question more generally, let $u(x)$ be smooth with $u_{xx}(x_j) \neq 0$. Show that the truncation error in the formula (3.38a) with $h_j = h$ and $h_{j-1} = h/2$ must decrease *linearly, and not faster*, as $h \to 0$.

3. Consider the following two methods for deriving an approximate formula for the second derivative $u_{xx}(x_j)$ of a smooth function $u(x)$ using three points x_{j-1}, $x_j = x_{j-1} + h_{j-1}$, and $x_{j+1} = x_j + h_j$, where $h_{j-1} \neq h_j$.

- Define $w(x) = u_x(x)$ and seek a *staggered mesh*, centered approximation

$$w_{j+1/2} = (u_{j+1} - u_j)/h_j; \quad w_{j-1/2} = (u_j - u_{j-1})/h_{j-1};$$

$$u_{xx}(x_j) \approx \frac{2(w_{j+1/2} - w_{j-1/2})}{h_{j-1} + h_j}.$$

The idea is that all the differences are short (i.e., no long differences for first derivatives) and centered.

- Using the second degree interpolating polynomial in Newton form, differentiated twice, define

$$u_{xx}(x_j) \approx 2u[x_{j-1}, x_j, x_{j+1}].$$

(a) Show that the above two methods are equivalent.

(b) Show that this method in general is only first order accurate.

(c) Run the method for the example of Exercise 2 (but for the second derivative of $u(x) = e^x$). What are your findings? Discuss.

4. Consider the boundary value ODE

$$-(au')' = q, \quad 0 < x < 1,$$
$$u(0) = 0, \quad u'(1) = 0,$$

where $a(x) > 0$ and $q(x)$ are known, smooth functions. It is well known (recall Section 3.1.3) that u also minimizes

$$T = \int_0^1 \left[a(u')^2 - 2uq \right] dx$$

over all functions with bounded first derivatives that satisfy the essential BC $u(0) = 0$. Consider next discretizing the integral on a nonuniform mesh,

$$0 = x_0 < x_1 < \cdots < x_J = 1,$$

as in Exercises 2 and 3.

(a) Show that, applying the midpoint rule for the first term and the trapezoidal rule for the second term in T for each subinterval, one obtains the problem of minimizing

$$T_h = \sum_{i=0}^{J-1} a(x_{i+1/2}) \frac{(v_{i+1} - v_i)^2}{h_i} - h_i \left(q(x_i) v_i + q(x_{i+1}) v_{i+1} \right)$$

with $v_0 = 0$. Thus, $T_h(u) = T(u) + O(h^2)$, where $h = \max_i h_i$.

(b) Obtain the necessary conditions

$$a(x_{j+1/2}) \frac{v_j - v_{j+1}}{h_j} + a(x_{j-1/2}) \frac{v_j - v_{j-1}}{h_{j-1}} = \frac{h_j + h_{j-1}}{2} q(x_j)$$

for $j = 1, \ldots, J$, where we set $v_{J+1} = v_J$.

(c) Show that upon writing the above as a linear system of equations $A\mathbf{v} = \mathbf{q}$, the matrix A is tridiagonal, symmetric, and positive definite despite the arbitrary nonuniformity of the mesh.

(d) Convince yourself by running a computational example that the solution $\mathbf{v}$ is second order accurate despite what Exercise 3 may suggest. Then try to prove it.

5. For the problem and notation of Exercise 4 define the **hat function** in one dimension:

$$\phi_i(x) = \begin{cases} \frac{x - x_{i-1}}{h_{i-1}}, & x_{i-1} \leq x < x_i, \\ \frac{x_{i+1} - x}{h_i}, & x_i \leq x < x_{i+1}, \\ 0 & \text{otherwise.} \end{cases}$$

3.3. Exercises

(a) Show that ϕ_i is piecewise linear and that any piecewise linear function w on this mesh that satisfies $w(0) = 0$ can be written as

$$w(x) = \sum_{i=1}^{J} w(x_i)\phi_i(x).$$

(b) Derive the Galerkin finite element method for the boundary value ODE. Show that the stiffness matrix A is tridiagonal, symmetric, and positive definite. How does this method relate to the method of Exercise 4?

6. Suppose that the problem of Example 3.4 has the BC $\frac{\partial u}{\partial n} = 0$, i.e., the normal derivative vanishes, at the wall $x = 0$. Derive the corresponding discrete boundary equations

$$\frac{dv_{0,j}}{dt} = h^{-2}[2a_{1/2,j}(u_{1,j} - u_{0,j})$$
$$+ a_{0,j+1/2}(u_{0,j+1} - u_{0,j}) - a_{0,j-1/2}(u_{0,j} - u_{0,j-1})] + q_{0,j}, \quad 1 \le j < J.$$

7. Consider a parabolic problem such as in Example 3.4, with Dirichlet BC $u|_{\partial\Omega} = g$, where $\partial\Omega$ is the boundary of a spatial, nonsquare domain Ω in 2D. Such a domain boundary is depicted in Figure 3.15.

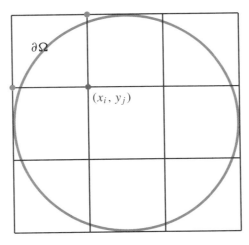

Figure 3.15. *A domain Ω with boundary $\partial\Omega$ that does not align with the mesh.*

Use nearby boundary conditions to eliminate the ghost unknowns marked in red for the difference scheme (3.15) applied at the marked mesh point (x_i, y_j) next to the boundary at the upper left corner.

8. Draw the Quadtree that corresponds to Figure 3.10.

9. (a) Suppose we are given values of the function $u(x, y)$ at the green nodes of Figure 3.16. It is desired to obtain approximations for u_x or u_y at the red points using *short differences*.

For each of these red points determine whether it is u_x or u_y that may be discretized and derive the corresponding formula.

(b) Now suppose that u is given at the blue points (cell centers). Derive an $O(h)$ formula for u_x at the two red points on the edges between the larger and the smaller cells.

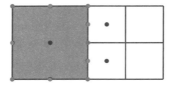

Figure 3.16. *Discretizing u derivatives on a Quadtree. Values of u_x or u_y at the red points should be obtained from values of u at the green points or at the blue points.*

10. Show the following:

    (a) The two-step BDF scheme formulated in Example 3.9 and Figure 3.12 can be written in the forms (3.27b)–(3.27d).

    (b) For this scheme Q_0 is diagonal, Q_1 is tridiagonal, and Q is in general a full matrix.

    (c) Both methods discussed in this example have order of accuracy $(2, 2)$.

11. Show that both the backward Euler scheme and the two-step BDF scheme applied to the heat equation are unconditionally stable.

12. For the Maxwell equations discussed in Example 3.6:

    (a) Derive all six equations for the staggered mesh semi-discretization.

    (b) Show that $\nabla \cdot \mathbf{E} = \nabla \cdot \mathbf{H} = 0$, provided that the initial value functions satisfy these identities.

    (c) Show that appropriate discretizations of $\nabla \cdot \mathbf{E}$ and $\nabla \cdot \mathbf{H}$ based on the same staggered mesh and short differences also vanish.

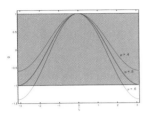

Chapter 4
Stability for Constant Coefficient Problems

This chapter continues from where Section 3.2.2 ends. We aim to establish conditions under which stability holds, i.e., to determine the stability function for a given difference scheme in time and space.

Rarely is the numerical solution of constant coefficient PDE problems of interest in itself. However, the stability analysis possible for such problems often gives a good indication of the performance of similar schemes in more complicated situations. The Fourier analysis developed in Sections 4.1.1 and 4.1.2 is, in fact, the only general tool known which provides an easy-to-check stability indication for full discretizations of time dependent PDEs. Most of this chapter is consequently devoted to it. In Sections 4.1.3 and 4.1.4 we make the connection between PDE stability and ODE absolute stability for the semi-discretization.

The eigenvalue analysis described in Section 4.2 is more limited in practice, but it sometimes works where the Fourier analysis method does not, e.g., in the presence of boundary conditions and variable coefficients for parabolic problems.

Much (but not all) of the material in this chapter and the next follows a combination of Kreiss [110], Gustafsson, Kreiss, and Oliger [76], Iserles [100], Morton and Mayers [137], and the true classic, Richtmyer and Morton [144].

4.1 Fourier analysis

Consider the difference scheme (3.27) which we rewrite here in general form,

$$v^{n+1} = Qv^n. \tag{4.1}$$

Recall the explicit difference scheme (3.27a) applied to a system of s PDEs,

$$\mathbf{v}_j^{n+1} = \sum_{i=-l}^{r} \beta_i \mathbf{v}_{j+i}^n. \tag{4.2}$$

We use boldface notation to emphasize that $\mathbf{v}_{j+i}^n$ are vectors of size s. Correspondingly, β_i are $s \times s$ matrices. We would like to apply to this scheme the same Fourier analysis that has been used in Section 1.2.1 for some simple examples. For this we need the same formula (4.2) to be applicable at all mesh points, with the coefficients β_i independent of the location j and n. Thus we require the following:

- The problem has constant coefficients.

- The problem is defined on an infinite mesh, i.e., it is a pure initial value problem. Alternatively, periodic boundary conditions are imposed, which allows a straightforward periodic extension into an infinite mesh for analysis purposes.

Under these conditions the trigonometric functions (1.26a) are eigenfunctions (or, discretely, eigenvectors), and the Fourier transform diagonalizes Q if $s = 1$ or block-diagonalizes it in case that $s > 1$, as we show next.

Let

$$\mathbf{v}(t, x) = \frac{1}{\sqrt{2\pi}} \int_{-\infty}^{\infty} e^{\iota \xi x} \hat{\mathbf{v}}(t, \xi) d\xi. \tag{4.3}$$

Then

$$\int_{-\infty}^{\infty} e^{\iota \xi x} \hat{\mathbf{v}}(t+k, \xi) d\xi = \sum_{i=-l}^{r} \beta_i \int_{-\infty}^{\infty} e^{\iota \xi (x+ih)} \hat{\mathbf{v}}(t, \xi) d\xi$$

$$= \int_{-\infty}^{\infty} \left(\sum_{i=-l}^{r} \beta_i e^{\iota \xi i h} \right) e^{\iota \xi x} \hat{\mathbf{v}}(t, \xi) d\xi.$$

Thus

$$\hat{\mathbf{v}}(t+k, \xi) = g(\zeta) \hat{\mathbf{v}}(t, \xi), \qquad \zeta = \xi h.$$

For a scalar PDE g is the **amplification factor**

$$g(\zeta) = \sum_{i=-l}^{r} \beta_i e^{\iota i \zeta}. \tag{4.4}$$

If the PDE has s components, then g is an $s \times s$ **amplification matrix**.

Now, for $t = nk$ with n a positive integer we can obviously write $\hat{\mathbf{v}}(t, \xi) = g(\zeta) \hat{\mathbf{v}}(t - k, \xi) = g(\zeta)^2 \hat{\mathbf{v}}(t - 2k, \xi) = \cdots$, so

$$\hat{\mathbf{v}}(t, \xi) = g(\zeta)^n \hat{\mathbf{v}}(0, \xi).$$

Recalling Parseval's equality (1.19), the stability condition of Section 3.2.1 becomes simply

$$\|g(\zeta)^n\| \leq \tilde{K} e^{\tilde{\alpha} n k}, \tag{4.5a}$$

which is satisfied if the following stability condition holds:

$$\|g(\zeta)\| \leq e^{\tilde{\alpha} k} \qquad \forall \, |\zeta| \leq \pi. \tag{4.5b}$$

4.1. Fourier Analysis

This corresponds to the stability condition (3.37), with the advantage that it is much easier to check in practice.

An analogous derivation can be applied for a general implicit difference scheme,

$$\sum_{i=-l}^{r} \gamma_i \mathbf{v}_{j+i}^{n+1} = \sum_{i=-l}^{r} \beta_i \mathbf{v}_{j+i}^{n}, \qquad (4.6)$$

where β_i and γ_i are $s \times s$ constant matrices. This yields the conditions (4.5) with

$$g(\zeta) = \left(\sum_{i=-l}^{r} \gamma_i e^{\iota i \zeta} \right)^{-1} \sum_{i=-l}^{r} \beta_i e^{\iota i \zeta}. \qquad (4.7)$$

The process just described can be viewed as a similarity transformation applied to the (infinite) matrix Q, whereby (4.1) is transformed into

$$\hat{v}^{n+1} = \hat{Q}\hat{v}^n$$

with

$$\hat{Q}(\zeta) = \begin{pmatrix} g(\zeta) & & & \\ & g(\zeta) & & \\ & & \ddots & \end{pmatrix}.$$

The integration process can be viewed as a summation (indeed it is, in the case of periodic boundary conditions; see Section 1.3.6) involving the trigonometric functions restricted to the mesh. These are eigenvectors when $s = 1$ (for which $\hat{Q}$ is diagonal); in any case the similarity transformation is orthogonal and hence norm preserving.

4.1.1 Stability for scalar equations

When $s = 1$ the stability condition (4.5b) simply reads

$$|g(\zeta)| \leq 1 + O(k), \qquad |\zeta| \leq \pi.$$

The $O(k)$ term in this bound is really used or needed only in cases where the exact solution to the homogeneous problem grows.

Example 4.1 For parabolic equations, as it turns out, a scalar PDE is not very different in principle from a system of PDEs in terms of stability conditions. Thus we return to the simple heat equation

$$u_t = u_{xx},$$

considered in Example 3.10. For the forward Euler method

$$v_j^{n+1} = v_j^n + \mu D_+ D_- v_j^n,$$

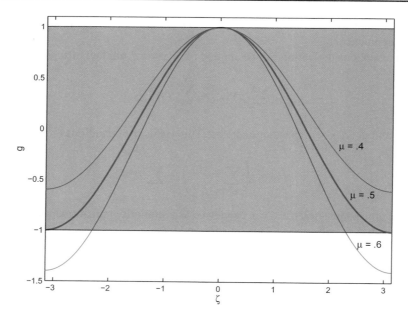

Figure 4.1. *Amplification factor for forward Euler applied to the heat equation. To be stable the curve must stay in the shaded area. Note that for $\mu = .6$ stability violation occurs at high wave numbers.*

where $\mu = k/h^2$, we have

$$D_+D_- e^{\iota \xi x} = (e^{\iota \zeta} - 2 + e^{-\iota \zeta})e^{\iota \xi x} = 2(\cos \zeta - 1)e^{\iota \xi x}$$
$$= -4\sin^2(\zeta/2)e^{\iota \xi x},$$

so

$$g(\zeta) = 1 - 4\mu \sin^2(\zeta/2).$$

Figure 4.1 depicts this amplification factor. Thus, to ensure that $|g(\zeta)| \leq 1$ we must require

$$\mu \leq 1/2, \quad \text{i.e.,} \quad k \leq h^2/2.$$

Moreover, it is the high wave numbers (ζ near $\pm \pi$) that may cause instability if $\mu > 1/2$.

The corresponding stability function $S(h)$ is depicted in Figure 3.14. This can be a very restrictive condition on the time step size k (corresponding to using a nonstiff method for a stiff ODE; see Sections 2.5 and 2.6). Note again that, although the order of accuracy $(1, 2)$ also suggests taking $\mu = O(1)$, here we *must* take $\mu \leq 1/2$ even if we integrate toward steady state and for accuracy reasons alone would expect to be able to take large time steps.

For the Crank–Nicolson method,

$$\left(1 - \frac{\mu}{2}D_+D_-\right)v_j^{n+1} = \left(1 + \frac{\mu}{2}D_+D_-\right)v_j^n,$$

4.1. Fourier Analysis

we get

$$g(\zeta) = \frac{1 - 2\mu \sin^2(\zeta/2)}{1 + 2\mu \sin^2(\zeta/2)}.$$

Here $|g(\zeta)| \leq 1$ for all $\mu > 0$, so the scheme is unconditionally stable.

Let $J = h^{-1}$. To advance from time level n to $n + 1$ using either the explicit Euler or the implicit Crank–Nicolson costs cJ operations with a modest constant c. If we take $h \sim k$, then it requires $O(J)$ steps to reach $t_f = 1$ for a total of $O(J^2)$ flops, whereas using the Euler scheme a similar task costs $O(J^3)$ flops.[18]

A slightly more involved parabolic PDE is considered in Exercise 1. ∎

When constructing amplification factors we often find ourselves combining contributions from standard difference operators such as D_+D_- and D_0. For convenience, they are listed in Table 4.1.

Table 4.1. *Difference operators and their effect in Fourier space.*

Difference operator	Contribution to $g(\zeta)$
D_+	$e^{\iota\zeta} - 1$
D_-	$1 - e^{-\iota\zeta}$
D_0	$2\iota \sin \zeta$
δ	$2\iota \sin(\zeta/2)$
D_+D_-	$-4 \sin^2(\zeta/2)$

4.1.2 Stability for systems of equations

For systems of PDEs the matrix $\hat{Q}$ is no longer diagonal, and $g(\zeta)$ is an $s \times s$ matrix. Let us emphasize this by denoting $g(\zeta) \equiv G(\zeta)$. The stability condition (4.5) still holds, with G replacing g.

Rather than checking the norm of $G(\zeta)$ for every $|\zeta| \leq \pi$, it is easier to require a similar condition on the *spectral radius*,

$$\rho(G(\zeta)) \leq e^{\tilde{\alpha}k} \qquad \forall \, |\zeta| \leq \pi. \tag{4.8}$$

This is the celebrated **von Neumann condition**. Recall from the review in Section 1.3.2 that $\rho(G) \leq \|G\|$ and that equality holds for the Euclidean norm in general only if G is normal.[19] Thus, the von Neumann condition (4.8) is in general only **necessary and not sufficient** for stability. It is sufficient if either of the following hold:

---

[18]Using a flop count to assess algorithm complexity should be frowned upon, strictly speaking, because it ignores other, crucial aspects affecting CPU time, such as fast memory access. The argument here is a classical one, though, and it often still gives the right flavor despite its imperfections.

[19]If G is normal, i.e., there is an orthogonal transformation T such that $T^T G T = \Lambda$ is diagonal, then $G = T \Lambda T^T$; hence $\|G\| \leq \rho(G)$ and equality follows: $\|G\| = \rho(G)$.

- If $G(\zeta)$ is *normal* $\forall\ |\zeta| \leq \pi$, then (4.8) implies (4.5b).

   The normality condition, however, often does not hold for hyperbolic problems.

- If $G(\zeta)$ is *uniformly diagonalizable* in ζ, i.e., there is a transformation $T(\zeta)$ such that $T^{-1}GT$ is diagonal and $\text{cond}(T) \leq \tilde{K}\ \forall\ |\zeta| \leq \pi$, then by (1.23) we obtain the stability condition (4.5a) from the von Neumann condition (4.8).

Example 4.2 Let **u** have two components and consider the trivial PDE $\mathbf{u}_t = \mathbf{0}$ and the (admittedly strange but) first order discretization

$$\mathbf{v}_j^{n+1} = \mathbf{v}_j^n - \begin{pmatrix} 0 & 1 \\ 0 & 0 \end{pmatrix} \mu^2 D_0^2 \mathbf{v}_j^n$$

with $\mu = k/h = 1$. This gives the amplification matrix

$$G(\zeta) = \begin{pmatrix} 1 & 4\sin^2 \zeta \\ 0 & 1 \end{pmatrix}.$$

So the von Neumann condition is fulfilled, and yet

$$G^n = I + n \begin{pmatrix} 0 & 4\sin^2 \zeta \\ 0 & 0 \end{pmatrix},$$

which grows unboundedly.

What happens here is that the matrix G is *defective* (i.e., not diagonalizable) and the off-diagonal element is not merely $O(k)$, so the eigenvalues do not tell the whole story about stability. ∎

Example 4.3 Consider the leapfrog scheme for a hyperbolic system,

$$\mathbf{u}_t + A\mathbf{u}_x = \mathbf{0}$$

with A constant and diagonalizable. To recall, the leapfrog scheme reads

$$\mathbf{v}_j^{n+1} = \mathbf{v}_j^{n-1} - \mu A(\mathbf{v}_{j+1}^n - \mathbf{v}_{j-1}^n),$$

which can be written as

$$\begin{pmatrix} \mathbf{v} \\ \mathbf{w} \end{pmatrix}_j^{n+1} = \begin{pmatrix} 0 & I \\ I & 0 \end{pmatrix} \begin{pmatrix} \mathbf{v} \\ \mathbf{w} \end{pmatrix}_j^n - \mu \begin{pmatrix} A & 0 \\ 0 & 0 \end{pmatrix} \begin{pmatrix} \mathbf{v} \\ \mathbf{w} \end{pmatrix}_{j+1}^n + \mu \begin{pmatrix} A & 0 \\ 0 & 0 \end{pmatrix} \begin{pmatrix} \mathbf{v} \\ \mathbf{w} \end{pmatrix}_{j-1}^n.$$

This is obviously in the form (4.2) with $2s \times 2s$ matrices β_i, $i = -1, 0, 1$.

For one PDE the amplification matrix reads

$$G(\zeta) = \begin{pmatrix} -2\iota a\mu \sin \zeta & 1 \\ 1 & 0 \end{pmatrix}.$$

4.1. Fourier Analysis

The eigenvalues are
$$\lambda = -\iota a\mu \sin \zeta \pm \sqrt{1 - a^2\mu^2 \sin^2 \zeta}.$$

Here, $|\lambda|^2 = a^2\mu^2 \sin^2 \zeta + (1 - a^2\mu^2 \sin^2 \zeta) = 1$ for both roots regardless of ζ as long as the square root expression stays real, so the von Neumann condition holds for

$$\mu|a| \leq 1$$

(the CFL condition). Now, if $\mu|a| < 1$, then the two roots differ: $\lambda_1 \neq \lambda_2$. Hence G is diagonalizable and the scheme is stable. But if $\mu|a| = 1$, then at $\zeta = \pi/2$ we have $1 - a^2\mu^2 \sin^2 \zeta = 0$; hence $\lambda_1 = \lambda_2$ and G is not diagonalizable. (Note that although G is symmetric, it is complex and not Hermitian! It is not normal, either.) Indeed, there is a mild instability in this case, and it can be easily seen that $\|G^n(\pi/2)\|$ grows linearly in n, as in the previous example.

This sort of instability is often tolerable in practice. It is very different from the exponential error blowup seen in Table 1.1 and Figure 1.12. But in any case, it is advisable in a more realistic model to keep $\mu|a| < 1$.

Returning to the case of a system of PDEs we have the amplification matrix

$$G(\zeta) = \begin{pmatrix} (-2\iota\mu \sin \zeta)A & I \\ I & 0 \end{pmatrix}.$$

The eigenvalues of G are pairs which obey the formula

$$\lambda_l^\pm = -\iota a_l\mu \sin \zeta \pm \sqrt{1 - a_l^2\mu^2 \sin^2 \zeta}$$

for each eigenvalue a_l of A. The stability condition is then

$$\mu < \frac{1}{\max_{1 \leq l \leq s} |a_l|}. \blacksquare$$

Kreiss gave, back in 1962, three equivalent conditions for the sufficiency of the von Neumann stability requirement [76]. Remember, however, that this analysis is only for the constant coefficient case.

Example 4.4 The von Neumann condition (4.8) is an easy to check, necessary and "almost sufficient" stability condition. But most problems encountered in practice are nonlinear, or at least have variable coefficients. Given such a practical problem to solve, the most popular basic search for a stability condition is to "freeze" its coefficients, ignore boundary conditions, and check (4.8).

For instance, the stability condition for the leapfrog discretization of a variable coefficient advection equation

$$u_t + a(x)u_x = 0$$

would read

$$\mu \max_x |a(x)| \leq 1 \quad (\mu = k/h).$$

Similarly, for the *anisotropic diffusion* equation

$$u_t = (a(u)u_x)_x, \tag{4.9a}$$

if we wish to apply the forward Euler method

$$v_j^{n+1} = v_j^n + \mu(a_{j+1/2}D_+ - a_{j-1/2}D_-)v_j^n \tag{4.9b}$$

(recall Examples 3.3 and 4.1), then the stability condition reads

$$\mu \max_u |a(u)| \leq 1/2 \quad (\mu = k/h^2). \tag{4.9c}$$

Of course, this gives rise to the question how to estimate $\max |a(u)|$ when u is our *unknown* function. But in practice such estimates are often known. Even if they are not, they can be obtained by solving the problem with rough k and h, for which computations and experimentations are cheap. The key observation is that (4.9c) is used only to limit the time step k, so with only a ballpark estimate $\tilde{a}$ of $\max |a(u)|$, we can "play safe" by simply setting k to, say, half its estimated maximum,

$$k = \frac{h^2}{4\tilde{a}}.$$

However, we caution that although such techniques are often satisfactory in practice, they do not always work. Example 5.1 documents a case of the latter type. Chapter 5 discusses the aftermath at length. ∎

4.1.3 Semi-discretization stability

For a semi-discretization (3.3) of an initial-value, homogeneous PDE with constant coefficients we write

$$v_t(t,x) = \sum_{i=-l}^{r} \alpha_i v(t, x+ih),$$

where the coefficients α_i are constant, independent of t and x.

Using the same Fourier transform as before we obtain

$$\hat{v}_t(t,\xi) = \tilde{g}(\zeta)\hat{v}(t,\xi), \tag{4.10a}$$

$$\tilde{g}(\zeta) = \sum_{i=-l}^{r} \alpha_i e^{ii\zeta}, \quad \zeta = \xi h. \tag{4.10b}$$

For each ξ the exact solution of this ODE in constant coefficients is

$$\hat{v}(t,\xi) = e^{\tilde{g}(\zeta)t}\hat{v}(0,\xi).$$

The semi-discretization is therefore stable iff there is a real constant $\hat{\alpha}$ independent of h and ξ such that

$$\|e^{\tilde{g}(\zeta)}\| \leq e^{\hat{\alpha}} \quad \forall \, |\zeta| \leq \pi. \tag{4.11}$$

4.1. Fourier Analysis

For a single PDE, $s = 1$, we obtain the simple stability condition

$$\mathcal{R}e\,\tilde{g}(\zeta) \leq \hat{\alpha} \quad \forall |\zeta| \leq \pi. \tag{4.12a}$$

For a system of PDEs let $\tilde{g}_1, \ldots, \tilde{g}_s$ be the eigenvalues of the matrix $\tilde{G}(\zeta)$. Then the (necessary but not always sufficient) von Neumann condition corresponds to

$$\max_{1 \leq l \leq s} \mathcal{R}e\,\tilde{g}_l(\zeta) \leq \hat{\alpha} \quad \forall |\zeta| \leq \pi. \tag{4.12b}$$

It is worthwhile at this point to compare the semi-discretization equations and conditions (4.10b), (4.11), and (4.12) with the corresponding full discretization expressions (4.4), (4.5b), and (4.8).

Example 4.5 For the PDE

$$u_t = \nu u_{xx} + a u_x + c u,$$

the semi-discretization

$$\frac{d}{dt} v_j(t) = \frac{\nu}{h^2}(v_{j+1} - 2v_j + v_{j-1}) + \frac{a}{2h}(v_{j+1} - v_{j-1}) + cv_j$$

yields the amplification factor

$$\tilde{g}(\zeta) = -\frac{4\nu}{h^2} \sin^2(\zeta/2) + c + \frac{\iota a}{h} \sin \zeta.$$

Thus, the stability condition (4.12a) holds, provided that $\nu \geq 0$, with $\hat{\alpha} = c$ if $c > 0$, and $\hat{\alpha} = 0$ otherwise.

There is no stability if $\nu < 0$, but then the initial value PDE is not well-posed either. Next, let $c = 0$ and assume $\nu > 0$. Recall from Chapter 1 that for this PDE

$$|e^{P(\iota \xi)}| = e^{-\nu \xi^2}$$

decreases as ξ increases. Here, similarly, $\mathcal{R}e\,\tilde{g}(\zeta)$ decreases toward its minimum as $|\zeta| \to \pi$. ∎

4.1.4 Fourier analysis and ODE absolute stability regions

For each fixed ξ the Fourier transform applied to the semi-discrete form yields a scalar ODE (4.10) with a constant, complex-valued coefficient. The stability condition (4.12) can be strengthened by requiring $\hat{\alpha} = 0$, i.e.,

$$\max_{1 \leq l \leq s} \mathcal{R}e\,\tilde{g}_l(\zeta) \leq 0 \quad \forall |\zeta| \leq \pi, \tag{4.13}$$

in cases where no solution growth is possible for the exact PDE for any initial value function $u_0 \in \mathcal{L}_2$. In other words, we require $\hat{\alpha} = 0$ when $\alpha \leq 0$ in (1.7) or (1.11).

Discretizing (3.3) in time yields a full discretization scheme where now we require that

$$\|v^{n+1}\| = \|\hat{v}^{n+1}\| \leq \|\hat{v}^n\| = \|v^n\|.$$

> **Note:** For a full discretization of the given PDE, which can be viewed as an ODE time-discretization applied to a semi-discretization in space, the semi-discretization can be seen to provide coefficients $\tilde{g}(\zeta)$ in an appropriate range satisfying (4.13), and the "subsequent" choice of time-discretization must then have suitable absolute stability and decay properties.

However, this condition coincides with the definition (2.21) which gives the region of absolute stability for the time-discretization method applied to (4.10)! Chapter 2 comes in handy at this point.

Example 4.6 Continuing Example 4.5 we now require $c \leq 0$ to allow no growth in the exact solution, i.e., $\alpha \leq 0$ in (1.11). The usual spatial discretizations of the terms νu_{xx} and cu contribute a real, nonpositive component to $\tilde{g}(\zeta)$ for any ζ, $|\zeta| \leq \pi$. The time-step restriction for the forward Euler method (and for any other explicit time discretization proposed in Chapter 2) in terms of $\mu = \frac{k}{h^2}$ follows directly.

From the "hyperbolic part" of the PDE, however, we obtain a purely imaginary eigenvalue. From Figure 2.6 it becomes clear why the method (1.15b) cannot be stable: there is no intersection between the absolute stability region of the forward Euler method and the imaginary axis for $h > 0$. On the other hand, the third and especially the fourth order explicit Runge–Kutta methods are good candidates for discretizing hyperbolic problems using the method of lines so long as the solution is smooth; see Section 5.3.2. ∎

4.2 Eigenvalue analysis

We can write a semi-discretization of an initial-value, homogeneous PDE with constant coefficients in the form (3.2),

$$\mathbf{v}_t = L\mathbf{v},$$

where L is a large, constant matrix. This matrix can be infinite or finite with some appropriate boundary conditions incorporated.

If L is diagonalizable, then there is a similarity transformation to a diagonal matrix

$$T^{-1}LT = \Lambda.$$

Consequently, for

$$\mathbf{w} = T^{-1}\mathbf{v}$$

we have the decoupled system

$$\mathbf{w}_t = \Lambda \mathbf{w}.$$

For each component of $\mathbf{w}$ we have a scalar equation,

$$\frac{dw_j}{dt} = \lambda_j w_j,$$

4.2. Eigenvalue Analysis

which can be analyzed as we have done with the Fourier technique. The system for **w** is then stable if there is a constant $\hat{\alpha}$ such that for all $h > 0$ small enough

$$\max_j \mathcal{R}e \, \lambda_j \leq \hat{\alpha}. \tag{4.14}$$

The big question is, Can we haul this stability result back to **v**? In general, all we have is

$$\|\mathbf{v}(t)\| \leq \text{cond}(T) e^{\max_j \mathcal{R}e \, \lambda_j t} \|\mathbf{v}(0)\|.$$

Since T is a very large matrix which does depend on h, *stability of* **w** *does not guarantee stability of* **v**! This eigenvalue analysis is worthwhile only if $\text{cond}(T)$ is *uniformly* bounded. Such boundedness would happen, e.g., if L is a normal matrix, in which case we can choose T orthogonal and hence $\text{cond}(T) = 1$.

An analogous treatment can be applied to the full discretization (3.27b) reproduced here:

$$v^{n+1} = Q v^n.$$

Viewing Q as a large matrix, we can deduce stability by considering its eigenvalues (assuming we can find them, or at least estimate them) using (4.14) and ODE absolute stability regions, provided Q is normal.

For hyperbolic systems Q may not be normal, or even diagonalizable (see Exercises 10–12); hence this approach requires much care. However, for certain parabolic problems it may be useful. Note that the approach can be applied if Q is symmetric, and moreover, boundary conditions can be incorporated and the PDE coefficients do not have to be constant.

Example 4.7 Consider the diffusion equation

$$u_t = (a(x) u_x)_x + q(t, x),$$

where $a(x) > 0$. Assume at first that a is constant and that the boundary condition $u = 0$ is prescribed at $x = 0$ and at $x = 1$ for $t > 0$.

Applying the same semi-discretization as in Examples 3.1, 3.2, and 4.5, and incorporating the Dirichlet boundary conditions, we obtain an ODE of the form (3.2) with the $J \times J$ matrix

$$L = \frac{a}{h^2} \begin{pmatrix} -2 & 1 & & & & & \\ 1 & -2 & 1 & & & & \\ & 1 & -2 & 1 & & & \\ & & \ddots & \ddots & \ddots & & \\ & & & & 1 & -2 & 1 \\ & & & & & 1 & -2 \end{pmatrix},$$

where $J + 1 = 1/h$. To find the eigenvalues of L, write

$$L \mathbf{v}^l = \lambda_l \mathbf{v}^l,$$

where for the lth eigenvector we guess

$$v_j^l = \sin(jlh\pi), \quad j = 1, \ldots, J.$$

This guess works (please check!) and we obtain the eigenvalues

$$\lambda_l = -\frac{4a}{h^2} \sin^2 \frac{lh\pi}{2}.$$

These eigenvalues are all nonpositive, hence stability follows. Note also that as lh increases λ_l gets smaller, which corresponds to the behavior of the PDE discussed in Example 4.5.

Turning to full discretizations, for the forward Euler method we have $Q = I + kL$, hence Q is also symmetric and its eigenvalues are

$$\lambda_l^{FE} = 1 - 4\mu a \sin^2 \frac{lh\pi}{2}.$$

To ensure that $\rho(Q) \leq 1$ we must require $\mu \leq 1/2$, which coincides with what the Fourier analysis gives in Example 4.1. For the trapeziodal method (Crank–Nicolson) we again obtain symmetric matrices and the eigenvalues

$$\lambda_l^{CN} = \frac{1 - 2\mu a \sin^2 \frac{lh\pi}{2}}{1 + 2\mu a \sin^2 \frac{lh\pi}{2}}.$$

Obviously, $|\lambda_l^{CN}| \leq 1$, so stability is unconditional. However, note that if μ is large (e.g., because we take $k = h \ll 1$), then $\lambda_J^{CN} \approx -1$, so high-frequency errors are kept in check but are not damped much. Finally, for the backward Euler method we obtain

$$\lambda_l^{BE} = \frac{1}{1 + 4\mu a \sin^2 \frac{lh\pi}{2}}.$$

Here, not only is stability unconditional but also the high-frequency eigenvectors are damped by the method, in parallel with the smoothing property of the differential operator and the semi-discretization.

The above results for the full discretizations can be obtained directly by ODE numerical stability analysis, using the concepts of absolute stability regions, A-stability and L-stability (see Sections 2.5 and 2.6).

For this simple equation we can also extend the results for a variable $a(x) > 0$ using the discretization of Example 3.2. It is easy to see using Gershgorin's theorem (Section 1.3.2) that the resulting matrix remains symmetric and diagonally dominant, and hence negative definite (see Exercise 3). Stability follows directly. ∎

4.3 Exercises

0. **Review questions**

    (a) Define amplification factor and amplification matrix, and explain their importance.

4.3. Exercises

   (b) Why can't the Fourier analysis be applied in general directly to nonlinear problems or to problems with Dirichlet BC?

   (c) Is the von Neumann condition applied to (4.2) sufficient for stability in the case of a scalar PDE? What about a system of PDEs?

   (d) What is the stability condition according to Fourier analysis for the semi-discretization case?

   (e) In what circumstances is the eigenvalue analysis preferred over the Fourier analysis?

1. Analyze the stability of the three discretizations discussed in Example 3.10 extended to the PDE
$$u_t = \nu u_{xx} + a u_x + c u$$
   with $\nu > 0$, a and c given constants.

2. Consider the advection equation
$$u_t + a u_x = 0,$$
   and recall that the consistent scheme (1.15b) is unconditionally unstable. The **Lax–Friedrichs** scheme is a variation:
$$v_j^{n+1} = \frac{1}{2}(v_{j-1}^n + v_{j+1}^n) - \frac{\mu a}{2}(v_{j+1}^n - v_{j-1}^n). \tag{4.15}$$
   Show that the Lax–Friedrichs scheme is stable, provided that the CFL condition holds.

3. Extend the argument of Example 4.7 for the case where $a(x)$ is varying and the discretization of Example 3.2 is used.

4. Show that a consistent difference scheme must satisfy $\rho(g(0)) = 1$.

5. For the scalar PDE
$$u_t = \nu u_{xx} + a u_x$$
   with $\nu > 0$, $a > 0$, consider discretizing the "parabolic part" implicitly and the "hyperbolic part" explicitly:
$$v_j^{n+1} = v_j^n + \frac{\nu k}{h^2} D_+ D_- v_j^{n+1} + \frac{ak}{h} D_+ v_j^n.$$
   What is the order of the method? Analyze its stability.

6. For the advection equation
$$u_t = a u_x,$$
   consider the corresponding semi-discretization obtained using a centered difference in space:
$$\frac{dv_j}{dt} = \frac{a}{2h} D_0 v_j.$$
   Analyze the stability of the full discretizations obtained using the following methods in time:

(a) Forward Euler.

(b) Trapezoidal method. Show that $|g(\zeta)| = 1\ \forall \zeta$. (Such a method is called *energy conserving*.)

(c) Backward Euler.

(d) The two-step Adams–Bashforth method:
$$v_j^{n+1} = v_j^n + \frac{a\mu}{4}(3D_0 v_j^n - D_0 v_j^{n-1}).$$

7. The leapfrog scheme for the heat equation $u_t = u_{xx}$ reads
$$v_j^{n+1} = v_j^{n-1} + 2\mu D_+ D_- v_j^n, \qquad \mu = \frac{k}{h^2}.$$

   Show that this scheme is unconditionally unstable.

8. Apply a Fourier stability analysis to the scheme
$$v_j^{n+1} = v_j^{n-1} - \mu D_0 \left[\rho + \frac{\nu}{h^2} D_+ D_-\right] v_j^n, \qquad \mu = \frac{k}{h},$$
   where ρ and ν are given parameters. This is a leapfrog discretization of the linearized KdV equation
$$u_t + \rho u_x + \nu u_{xxx} = 0.$$

9. The *Dufort–Frankel* scheme has to be the quirkiest of all difference schemes which bear someone's name. It alleviates the leapfrog instability for the heat equation (Exercise 7) by replacing v_j^n with the average $(v_j^{n+1} + v_j^{n-1})/2$. This yields (please check)
$$(1 + 2\mu)v_j^{n+1} = (1 - 2\mu)v_j^{n-1} + 2\mu(v_{j+1}^n + v_{j-1}^n). \qquad (4.16)$$

   (a) Show that this scheme is unconditionally stable.

   (b) Dividing (4.16) by $1 + 2\mu$ we obtain what looks like an explicit scheme. Explain why the scheme is nonetheless implicit, in general.

   (c) Show that the scheme is generally inconsistent! Specifically, the local truncation error does not shrink to 0 for *all* families of steps $k \to 0$ and $h \to 0$.

   (d) On the other hand, the scheme is *conditionally consistent*: show that if $k \to 0$ and $h \to 0$ such that μ remains constant, then the scheme has order $(1, 2)$.

So there is really no stability condition as such, but if we keep increasing μ, then accuracy deteriorates more severely than for Crank–Nicolson, say. In particular, for $k = h$, which is a good choice for Crank–Nicolson, the Dufort–Frankel scheme is only zero-order accurate.

4.3. Exercises

10. For the advection equation
$$u_t + au_x = 0$$
subject to periodic boundary conditions, consider the semi-discretization obtained using a centered difference in space:
$$\frac{dv_j}{dt} + \frac{a}{2h} D_0 v_j = 0.$$
Writing the result in the form of an ODE system,
$$\mathbf{v}_t = L\mathbf{v},$$
what properties does the constant matrix L have? Specifically, is it symmetric, skew-symmetric, or neither? Does it have real, imaginary, or generally complex eigenvalues? Under what conditions are we guaranteed that the real parts of these eigenvalues stay nonpositive?

11. Consider the same setting and answer the same questions as in Exercise 10 but now with the boundary condition $u(t, 1) = 0$ and the one-sided difference
$$\frac{dv_j}{dt} + \frac{a}{h} D_+ v_j = 0.$$

12. Consider the initial-boundary value problem for the advection equation $u_t - u_x = 0$ for $t \geq 0$, $0 \leq x \leq 1$, with two types of BCs:
    - Dirichlet $u(t, 1) = 0$ and
    - periodic $u(t, 0) = u(t, 1)$.

    (a) For each BC write the one-sided discretization (1.15a) as $v^{n+1} = Qv^n$ for a finite matrix Q. Is Q diagonalizable? Is it normal?

    (b) Repeat the tasks above for the scheme (1.15b) which sports a symmetric spatial discretization.

Chapter 5
Variable Coefficient and Nonlinear Problems

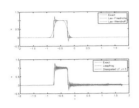

The previous chapter provides us with a useful practical tool for checking stability for constant coefficient problems, determining conditions under which a given difference method is stable. The next task is naturally to investigate the application and applicability of this tool in a more general, realistic context. Treatment of stability effects of boundary conditions, especially nontrivial ones, is still delayed and will be considered in Chapter 8. Here we consider extensions to problems with variable coefficients and ultimately to nonlinear problems.

Example 5.1 (An unexpected instability) The purpose of this example is to motivate the entire present chapter.

The *Korteweg–de Vries* (KdV) equation is a nonlinear PDE which arises in many applications. We write it as

$$u_t + (\alpha u^2 + \nu u_{xx})_x = 0, \tag{5.1}$$

where ν and α are known constants. The equation has a long history and admits, in particular, *solitary wave solutions* [58]. We shall return to it later in this chapter as well as in Section 7.3. Right now, let us simply state that the linearization of this equation is a hyperbolic PDE, and that the initial value problem is well-posed with a bounded solution existing for all $t \geq 0$.

The following explicit, leapfrog scheme was proposed in the 1960s [187]. With $\mu = \frac{k}{h}$ the scheme reads

$$v_j^{n+1} = v_j^{n-1} - \frac{2\alpha\mu}{3}(v_{j-1}^n + v_j^n + v_{j+1}^n)(v_{j+1}^n - v_{j-1}^n)$$
$$- \frac{\nu\mu}{h^2}(v_{j+2}^n - 2v_{j+1}^n + 2v_{j-1}^n - v_{j-2}^n).$$

It is accurate of order $(2, 2)$. The discretization of the nonlinear term is reasonable and will be discussed in Section 5.3.

Upon freezing coefficients and applying von Neumann's constant coefficient stability analysis (Exercise 4.8) we obtain that the time step must be restricted to satisfy

$$k < h / \left[\frac{4|\nu|}{h^2} + 2|\alpha u_{\max}| \right],$$

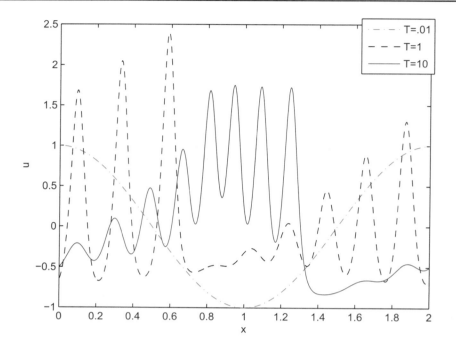

Figure 5.1. *Snapshots of the solution for Example 5.1.*

where $u_{\max}$ is the least upper bound on $|u(t, x)|$. Following [188, 12] we choose the data

$$\nu = 0.022^2, \quad \alpha = 0.5,$$
$$u_0(x) = \cos(\pi x), \quad u(0, t) = u(2, t).$$

The solution profile at different times is displayed in Figure 5.1.

Thus, we can generously estimate $u_{\max} \leq 5$ and conclude that it would be wise to select

$$k < h/(4 \times .022^2/h^2 + 5).$$

For $h = .01$ this is satisfied if $k < .0004$.

And yet, here is what happens: the solution calculated using MATLAB standard arithmetic for this explicit scheme with $h = .01$ and $k = .0001$ is qualitatively correct for a while. However, around $t = 5$, after about 50,000 time steps, the numerical solution blows up![20] Reducing the time step further does not help either.

What happens is that enough high wave number error accumulates in the solution after many time steps so that the linear stability bound becomes violated. The primary lesson for us now is that the constant coefficient Fourier analysis is not always sufficient in practice. ∎

---

[20]It is conceivable that with the computing equipment available in the early 1960s, calculating 50,000 time steps of *anything* was considered an unaffordable luxury.

Despite Example 5.1 it turns out that extension of the constant coefficient analysis is possible if we require that high-frequency amplifications are attenuated and not merely bounded by 1. This leads to the concept of *dissipativity*, considered in Section 5.1.

Dissipativity is obtained naturally when consistent discretizations are designed for problems that contain *dissipation*. But for hyperbolic equations, which are not dissipative, the design of dissipative methods is tricky. This is the subject of Section 5.2, and it arises in future chapters as well. We use the opportunity to introduce several schemes for hyperbolic PDE systems in one spatial variable and to demonstrate their performance.

Extensions of the Fourier and eigenvalue techniques are not always possible. An alternative approach, the *energy method*, for ascertaining stability in the nonlinear setting is discussed in Section 5.3. The approach is more ad hoc, but sometimes it works beautifully. In Section 5.3.2, in particular, we consider the popular classical fourth order Runge–Kutta time discretization and show its dissipation properties by "energy" estimates.

5.1 Freezing coefficients and dissipativity

We consider a general difference scheme for a linear initial value problem. A natural idea is to "freeze" the coefficients, i.e., assume that the (variable) coefficients of the PDE are constant, with constant values in the range of the actual coefficient functions, and check stability using the methods of the previous chapter.

The question is, suppose that all these constant-coefficient schemes come out stable—is the method for the variable-coefficient problem stable as well? As it turns out, the answer is *not necessarily*, even though in practice usually the technique of freezing coefficients works well, especially in giving a basic feeling on what should work when. However, some unusual examples exist to show that, strictly speaking, the stability of the method for the frozen problem is neither sufficient nor necessary for stability in the variable coefficient case.

Basically, the difficulty is as follows: if we allow the eigenvalues of the amplification matrix G defined in Section 4.1.2 to be on the unit circle (rather than inside), then some perturbation, which cannot always be controlled, may push them in the "wrong" direction and produce an instability. Now, by Exercise 4.4 we cannot require $\max_\zeta \rho(G(\zeta)) < 1$, because consistency implies that $\rho(G(0)) = 1$. See also Figure 4.1. But our concern is not low wave numbers anyway—it is the high wave numbers $|\zeta| \lesssim \pi$, i.e., $|\xi| \lesssim \pi/h$ where h is small, whose growth may cause instability. So, we require damping of the high wave numbers.

The difference scheme is called **dissipative**[21] of order $2r$ if there is a constant $\delta > 0$ such that for all the associated frozen coefficient schemes the eigenvalues of the amplification matrix $G(\zeta)$ satisfy

$$\rho(G(\zeta)) \leq e^{\tilde{\alpha} k}(1 - \delta|\zeta|^{2r}) \quad \forall |\zeta| \leq \pi. \tag{5.2}$$

Thus, at $\zeta = 0$ the von Neumann condition is retrieved while for higher wave numbers the present requirement is more demanding. The dissipativity condition (5.2) just defined is *not necessary* for the "good behavior" of a scheme, essentially because it is not necessary

---
[21]Sometimes, the term *strictly dissipative* is used.

for producing a smooth approximate solution. But it allowed Kreiss to prove the following useful results [76, 110].

Theorem 5.2. *If the linear evolutionary PDE* (3.1) *is well-posed and the difference method is consistent, dissipative of order* $2r$, *and has coefficients that are Lipschitz continuous in x and t, then the scheme is stable in each of the following cases:*

- *The PDE is strictly hyperbolic (i.e., $\mathbf{u}_t + A\mathbf{u}_x = \mathbf{0}$ with the eigenvalues of A real and distinct).*

- *The PDE is symmetric hyperbolic and the difference scheme is symmetric and accurate of order at least $2r - 2$.*

- *The PDE is parabolic, with μ bounded by a constant.*

Example 5.3 For parabolic problems, with sufficiently moderate values of μ all the schemes that we have seen are dissipative in the sense of (5.2); see Figures 4.1, 5.2, and 5.3. Note that even forward Euler is dissipative for $\mu < .5$. When μ is raised beyond the stability restriction of the explicit scheme but remains moderate, then the Crank–Nicolson scheme is still dissipative; see Figure 5.2. This is not surprising since the schemes, being consistent, inherit dissipation from the PDE. Thus, the von Neumann analysis is usually indicative of what happens for general parabolic problems. We have seen some of this already in Example 4.7 and Exercise 4.3. As the time step k is increased further so that μ becomes large, the Crank–Nicolson scheme loses dissipation, leaving the backward Euler scheme as the only strongly dissipative one in the limit of large time steps; see Figure 5.3. This corresponds to the stiff decay property in numerical ODEs. ∎

But the real interest in Theorem 5.2 is what it says about schemes for hyperbolic, not parabolic, problems. Let us consider this next.

5.2 Schemes for hyperbolic systems in one dimension

For a hyperbolic $s \times s$ system

$$\mathbf{u}_t + A\mathbf{u}_x = \mathbf{0},$$

i.e., A has a complete set of eigenvectors and real eigenvalues, recall from Chapter 1 that there is no attenuation in high wave number components of the *symbol* $P(\iota\xi)$. Thus, dissipation in the numerical scheme is in a sense an artificial property, and its amount must be carefully controlled, lest the numerical solution end up displaying different qualitative features from those of the exact solution. As usual, we set $\mu = k/h$.

5.2. Schemes for Hyperbolic Systems in One Dimension

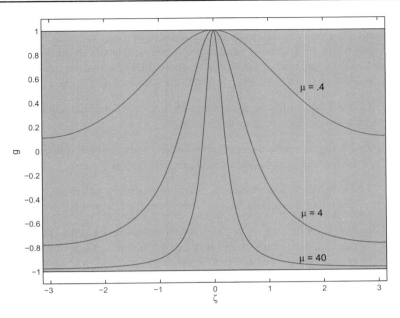

Figure 5.2. *Amplification factors for the Crank–Nicolson scheme applied to the heat equation. The scheme remains dissipative for all μ, although as μ grows the amplification factor inches toward -1 at high wave numbers.*

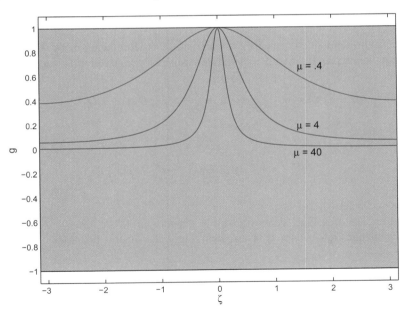

Figure 5.3. *Amplification factors for the backward Euler scheme applied to the heat equation. The scheme remains dissipative for all μ. As μ grows the amplification factor inches toward 0 at high wave numbers, as it should.*

> **Note:** There are several issues that arise for hyperbolic-type PDEs which do not occur for parabolic or elliptic problems. These can all be traced back to the conservative nature of hyperbolic PDEs, where the linear PDE $u_t = P(\partial_x)u$ satisfies $|e^{P(\iota\xi)t}| = 1 \ \forall \xi$; see Section 1.1.
>
> 1. The PDE admits no dissipation. Dissipative numerical schemes are considered in the present chapter.
>
> 2. There is numerical dispersion whether or not the PDE is dispersive. Numerical dispersion is considered in Chapter 7.
>
> 3. There are delicate issues concerning boundary conditions (BCs). These and numerical BCs are considered in Chapter 8.
>
> 4. The solution may have discontinuities inside the time–space domain. Discontinuities and methods for handling them are considered in Section 5.2 and in Chapter 10.

5.2.1 Lax–Wendroff and variants for conservation laws

Let us derive this scheme for the constant coefficient case first. By Taylor's expansion for one equation,

$$u(t+k, x) = u + ku_t + \frac{k^2}{2}u_{tt} + O(k^3),$$

where the right-hand-side terms are evaluated at (t, x). Replacing u_t by $-au_x$ and u_{tt} by $a^2 u_{xx}$, and then approximating the spatial derivatives by centered differences, we obtain

$$u(t+k, x) = u - \frac{k}{2h}aD_0 u + \frac{k^2}{2h^2}a^2 D_+ D_- u + k\,O(k^2 + h^2).$$

Hence the scheme

$$v_j^{n+1} = \left(I - \frac{\mu}{2}aD_0 + \frac{\mu^2}{2}a^2 D_+ D_-\right) v_j^n \tag{5.3a}$$

is accurate of order (2, 2).[22]

In the case of a PDE system where A depends on x the term $A^2 \mathbf{u}_{xx}$ in the above derivation changes into $A(A\mathbf{u}_x)_x$. Correspondingly, the scheme becomes

$$\mathbf{v}_j^{n+1} = \left[I - \frac{\mu}{2}A_j D_0 + \frac{\mu^2}{2}A_j \delta(A_j \delta)\right] \mathbf{v}_j^n. \tag{5.3b}$$

For an even more general A we must calculate $A^{n+1/2}$, which may prove to be a drag, especially in the nonlinear case.

---
[22] Note that (5.3a) is *not* naturally derived from a time differencing of a semi-discretization.

5.2. Schemes for Hyperbolic Systems in One Dimension

Fortunately, many nonlinear problems arising in practice appear in **conservation form**:

$$\mathbf{u}_t + \mathbf{f}(\mathbf{u})_x = \mathbf{0}. \tag{5.4}$$

This is the case for many flow problems, where the equations in (5.4) express conservation of various physical quantities such as mass, energy, and momentum.

Example 5.4 The inviscid Burgers equation

$$u_t + uu_x = 0$$

can be expressed in the conservation form

$$u_t + \frac{1}{2}(u^2)_x = 0.$$

More generally, the Navier–Stokes equations describing fluid flow in three space variables contain the material derivative (for the fluid velocity components u, v, and w),

$$\frac{Du}{Dt} = u_t + uu_x + vu_y + wu_z,$$

$$\frac{Dv}{Dt} = v_t + uv_x + vv_y + wv_z,$$

$$\frac{Dw}{Dt} = w_t + uw_x + vw_y + ww_z,$$

which does not appear to be in conservation form. However, in the case where the fluid is incompressible, the incompressibility condition

$$u_x + v_y + w_z = 0$$

can be used to write the material derivative in the conservation form,

$$\frac{Du}{Dt} = u_t + (u^2)_x + (vu)_y + (wu)_z,$$

$$\frac{Dv}{Dt} = v_t + (uv)_x + (v^2)_y + (wv)_z,$$

$$\frac{Dw}{Dt} = w_t + (uw)_x + (vw)_y + (w^2)_z. \quad \blacksquare$$

The conservation form (5.4) can be written as $\mathbf{u}_t + A\mathbf{u}_x = \mathbf{0}$, where $A = \frac{\partial \mathbf{f}}{\partial \mathbf{u}}$ is the Jacobian matrix which generally depends on $\mathbf{u}$. But in the derivation of the *Lax–Wendroff* scheme we can use $\mathbf{f}$ directly to obtain the variant

$$\mathbf{v}_j^{n+1} = \mathbf{v}_j^n - \frac{\mu}{2}(\mathbf{f}_{j+1}^n - \mathbf{f}_{j-1}^n)$$

$$+ \frac{\mu^2}{2}[A_{j+1/2}^n(\mathbf{f}_{j+1}^n - \mathbf{f}_j^n) - A_{j-1/2}^n(\mathbf{f}_j^n - \mathbf{f}_{j-1}^n)]. \tag{5.5a}$$

This is an explicit, second order formula for nonlinear systems of conservation laws for which we apply a stability analysis below. Before that, though, we mention two other variants due to *Richtmyer* and *MacCormack*, respectively, which avoid using the Jacobian matrix altogether and reduce to the Lax–Wendroff scheme in the constant-coefficient case [144, 133, 92]:

$$\bar{\mathbf{v}}_j = \frac{1}{2}(\mathbf{v}_j^n + \mathbf{v}_{j+1}^n) - \frac{1}{2}\mu(\mathbf{f}_{j+1}^n - \mathbf{f}_j^n),$$
$$\mathbf{v}_j^{n+1} = \mathbf{v}_j^n - \mu(\mathbf{f}(\bar{\mathbf{v}}_j) - \mathbf{f}(\bar{\mathbf{v}}_{j-1})) \tag{5.5b}$$

(see Exercise 5) and

$$\bar{\mathbf{v}}_j = \mathbf{v}_j^n - \mu(\mathbf{f}_j^n - \mathbf{f}_{j-1}^n),$$
$$\mathbf{v}_j^{n+1} = \frac{1}{2}(\mathbf{v}_j^n + \bar{\mathbf{v}}_j^n) - \frac{1}{2}\mu(\mathbf{f}(\bar{\mathbf{v}}_{j+1}) - \mathbf{f}(\bar{\mathbf{v}}_j)). \tag{5.5c}$$

Let us consider the stability properties of the Lax–Wendroff scheme. Note that for constant coefficients all the above variants coincide and reduce to the system version of (5.3a). For each eigenvalue a_l of A there corresponds an eigenvalue of the amplification matrix $G(\zeta)$:

$$\lambda_l(\zeta) = 1 - \iota\mu a_l \sin\zeta - 2\mu^2 a_l^2 \sin^2(\zeta/2).$$

From Exercise 1 we learn that

$$|\lambda_l(\zeta)| \leq 1 - \delta|\zeta|^4,$$

where $\delta > 0$ so long as the strict CFL condition, $\mu \max |a_l| < 1$, holds. Thus, the various Lax–Wendroff schemes are strictly dissipative and hence stable as per the stipulations of Theorem 5.2.

Note that the amount of dissipation in the Lax–Wendroff scheme is not controlled: the high wave number components get damped more than the low ones and the solution is smoothed. Thus, sharp solution features may be smeared and integration over a long time could yield undesirable results. It is important then to introduce as little dissipation as possible—just enough to prevent instability.

5.2.2 Leapfrog and Lax–Friedrichs

The *leapfrog* scheme can be written as

$$\mathbf{v}_j^{n+1} = \mathbf{v}_j^{n-1} - \mu A_j^n(\mathbf{v}_{j+1}^n - \mathbf{v}_{j-1}^n), \tag{5.6a}$$

extending to conservation laws in an obvious way:

$$\mathbf{v}_j^{n+1} = \mathbf{v}_j^{n-1} - \mu(\mathbf{f}_{j+1}^n - \mathbf{f}_{j-1}^n). \tag{5.6b}$$

Like the Lax–Wendroff scheme, it has order (2, 2). We have already seen in Example 4.3 that, provided the CFL condition holds, the eigenvalues of the amplification matrix satisfy

5.2. Schemes for Hyperbolic Systems in One Dimension

$|\lambda_l| = 1$ for all relevant l and ζ, hence the scheme is **energy-conserving** and *nondissipative*. To avoid instabilities, which may arise especially in nonlinear problems, we can modify the scheme as in [110]:

$$\mathbf{v}_j^{n+1} = \left(I - \frac{\varepsilon}{16} D_+^2 D_-^2\right) \mathbf{v}_j^{n-1} - \mu A_j^n D_0 \mathbf{v}_j^n, \tag{5.7}$$

where $\varepsilon < 1$ is a parameter. Then

$$|\lambda_l| = 1 - \varepsilon \sin^4(\zeta/2) \quad \text{provided } \mu|a_l| < 1 - \varepsilon.$$

So, we get a dissipativity of order 4, with ε controlling the amount of dissipation. There is no loss of accuracy order in the modified scheme.

Next, we recall the *Lax–Friedrichs scheme* (4.15), to which Exercise 4.2 is devoted. This scheme extends to systems in an obvious way, with a replaced by A_j^n in (4.15). It also immediately extends to nonlinear conservation laws. The order of accuracy is (1, 1). The scheme can further be written as

$$\mathbf{v}_j^{n+1} = \mathbf{v}_j^n - \frac{\mu}{2} A_j^n (\mathbf{v}_{j+1}^n - \mathbf{v}_{j-1}^n) + \frac{1}{2}(\mathbf{v}_{j-1}^n - 2\mathbf{v}_j^n + \mathbf{v}_{j+1}^n). \tag{5.8}$$

Note that this scheme is not strictly dissipative; i.e., it does not satisfy (5.2); see Exercise 2. Yet we show next that the Lax–Friedrichs scheme is certainly stable for variable coefficient problems nonetheless, provided that the CFL condition holds. Let us rewrite the one-equation version of (5.8) as

$$v_j^{n+1} = \frac{1}{2}(1 - \mu a_j^n) v_{j+1}^n + \frac{1}{2}(1 + \mu a_j^n) v_{j-1}^n.$$

If $\mu \max |a| \leq 1$, then the scheme has nonnegative coefficients multiplying the solution values on the right-hand side. Schemes with this property are called **monotone**. We encounter this property again in Chapter 10. Since the sum of the coefficients is 1 (this must be so by consistency) we get $|v_j^{n+1}| \leq \max\{|v_{j+1}^n|, |v_{j-1}^n|\}$, and thus also

$$\max_j |v_j^{n+1}| \leq \max_j |v_j^n|. \tag{5.9}$$

This is a strong stability bound in the maximum (or l_∞-) norm. The scheme does not require additional artificial dissipativity terms! In fact, (5.9) guarantees more than just stability, as is discussed in Example 5.6 below.

Although not derived in that way, both Lax–Wendroff and Lax–Friedrichs schemes can be viewed as a forward Euler discretization of a large ODE system in time. Exercise 6 shows that the eigenvalues of the corresponding Jacobian matrix have negative real parts and are no longer purely imaginary, unlike for conservative schemes such as leapfrog and (1.15b).

Finally, we mention here that another way to "correct" the instability of the scheme (1.15b) is to replace the forward Euler time discretization by the classical Runge–Kutta method of order 4 (*RK4*). Figure 2.6 suggests that RK4 has the stability that forward Euler lacks along the imaginary axis. We delay further discussion of this scheme to Section 5.3.2, where a numerical example is given as well.

Table 5.1. *Maximum errors at $t = 1$ for three difference schemes applied to the equation $u_t - u_x = 0$ with $u_0(x) = \sin \eta x$ and periodic BC. Error blowup is denoted by *. Note that $k = \mu h$.*

η	h	μ	Lax–Wendroff	Lax–Friedrichs	Box
1	$.01\pi$	0.5	1.2e-4	2.3e-2	6.1e-5
	$.001\pi$	0.5	1.2e-6	2.4e-3	6.2e-7
	$.001\pi$	5.0	*	*	2.0e-5
10	$.01\pi$	0.5	1.2e-1	9.0e-1	6.1e-2
	$.001\pi$	0.5	1.2e-3	2.1e-1	6.2e-4
	$.0005\pi$	0.5	3.1e-4	1.1e-1	1.5e-4

Example 5.5 In Figure 1.9 we displayed results using the leapfrog and other schemes for the simple advection equation $u_t - u_x = 0$. For the smooth initial data $u_0(x) = \sin \eta x$ the stable schemes appear to perform well. Errors for the various schemes presented in this section are recorded in Table 5.1, which is to be compared with Table 1.1. From these results it is clear that the Lax–Wendroff scheme is second order; indeed the errors are very similar to those of the leapfrog scheme. On the other hand, the significantly less accurate Lax–Friedrichs is only first order in both x and t, and the error magnitudes are about twice those recorded in Table 1.1 for the upwind scheme (1.15a). ∎

Example 5.6 For the advection equation Examples 5.5 and 1.2 show that if the initial data are smooth, then we may expect decent results from all schemes considered here, essentially reflecting their order, provided that the CFL condition holds. But all this changes dramatically if the initial data function is discontinuous. Here we run the schemes discussed so far in the present section for the same problem but for initial data with discontinuities. Thus we choose $u_0 = 0$, except when $.25 \leq x < .75$, where we set $u_0 = 1$. The exact solution for $u_t - u_x = 0$ then has this square wave moving left at speed $\frac{dx}{dt} = -1$. Figures 5.4 and 5.5 display results at $t = 1$ using the Lax–Friedrichs scheme, the Lax–Wendroff scheme, the leapfrog scheme, and the dissipated leapfrog (5.7) with $\varepsilon = 0.5$.

Unlike the Fourier transform of the smooth initial function from Chapter 1, here the Fourier transform of the square wave is rich in all wave numbers. When these Fourier modes are heavily damped, as in the Lax–Friedrichs scheme, we obtain a solution which certainly looks stable but is simply too smooth and too smeared out. It does satisfy (5.9), though, and no spurious oscillations arise. The Lax–Wendroff scheme also smooths out the solution, but much less so. The leapfrog scheme has annoying oscillations which spread also away from the square wave. This is improved upon by the dissipative term, and the modified leapfrog performs almost as well as the Lax–Wendroff scheme in this case.

The display in Figure 5.5 indicates that even for a fine discretization (finer than is typically affordable in complex applications) there is excessive smoothing in the Lax–Friedrichs scheme, and the **Gibbs phenomenon** (i.e., the oscillations in the approximate

5.2. Schemes for Hyperbolic Systems in One Dimension 161

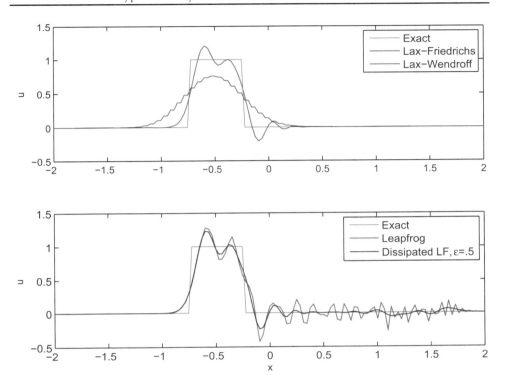

Figure 5.4. *Solving $u_t - u_x = 0$ starting from a square wave, using $\mu = 0.5$ and $h = 0.01\pi$. The exact solution at $t = 1$ is plotted, together with the approximate results. The usual piecewise linear interpolation is employed between mesh values: this does not generate any spurious maxima or minima.*

solution near the discontinuities) in the other schemes does not disappear. The leapfrog scheme is bad again, and the added dissipativity improves it, but not beyond what is achieved by the Lax–Wendroff scheme. ∎

The solution profile for the Lax–Friedrichs scheme in Figure 5.4 features an annoying staircasing effect which is the result of the fact that this scheme can be split into odd and even parts: only solution variables v_l^n with even l participate in forming v_j^{n+1} for odd j, and vice versa. There are ways to fix this, though. For instance, a smoother curve of the same basic accuracy can be obtained by simple postprocessing: at t_N where we wish to display the results, replace v^N by $\tilde{v}^N$, where

$$\tilde{v}_j^N = \frac{1}{4}[v_{j-1}^N + 2v_j^N + v_{j+1}^N]. \tag{5.10}$$

Although this may smooth true discontinuities somewhat, there is nothing close to discontinuity in the case of the Lax–Friedrichs scheme anyway.

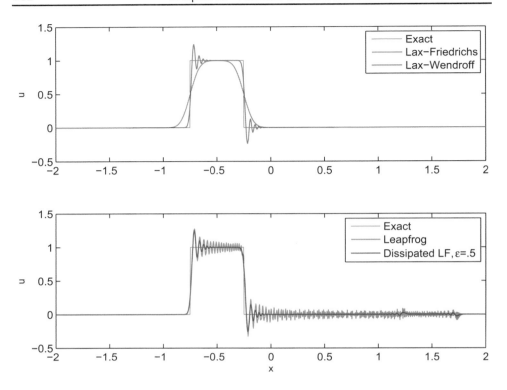

Figure 5.5. *Solving $u_t - u_x = 0$ starting from a square wave, using $\mu = 0.5$ and $h = 0.001\pi$. The exact solution at $t = 1$ is plotted, together with the approximate results.*

5.2.3 Upwind scheme and the modified PDE

The results in Example 5.6 are typical for problems with discontinuous solutions. Generally, the schemes we have seen so far are simply not sufficiently good for such problems. Note that for nonlinear hyperbolic PDEs discontinuities may develop in the solution even if the initial and boundary data are smooth! Essentially, methods that are based on a *fixed* spatial discretization coupled with a time discretization are adequate for parabolic problems and for hyperbolic problems with smooth solutions. But when faced with a hyperbolic problem with some solution discontinuities, the problem's characteristic directions must be taken more into account. This implies that the spatial discretization requires "individual attention" at each mesh point.

The *upwind scheme* featured in Exercise 1.5 is a discretization of this more elaborate sort. For the advection equation $u_t + au_x = 0$ it is defined by

$$v_j^{n+1} = v_j^n - \mu a \begin{cases} (v_{j+1}^n - v_j^n), & a < 0, \\ (v_j^n - v_{j-1}^n), & a \geq 0. \end{cases} \tag{5.11}$$

The name "upwind" becomes more intuitive if we imagine a variable coefficient $a(t, x)$ that may change sign. The discretization stencil then depends on the location in time and space.

5.2. Schemes for Hyperbolic Systems in One Dimension

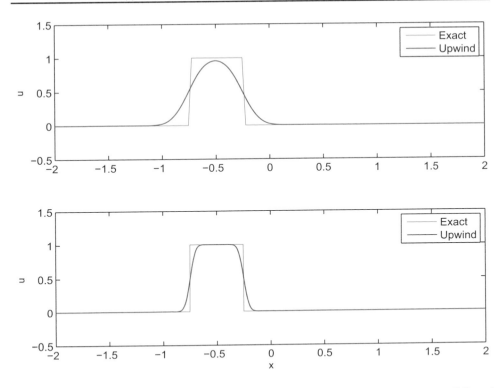

Figure 5.6. *Solving $u_t - u_x = 0$ starting from a square wave, using $\mu = 0.5$ and either* (i) $h = 0.01\pi$ *(upper plate) or* (ii) $h = 0.001\pi$ *(lower plate). The exact square wave solution at $t = 1$ is plotted, together with the approximate solution.*

Example 5.7 Repeating the experiments of Example 5.6 using (5.11) yields the results displayed in Figure 5.6.

We see that this upwind scheme exhibits a desirable behavior similar to the Lax–Friedrichs one, with the obtained profiles being sharper. These are arguably the best results we have in the present chapter for problems with solution discontinuities. ∎

Indeed, the upwind and the Lax–Friedrichs schemes have a number of properties in common. The upwind scheme has order $(1, 1)$ and, writing it as

$$v_j^{n+1} = v_j^n - \mu|a|v_j^n + \mu|a|v_{j-\text{sign}(a)}^n \tag{5.12a}$$

$$= v_j^n - \frac{\mu a}{2}(v_{j+1}^n - v_{j-1}^n) + \frac{\mu|a|}{2}(v_{j+1}^n - 2v_j^n + v_{j-1}^n), \tag{5.12b}$$

we see from (5.12a) that it is monotone (has nonnegative coefficients) provided the CFL condition holds. This yields the bound (5.9), which guarantees no spurious oscillations in the computed approximate solution.

Additional insight may be obtained by considering the modified PDE, developed next.

Modified PDE

Let us recall the definitions of local truncation error (3.29) and order of accuracy (3.30). To derive or show the order of a given method one uses Taylor expansions, as in Example 3.10 and many exercises. Thus, the truncation error is typically written as a sum of two leading terms of size $O(k^{p_1})$ and $O(h^{p_2})$, plus higher order terms. These leading order terms involve solution derivatives, of course, so incorporating them may suggest a *modified PDE* that the given discretization approximates more closely than it does the intended one. The properties of such a modified PDE may then shed light on our numerical method.

To see how this works consider the upwind scheme for the advection equation with $a < 0$. Then the scheme (5.11) boils down to (1.15a). We next write

$$u(t+k, x) = u + ku_t + \frac{k^2}{2}u_{tt} + \frac{k^3}{6}u_{ttt} + \cdots,$$

$$u(t, x+h) = u + hu_x + \frac{h^2}{2}u_{xx} + \frac{h^3}{6}u_{xxx} + \cdots,$$

$$\tau = \frac{u(t+k,x) - u}{k} + a\frac{u(t,x+h) - u}{h}$$

$$= \left[u_t + \frac{k}{2}u_{tt} + \frac{k^2}{6}u_{ttt} + \cdots\right] + a\left[u_x + \frac{h}{2}u_{xx} + \frac{h^2}{6}u_{xxx} + \cdots\right].$$

The expansions are about the point (t, x), and we omit this argument when it arises above. Now, canceling out $u_t = -au_x$ yields the usual order $(1, 1)$ for this upwind method. Further, however, we apply the differential equation again to obtain $u_{tt} = a^2 u_{xx}$. The leading terms of the local truncation error can therefore be written as $-\nu u_{xx}$, where $\nu = -\frac{1}{2}[ka^2 + ha]$, or

$$\nu = \frac{h}{2\mu}(\mu|a| - \mu^2 a^2). \tag{5.13a}$$

Since $\tau + \nu u_{xx} = O(k^2) + O(h^2)$, *the same* upwind method may be viewed as approximating to a higher order the advection-diffusion equation

$$u_t + au_x = \nu u_{xx}. \tag{5.13b}$$

There is dissipation in the modified, parabolic PDE (5.13b), and correspondingly there is attenuation of high wave number amplification factors in this scheme.

Exercise 7 shows that a similar modified PDE but with the larger value $\nu = \frac{h}{2\mu}(1 - \mu^2 a^2)$ is obtained for the Lax–Friedrichs scheme. Thus, both these schemes can be seen as simple discretizations resulting from adding an **artificial viscosity**, or **artificial diffusion** term to the hyperbolic PDE which is the advection equation.

Whereas the Lax–Friedrichs scheme generalizes automatically to systems of conservation laws such as (9.2), the generalization of the upwind scheme requires a lot of attention. We leave this to Chapter 10. But note that when it is applicable, the upwind scheme adds less artificial diffusion (less since the CFL condition $\mu|a| \leq 1$ must hold in any event) and produces sharper, nonoscillatory solution profiles. Note that for the Lax–Friedrichs scheme there is relatively more artificial diffusion as k gets smaller with h fixed. Its corresponding results in Figures 5.4 and 5.5 could be improved by selecting $\mu = 0.9$ instead of $\mu = 0.5$. But this is not always possible in more complex situations; see Example 10.10.

The Lax–Wendroff scheme (5.3a) can also be seen as a higher order (specifically, third order) approximation to a modified PDE (5.26b) (see Exercise 7), but that modified equation involves u_{xxx}, not u_{xx}, and the resulting PDE is a dispersive, hyperbolic equation which does not "control" the oscillations.

For more on modified PDEs see, e.g., [137, 134]. Care must be taken to use this tool rigorously. We have not done this here and are using modified PDEs only to get a feel for the relative performance of different methods.

5.2.4 Box and Crank–Nicolson

Next we introduce two schemes that are not dissipative but are energy conserving (i.e., the amplification factors equal 1 in magnitude for all wave numbers). They are also both implicit and thus need not obey the CFL condition.

The *Crank–Nicolson* scheme for the hyperbolic system is obtained as a limit of the same scheme applied to the parabolic problem, i.e., the *trapezoidal* method (2.8) is applied in time. This yields

$$\left(I + \frac{1}{4}\mu A D_0\right) \mathbf{v}_j^{n+1} = \left(I - \frac{1}{4}\mu A D_0\right) \mathbf{v}_j^n, \qquad (5.14)$$

with $\mu = k/h$. The matrix A should be evaluated as A_j^{n+1} in the left-hand term and as A_j^n in the right-hand term. Alternatively, we can apply the midpoint rule (2.12) instead and evaluate $A = A_j^{n+1/2}$ in both sides of the equation; see Chapter 2. Some prefer the latter variant, although there is no huge difference. A variant for conservation laws can again be written down naturally.

The two variants of Crank–Nicolson coincide in the constant coefficient case and the scheme is again energy conserving. It is possible to add a dissipative term to the Crank–Nicolson scheme by modifying (5.14) into

$$\left(I + \frac{1+\varepsilon}{4}\mu A D_0\right) \mathbf{v}_j^{n+1} = \left(I - \frac{1-\varepsilon}{4}\mu A D_0\right) \mathbf{v}_j^n, \qquad (5.15)$$

where $0 \leq \varepsilon = O(kh)$. This relates to the θ-method of Exercise 2.2. Moreover, stability for variable coefficient problems is unconditional. However, the scheme is also implicit, unlike the modified leapfrog, which is explicit.

Example 5.8 Consider the problem

$$u_t + au_x = \sigma u_{xx},$$

$$u(0, x) = \begin{cases} 1 & x \leq 0, \\ 0 & x > 0, \end{cases} \quad u(t, -\pi) = 1, \quad u(t, \pi) = 0,$$

which we wish to integrate from $t = 0$ to $t = 1$.

This is a parabolic problem for $\sigma > 0$, but if $\sigma \ll 1$, then the PDE is "almost hyperbolic." Specifically, with $a = -1$ it is "almost" the same advection equation that gives rise to Figure 1.3. Thus, we expect the solution to look more like the latter than like the one depicted in Figure 1.4. Let us then set $a = -1$, $\sigma = 10^{-3}$.

The familiar Crank–Nicolson scheme reads

$$\left(1 + \frac{ka}{4h}D_0 - \frac{k\sigma}{2h^2}D_+D_-\right)v_j^{n+1} = \left(1 - \frac{ka}{4h}D_0 + \frac{k\sigma}{2h^2}D_+D_-\right)v_j^n, \quad 1 \le j \le J,$$

and it boils down to the scheme (5.14) introduced above when $\sigma = 0$.

Using the boundary conditions to close off the system in space, we can propagate it in time. The tridiagonal matrix that appears at the left-hand side for the implicit method can be assembled and decomposed once, and then for each time step only a quick forward-backward substitution is required. We first calculate an accurate solution by employing a fine discretization with step sizes $k = .0001$, $h = .0001\pi$. The result is plotted in Figure 5.7(a). Indeed, the solution resembles the step function of Figure 1.3.

But taking very small, and hence many, steps in x and t does not seem right for such a simple looking solution. Hence we next try increasing the spatial step size to $h = .01\pi$ and the time step size to $k = .01$. Note that both these steps are larger than σ; hence we are really looking in Figure 5.7(b) basically at what the Crank–Nicolson scheme does for the advection equation. It is not surprising to observe again the artificial wiggles near the somewhat smoothed discontinuity, because the scheme has no dissipativity when $\sigma = 0$.

Next consider again the upwind scheme for the advection equation. To its right-hand side we add $\frac{k\sigma}{h^2}D_+D_-v_j^n$, discretizing the term σu_{xx} in a forward Euler fashion. This is easy and natural with the form (5.12b)! We obtain the scheme

$$v_j^{n+1} = v_j^n - \frac{ka}{2h}\left(v_{j+1}^n - v_{j-1}^n\right) + \left[\frac{k\sigma}{h^2} + \frac{k|a|}{2h}\right]\left(v_{j+1}^n - 2v_j^n + v_{j-1}^n\right).$$

Thus, the actual diffusion is added to the artificial diffusion. The amplification factor of this scheme can readily be calculated to yield

$$g(\zeta) = 1 - \iota\frac{k}{h}a\sin(\zeta) - 4\frac{k}{h}|a|b\sin^2(\zeta/2), \quad b = \left[\frac{\sigma}{|a|h} + \frac{1}{2}\right],$$

and for parameters like the ones we are using stability is ensured if all is well at $\zeta = \pi$, i.e., if we restrict k to satisfy

$$k \le \frac{h}{2|a|b}.$$

This restriction is only slightly worse than the CFL condition for the case $\sigma = 0$, and it holds for the above choices of σ, a, h, and k.

The result in depicted in Figure 5.7(c). Although the sharp front is smoothed somewhat, it looks much more pleasing than Figure 5.7(b).

It can be argued that instead of $\frac{k\sigma}{h^2} + \frac{k|a|}{2h}$ we ought to have taken $\max\{\frac{k\sigma}{h^2}, \frac{k|a|}{2h}\}$. The difference in the result here is not huge, but the latter expression yields a smoother transition from artificial to real diffusion as σ grows. We defer further discussion and demonstration to Section 10.6. ∎

5.2. Schemes for Hyperbolic Systems in One Dimension

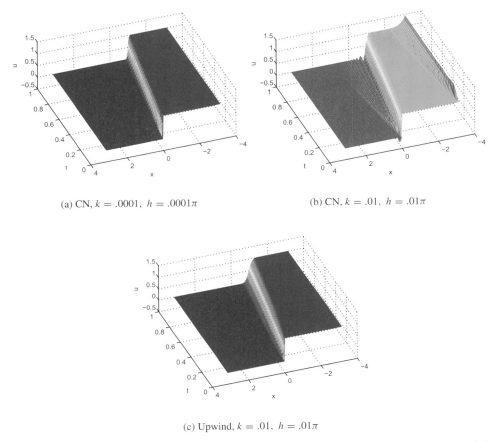

(a) CN, $k = .0001$, $h = .0001\pi$

(b) CN, $k = .01$, $h = .01\pi$

(c) Upwind, $k = .01$, $h = .01\pi$

Figure 5.7. *Solving an "almost hyperbolic" parabolic problem, Example 5.8. When the spatial step size is large enough so that $h \gg \sigma$, the Crank–Nicolson scheme develops wiggles while the modified upwind scheme does not.*

And now on to something different. Let us introduce the only symmetric scheme in this chapter that is *not* based on the noncompact, long centered difference D_0. The *box scheme* is described in [137], although its roots are much older [143, 107, 165]. This scheme treats time and space locally in a similar fashion (as does the leapfrog scheme, incidentally), applying **short** differences to both over a **control volume**.

Indeed, each conservation law (5.4) can be considered as the divergence in (t, x) of the "flux" vector

$$\begin{pmatrix} u \\ f \end{pmatrix}$$

equated to 0. Integrating (5.4) over the control volume $V = [t_n, t_{n+1}] \times [x_j, x_{j+1}]$ (from where the name "box" derives its origin, especially when envisioned in more than one space dimension), and applying the *Gauss divergence formula*, we obtain as in Example 3.4

$$\mathbf{u}_{j+1/2}^{n+1} - \mathbf{u}_{j+1/2}^{n} + \mathbf{f}_{j+1}^{n+1/2} - \mathbf{f}_{j}^{n+1/2} = \mathbf{0}, \tag{5.16a}$$

where, e.g.,

$$\mathbf{u}_{j+1/2}^{n} = \int_{x_j}^{x_{j+1}} \mathbf{u}(t_n, x) dx, \quad \mathbf{f}_{j}^{n+1/2} = \int_{t_n}^{t_{n+1}} \mathbf{f}(\mathbf{u}(t, x_j)) dt. \tag{5.16b}$$

For the integrals in (5.16b) we apply the trapezoidal rule, which introduces a truncation error of order $(2, 2)$ and yields the formula

$$\mathbf{v}_{j+1}^{n+1} + \mathbf{v}_{j}^{n+1} - \mathbf{v}_{j+1}^{n} - \mathbf{v}_{j}^{n} + \mu \left(\mathbf{f}_{j+1}^{n+1} + \mathbf{f}_{j+1}^{n} - \mathbf{f}_{j}^{n+1} - \mathbf{f}_{j}^{n} \right) = \mathbf{0}. \tag{5.17}$$

The formula (5.17) can be written neatly using the short difference and averaging operators (see Exercise 8).

Like Crank–Nicolson this method is **centered, second order accurate, energy conserving, unconditionally stable**, and (alas) **implicit**. Unlike Crank–Nicolson and leapfrog, it is **compact**, avoiding the long difference D_0. We return to this point in Chapter 7.

Example 5.9 We apply the box scheme to the simple advection equation with initial conditions as in Examples 5.5 and 5.6. The results for the Crank–Nicolson scheme are similar and not better, so we omit them. Results for the smooth initial conditions appear in Table 5.1. For $\mu < 1$ they are comparable to (and slightly better than) the results for the Lax–Wendroff or leapfrog schemes. The latter are therefore preferable, because they are cheaper per step, unless there is reason to want the time step to be significantly larger than the spatial step: as can be seen in the table, for the choice $\mu = 5.0$ the box scheme alone produces decent results.

When applying the box scheme to the same problem with the square wave as initial conditions, the results resemble qualitatively those for the leapfrog scheme (see Figures 5.4 and 5.5)—so much so that they are not reproduced here. It should come as no surprise at this point that conservative methods like these do not perform well in the presence of discontinuities, as they contain no damping mechanism for high wave number modes. ∎

5.3 Nonlinear stability and energy methods

Strictly speaking, the techniques introduced hitherto for the stability analysis of difference schemes do not extend to nonlinear problems.

Practically, we can still freeze coefficients and check stability. For instance, for the *inviscid Burgers's equation* $u_t + \frac{1}{2}(u^2)_x = u_t + uu_x = 0$, we "freeze" $u = \tilde{u}$, where $\tilde{u}$ takes on various values from the (somehow estimated) range of u, and check stability for the resulting advection equation $u_t + \tilde{u}u_x = 0$. A stability restriction of the form

5.3. Nonlinear Stability and Energy Methods

$$k < \frac{h}{\max |u|}$$

is sure to follow, upon using one of the explicit, conditionally stable methods described so far.

While freezing coefficients in this manner provides a useful practical check, securing linear stability generally *does not guarantee* stability for the nonlinear problem. The discrepancy between the linear and nonlinear situations is of importance especially when the computed solution is not smooth, as in the case of the leapfrog method in Examples 5.1 and 5.6. Then some nonlinear instability phenomena may arise, as we have seen in Example 5.1.

For nonlinear problems we may have to directly check boundedness of the approximate solution to guarantee stability, which is occasionally possible using an *energy method*.

5.3.1 Energy method

This is a name for a loosely defined collection of techniques. The objective is to show stability directly, by ensuring that

$$\|v(nk, \cdot)\|$$

does not increase too much (in the l_2-norm) with n.

In Chapter 1 we note that the stability requirement of a numerical scheme corresponds to the well-posedness of the differential problem. Here, too, let us discuss the differential problem first, as in Morton and Mayers [137].

The basic observation is very simple: upon multiplication of u_t by $2u$ and integration we have

$$\int_{-\infty}^{\infty} 2u u_t \, dx = \frac{\partial}{\partial t} \int_{-\infty}^{\infty} u^2(t, x) \, dx.$$

Therefore, applying the same procedure for the general, possibly nonlinear initial value problem

$$u_t = f(t, x, u_x, u_{xx}, u_{xxx}, \ldots), \qquad -\infty < x < \infty, \ t > 0, \qquad (5.18a)$$

we have that if

$$\int_{-\infty}^{\infty} u f(t, x, u_x, u_{xx}, u_{xxx}, \ldots) \, dx \leq 0, \qquad (5.18b)$$

then $\|u(t)\| \leq \|u(0)\|$ for all $t \geq 0$, and **stability** in the $\mathcal{L}_2$-norm follows. Moreover, if it turns out that

$$\int_{-\infty}^{\infty} u f(t, x, u_x, u_{xx}, u_{xxx}, \ldots) \, dx = 0, \qquad (5.18c)$$

then we have the **conservation property** that $\|u(t)\| = \|u(0)\|$ for all $t \geq 0$. The basic tool for obtaining such estimates is **integration by parts**.[23] This involves BCs, in general, but for the pure IVP we may assume that u and derivatives as required vanish at $\pm\infty$ or are periodic. The above can certainly be developed in a similar way for finite intervals in x.

Example 5.10 For the heat equation (yes, again) $u_t = u_{xx}$, we note that $(uu_x)_x = uu_{xx} + (u_x)^2$ and obtain

$$\int_{-\infty}^{\infty} uu_{xx}dx = -\int_{-\infty}^{\infty} (u_x(t,x))^2 dx \leq 0.$$

Therefore (5.18b) holds. Indeed we can see that $\|u(t)\|$ actually decreases with t barring trivial cases.

Next, let us turn to the nonlinear KdV equation (5.1). Using integration by parts once again (NB $(u^3)_x = u(u^2)_x + 2u^2 u_x$), we have

$$\int u(2uu_x)dx = \int u(u^2)_x dx = -\int 2u^2 u_x dx.$$

This integral equals its negative and hence must vanish, so (5.18c) holds for the case $\nu = 0$. Likewise, we can show that

$$\int uu_{xxx}dx = 0,$$

hence (5.18c) holds for $\nu \neq 0$ as well. The KdV solution therefore satisfies norm conservation

$$\|u(t)\| = \|u(0)\| \quad \forall t.$$

Note that for $\nu = 0$ and $\alpha = 1/2$ we have the Burgers equation. This equation also conserves the spatial $\mathcal{L}_2$-norm so long as the solution is smooth. In general, if the solution has (or develops) discontinuities, then of course the above manipulations cannot be carried out and all bets are off. As it turns out the KdV solution exists for all times but the Burgers solution typically does not.

Finally, consider the nonlinear *Schrödinger equation* from Exercise 1.7,

$$\iota \psi_t = -\psi_{xx} + V'(|\psi|)\psi.$$

Here, $\psi(t, x)$ is complex valued, $|\psi|^2 = \bar{\psi}\psi$, and the $\mathcal{L}_2$-norm is defined by $\|\psi(t)\|^2 = \int \bar{\psi}(t,x)\psi(t,x)dx$. Therefore

$$\frac{\partial}{\partial t}\|\psi\|^2 = \int \bar{\psi}\psi_t + \bar{\psi}_t \psi = 2\mathcal{R}e \int \bar{\psi}\psi_t.$$

---

[23] If $f(x)$ and $g(x)$ are differentiable on an interval $[a, b]$, then

$$f(b)g(b) - f(a)g(a) = \int_a^b (fg)'dx = \int_a^b f'gdx + \int_a^b fg'dx;$$

hence

$$\int_a^b f'gdx = -\int_a^b fg'dx + f(b)g(b) - f(a)g(a). \tag{5.19}$$

If $f(a)g(a) = f(b)g(b)$, then the boundary terms disappear in (5.19), and we can write it as

$$(f', g) = -(f, g').$$

5.3. Nonlinear Stability and Energy Methods

So, we must check $\mathcal{R}\text{e}\,\{\imath \int \bar{\psi}(\psi_{xx} - V'(|\psi|)\psi)\}$. By Exercise 9, however, this quantity vanishes. Thus, norm conservation (5.18c) holds here as well. ∎

Moving on to the discrete case, let v and w be mesh functions (which can be viewed as possibly infinite vectors). Their components may be complex numbers. Define the inner product and norm as in (1.14):

$$(v, w) = h \sum_{j=-\infty}^{\infty} v_j \bar{w}_j, \qquad \|v\|^2 = (v, v). \tag{5.20}$$

Note that the infinite summation range can become finite upon use of periodic or other boundary conditions. The factor h scales the norm properly in the case that the interval of integration in x is of length ≈ 1.

Now, for a semi-discretization of a PDE whose solution does not grow in time we obtain stability in the strict form $\|v(t_{n+1})\| \leq \|v(t_n)\|$ if we can show that

$$\frac{\partial}{\partial t}(\|v\|^2) \leq 0.$$

The main tool is **summation by parts**. The method works for nonlinear problems and in the presence of boundary conditions. But it is not a simple, general prescription for stability checking that can be automated. Below we proceed with an example and a few additional comments.

We can establish the following identities involving inner products:

1. For the time derivative

$$\frac{\partial}{\partial t}(\|v\|^2) = (v, v_t) + (v_t, v) = (v, v_t) + \overline{(v, v_t)} = 2\mathcal{R}\text{e}\,(v, v_t).$$

So

$$\frac{\partial}{\partial t}(\|v\|^2) = 2\mathcal{R}\text{e}\,(v, v_t). \tag{5.21a}$$

2. For the centered difference operator

$$(v, D_0 w) = h \sum v_j (\bar{w}_{j+1} - \bar{w}_{j-1}) = h \sum \bar{w}_j (v_{j-1} - v_{j+1}).$$

(It should be clear how the ends of a finite sum would have to be carried along for nonzero BC.) So

$$(v, D_0 w) = -(D_0 v, w); \qquad \text{hence } \mathcal{R}\text{e}\,(v, D_0 v) = 0. \tag{5.21b}$$

3. For one-sided operators we can show in a similar fashion that

$$(v, D_+ w) = -(D_- v, w). \tag{5.21c}$$

4. For any difference operator R

$$\mathcal{R}e\,(v,\,Rv) = 0 \iff \mathcal{R}e\,(v,\,Rw) = -\mathcal{R}e\,(Rw,\,v)\ \forall\ v,w \in l_2. \quad (5.21d)$$

We leave the proof to Exercise 10.

Example 5.11 Consider again the inviscid Burgers equation where it has a differentiable solution. We saw in Example 5.4 that it can be written in two forms, conservative and nonconservative. We now write a general combination of the two,

$$u_t + \frac{\theta}{2}(u^2)_x + (1-\theta)u u_x = 0,$$

where θ is a weight parameter, $0 \leq \theta \leq 1$. A centered semi-discretization of the latter form reads

$$v_t + \frac{1}{2h}\left[\frac{\theta}{2}D_0(v^2) + (1-\theta)v D_0 v\right] = 0.$$

The semi-discretization solution, unlike that of the PDE, does depend on the value of the parameter θ. We now take the inner product of the latter equation (which holds at each mesh point) with the mesh function v. This gives

$$(v,\,v_t) + \frac{1}{2h}\left[\frac{\theta}{2}(v,\,D_0(v^2)) + (1-\theta)(v,\,v D_0 v)\right] = 0.$$

By (5.21b), $(v,\,D_0(v^2)) = -(D_0 v,\,v^2)$, so

$$(v,\,v D_0 v) = (v^2,\,D_0 v) = (D_0 v,\,v^2) = -(v,\,D_0(v^2)).$$

Taking

$$\theta = 2/3$$

gives $\frac{\theta}{2} = 1 - \theta$, which therefore yields $(v,\,v_t) = 0$; hence by (5.21a)

$$\frac{\partial}{\partial t}(\|v\|^2) = 0.$$

We have obtained nonlinear stability (indeed, the conservation of $\|v(t)\|$!) for the semi-discretization for Burgers's equation, so long as the exact solution is smooth, upon using a seemingly unlikely combination of conservative and nonconservative forms!

It was further shown [76] that the semi-discretization with $\theta = 0$ can be unstable. Moreover, passing to the full discretization, it can be shown that using the leapfrog scheme may still yield some instability, as in Example 5.1.

However, for Crank–Nicolson with $\theta = 2/3$ stability in the l_2-sense is guaranteed. Consider the midpoint version

$$\frac{1}{k}(v_j^{n+1} - v_j^n) + \frac{1}{2h}\left[\frac{\theta}{2}D_0(v_j^{n+1/2})^2 + (1-\theta)v_j^{n+1/2}D_0 v_j^{n+1/2}\right] = 0,$$

5.3. Nonlinear Stability and Energy Methods

where
$$v_j^{n+1/2} = \frac{1}{2}(v_j^n + v_j^{n+1}).$$

Multiplying through by $2kv_j^{n+1/2}$ and summing up on j we have

$$\|v^{n+1}\|^2 - \|v^n\|^2 + \mu\left[\frac{\theta}{2}\left(v_j^{n+1/2}, D_0(v_j^{n+1/2})^2\right) + (1-\theta)\left(v_j^{n+1/2}, v_j^{n+1/2}D_0v_j^{n+1/2}\right)\right] = 0.$$

Repeating the argument above, we obtain that the term that μ multiplies vanishes when $\theta = 2/3$, whence
$$\|v^{n+1}\| = \|v^n\| \quad \forall\, n \geq 0. \quad \blacksquare$$

A class of schemes of the form

$$(I - kR_1)v^{n+1} = 2kR_0 v^n + (I + kR_1)v^{n-1} \tag{5.22}$$

is considered in [110, 76], where R_0 stands for a symmetric difference operator such as D_0 and the part involving R_1 corresponds to an implicit discretization. Thus we have in (5.22) a family of implicit-explicit discretizations. However, the form involving the time level $n-1$ is not always natural, so we refer the reader to the cited literature for further results; see also Section 7.3 for some numerical experiments.

5.3.2 Runge–Kutta for skew-symmetric semi-discretizations

In this subsection we apply simple energy estimates to investigate the properties of time-differencing methods for the semi-discretization

$$\mathbf{v}' = \mathbf{v}_t = L\mathbf{v}, \tag{5.23}$$

where L is a real $J \times J$, possibly large, skew-symmetric matrix

$$L^T = -L.$$

The matrix L may arise as a result of a spatial discretization of a hyperbolic PDE system with a symmetric coefficient matrix; see Exercise 4.10. A relevant reference is Levy and Tadmor [118], although we are really following no one in particular here.

It is not difficult to verify that the eigenvalues of L all satisfy $\mathcal{R}e\,\lambda_i = 0$, and moreover

$$\|L\| = \rho(L) = \max_i |\lambda_i|.$$

Note also that for any real vector $\mathbf{x} \in \mathbb{R}^J$

$$\mathbf{x}^T L \mathbf{x} = (L\mathbf{x}, \mathbf{x}) = -(\mathbf{x}, L\mathbf{x}) = -\mathbf{x}^T L \mathbf{x} = 0.$$

Finally, it also holds that

$$\|\mathbf{v}(t)\| = \|\mathbf{v}(0)\| \quad \forall\, t \geq 0.$$

These properties follow directly from the skew-symmetry of L and hold even if $L = L(t, \mathbf{v})$.
Next consider the **implicit midpoint** method (2.12) for the time discretization

$$\frac{\mathbf{v}^{n+1} - \mathbf{v}^n}{k} = L \frac{\mathbf{v}^{n+1} + \mathbf{v}^n}{2}. \tag{5.24}$$

(Evaluate $L = L(t^{n+1/2}, \frac{\mathbf{v}^{n+1}+\mathbf{v}^n}{2})$ if it is not constant.) Multiplying both sides of (5.24) by $(\mathbf{v}^{n+1} + \mathbf{v}^n)^T$ gives

$$\|\mathbf{v}^{n+1}\|^2 - \|\mathbf{v}^n\|^2 = \frac{k}{2}(\mathbf{v}^{n+1} + \mathbf{v}^n)^T L(\mathbf{v}^{n+1} + \mathbf{v}^n) = 0.$$

Hence we obtain not only unconditional stability (even for the nonlinear problem!) but also "energy" conservation

$$\|\mathbf{v}^n\| = \|\mathbf{v}^0\| \quad \forall\, n \geq 0.$$

The present paragraph makes a point that is a bit more technical but also elegant. The result for the midpoint method can also be extended for higher order *implicit Runge–Kutta* methods based on **Gauss points** (zeros of the Legendre polynomial). These methods are equivalent to **piecewise polynomial collocation** and have order $p = 2l$ when l collocation points are used (see Sections 2.2 and 6.2.1), as well as [14]. Denoting the collocation solution by $\mathbf{v}_\Delta(t)$ (it is a polynomial on each time-interval $[t_n, t_{n+1}]$ and it satisfies $\mathbf{v}_\Delta(t_n) = \mathbf{v}^n\ \forall n$) and the collocation points by $t_{nj},\ j = 1, \ldots, l$, we have

$$\|\mathbf{v}^{n+1}\|^2 - \|\mathbf{v}^n\|^2 = 2\int_{t_n}^{t_{n+1}} \mathbf{v}_\Delta^T(t)\mathbf{v}_\Delta'(t)dt = 2k\sum_{j=1}^{l} b_j \mathbf{v}_\Delta^T(t_{nj})\mathbf{v}_\Delta'(t_{nj})$$

$$= 2k\sum_{j=1}^{l} b_j \mathbf{v}_\Delta^T(t_{nj}) L \mathbf{v}_\Delta(t_{nj}) = 0.$$

The crucial step in the above derivation is the equality of the integral to the sum (b_j are quadrature weights), and this holds because $\mathbf{v}_\Delta^T \mathbf{v}_\Delta'$ is a polynomial of degree (at most) $2l - 1$ and the points are Gaussian; see Section 2.12.4 and [14].

The big disadvantage of implicit Runge–Kutta methods is that they are implicit. The time step typically allowed by explicit methods may not be unbearably small as it is for parabolic problems, and the matrix L is not symmetric positive definite (which makes solving linear systems of equations harder); hence there is an incentive to stay with explicit methods. An inspection of absolute stability regions for such methods (see, e.g., Figure 2.6) suggests that among p-stage methods of order p only third and fourth order methods have a chance to be suitable, because their region of absolute stability contains a segment of the imaginary axis.

Let L be constant. Then for the explicit Runge–Kutta methods considered here we can write

$$\mathbf{v}^{n+1} = \sum_{i=0}^{p} \frac{k^i}{i!} L^i \mathbf{v}^n.$$

5.3. Nonlinear Stability and Energy Methods

See, e.g., [14, 118]. For instance, all four-stage, fourth order Runge–Kutta methods coincide for this constant coefficient case. Thus

$$\|\mathbf{v}^{n+1}\|^2 - \|\mathbf{v}^n\|^2 = \left(\sum_{i=0}^{p} \frac{k^i}{i!} L^i \mathbf{v}^n, \sum_{i=0}^{p} \frac{k^i}{i!} L^i \mathbf{v}^n \right) - (\mathbf{v}^n, \mathbf{v}^n).$$

Now, notice that

$$(L^i \mathbf{v}, L^j \mathbf{v}) = \begin{cases} 0, & |i-j| \text{ odd}, \\ (-1)^{|i-j|/2} \|L^{(i+j)/2} \mathbf{v}\|^2, & |i-j| \text{ even}. \end{cases}$$

For the forward Euler method we have

$$\|\mathbf{v}^{n+1}\|^2 - \|\mathbf{v}^n\|^2 = k^2 \|L\mathbf{v}^n\|^2,$$

and no stability is possible for any $k > 0$. Indeed, this agrees with the fact that the absolute stability region of the forward Euler method does not intersect the imaginary axis.

For the fourth order, "classical" Runge–Kutta method **RK4**, however, there is more hope. We obtain

$$\|\mathbf{v}^{n+1}\|^2 - \|\mathbf{v}^n\|^2 = k^2 \|L\mathbf{v}^n\|^2 + \frac{k^4}{4} \|L^2\mathbf{v}^n\|^2 + \frac{k^6}{36} \|L^3\mathbf{v}^n\|^2 + \frac{k^8}{576} \|L^4\mathbf{v}^n\|^2$$
$$- \frac{k^2}{2} \|L\mathbf{v}^n\|^2 + \frac{k^4}{24} \|L^2\mathbf{v}^n\|^2 - \frac{k^4}{6} \|L^2\mathbf{v}^n\|^2$$
$$- \frac{k^2}{2} \|L\mathbf{v}^n\|^2 - \frac{k^6}{48} \|L^3\mathbf{v}^n\|^2 - \frac{k^4}{6} \|L^2\mathbf{v}^n\|^2 + \frac{k^4}{24} \|L^2\mathbf{v}^n\|^2$$
$$- \frac{k^6}{48} \|L^3\mathbf{v}^n\|^2 = \frac{k^6}{72} \left(-\|L^3\mathbf{v}^n\|^2 + \frac{k^2}{8} \|L^4\mathbf{v}^n\|^2 \right).$$

Thus, we have stability by the energy method

$$\|\mathbf{v}^{n+1}\|^2 \leq \|\mathbf{v}^n\|^2,$$

provided that

$$k \leq \frac{2\sqrt{2}}{\rho(L)}. \tag{5.25}$$

Example 5.12 We return to the simple advection equation

$$u_t - u_x = 0$$

with periodic boundary conditions $u(t, \pi) = u(t, -\pi)$ and employ a centered difference method in space and the fourth order classical Runge–Kutta method in time. In space we use two discretizations:

i. The second order approximation D_0 used so far almost exclusively,

$$\frac{D_0 v_j}{2h} = \frac{v_{j+1} - v_{j-1}}{2h},$$

Table 5.2. *Errors using the fourth order classical Runge–Kutta method in time and two centered discretizations in space, applied to the equation $u_t - u_x = 0$ with $u_0(x) = \sin \eta x$. Note that $k = \mu h$.*

η	h	μ	Error using D_0	Error using $\tilde{D}_0$
1	$.01\pi$	0.5	1.6e-4	3.3e-8
	$.001\pi$	0.5	1.6e-6	3.3e-12
	$.001\pi$	2.0	1.6e-6	1.6e-11
10	$.01\pi$	0.5	1.6e-1	3.2e-3
	$.001\pi$	0.5	1.6e-3	3.3e-7
	$.001\pi$	2.0	1.6e-3	1.6e-6

ii. The fourth order approximation $\tilde{D}_0 = D_0(1 - \frac{1}{6} D_+ D_-)$,

$$\frac{\tilde{D}_0 v_j}{2h} = \frac{1}{12h}(-v_{j+2} + 8v_{j+1} - 8v_{j-1} + v_{j-2}).$$

Handling the extra neighbors when using $\tilde{D}_0$ near the boundaries poses no problem here because of the periodic boundary conditions which imply that $u(t, \pi + \delta) = u(t, -\pi + \delta)$, and we can take $\delta = \pm h$; see Figure 1.11.

We try three types of initial conditions:

1. For the smooth initial data
$$u_0(x) = \sin \eta x,$$
Table 5.2 displays results to be compared to those in Tables 1.1 and 5.1. Note that the error is only second order when using D_0: indeed, we have a local truncation error $O(k^4) + O(h^2)$ with $k = O(h)$ and solution derivatives of a similar size in x and t, so it is not surprising that the $O(h^2)$ term dominates the error. The results are similar to those obtained in Table 1.1 with the leapfrog scheme. Note, though, that μ is allowed to climb up to ≈ 2.8, and the results for $k = 2h$ reflect a stable scheme. When switching to $\tilde{D}_0$ we have a scheme of order $(4, 4)$, and indeed the error decreases like k^4 as the mesh is refined.

2. Upon using the square wave of Example 5.6 and $\mu = 0.5$ we obtain results which are qualitatively similar to those found using the leapfrog scheme; see Figures 5.4 and 5.5. Choosing D_0 or $\tilde{D}_0$ does not make much difference here. Increasing μ to about $\mu = 2$ amounts to adding dissipation, in view of the absolute stability region of the Runge–Kutta method (see Figure 2.6 or Figure 6.11). Indeed, the resulting plots look somewhat like the leapfrog scheme with dissipation added. The results are not better, however, than those displayed in Figures 5.4 and 5.5 for the Lax–Wendroff scheme, so they are omitted.

3. We also tried
$$u_0(x) = \sin(50x)e^{-50x^2},$$
as discussed further in Example 7.1. The results are qualitatively similar to those obtained for the square wave in relation to other schemes. This is true so long as the

spatial discretization is too coarse to properly resolve the initial solution profile—see Figures 7.2 and 7.3. However, upon choosing $h = .001\pi$ the solution profile is sufficiently resolved and the maximum errors at $t = 1$ using $\mu = 2$ are 2.3e-1 using D_0 and 7.1e-3 upon using the more accurate $\tilde{D}_0$. ■

5.4 Exercises

0. **Review questions**

   (a) Why is the instability in Example 5.1 "unexpected"?

   (b) Define (strict) dissipativity. Is it necessary for stability for strictly hyperbolic PDEs? Is it sufficient?

   (c) What is a conservation form? Is the PDE $u_t + a(u)u_x = 0$ in conservation form?

   (d) The Lax–Wendroff and leapfrog schemes are both second order accurate in time and space, and both require the CFL condition to hold in order to produce meaningful results. What is the basic difference between them in terms of stability?

   (e) What is a monotone scheme, and why is this property desirable?

   (f) Name one advantage and one disadvantage of the Lax–Friedrichs scheme in comparison to the Lax–Wendroff scheme for hyperbolic conservation laws.

   (g) Define the upwind scheme for the nonlinear advection equation $u_t + a(u)u_x = 0$. Would you expect the cost of carrying out a step of the upwind scheme to be lower or higher than that of the leapfrog scheme? Explain.

   (h) When simulating fluid flow in computer graphics people usually apply some variant of the upwind scheme. Why?

   (i) What is a modified PDE? Illustrate its use.

   (j) What is different in the box scheme from all other schemes proposed in this chapter? Under what circumstances can this scheme be advantageous?

   (k) What is an "energy method" for proving stability? Can you think of an instance in which such a method would help where the usual stability analysis is unsatisfactory or incomplete?

   (l) Describe the method of collocation at Gaussian points. How does it relate to Runge–Kutta methods?

   (m) Describe two advantages that RK4 has over forward Euler when applied to the semi-discretization $\mathbf{v}_t = L\mathbf{v}$ with L skew-symmetric.

1. Prove that the Lax–Wendroff scheme (5.5) is dissipative by showing that
$$|\mu_l(\zeta)| \leq 1 - \delta_l|\zeta|^4,$$
   where
$$\delta_l = \frac{1}{4}\mu^2|a_l|^2(1 - \mu^2|a_l|^2).$$
   Try to do this without consulting [133].

2. Show that the Lax–Friedrichs scheme is *not* dissipative in the sense that (5.2) does not hold for any $\delta > 0$. However, many high-frequency amplification components are damped nonetheless.

3. Show that the backward Euler scheme for a hyperbolic system

$$\mathbf{v}_j^{n+1} = \mathbf{v}_j^n - \frac{1}{2}\mu A_j^{n+1}(\mathbf{v}_{j+1}^{n+1} - \mathbf{v}_{j-1}^{n+1})$$

satisfies the von Neumann condition for any $\mu \geq 0$. And yet, like the Lax–Friedrichs scheme, it is not strictly dissipative and it is not "energy conserving."

4. Write a program that solves the advection problem $u_t - u_x = 0$ using the Lax–Friedrichs, Lax–Wendroff, and box schemes. Try the initial data

$$u_0(x) = \begin{cases} 1 - |x|, & |x| \leq 1, \\ 0, & |x| > 1 \end{cases}$$

(cf. [159]). Use periodic boundary conditions on $[-\pi, \pi]$. Keeping $\mu = 0.5$ fixed, run for $h = .01\pi$ and $h = .001\pi$. Plot the resulting solutions at $t = 1$ and tabulate maximum errors. What are your observations?

5. Show that the Richtmyer variant (5.5b) of the Lax–Wendroff method can be written as an explicit midpoint method,

$$\mathbf{v}_j^{n+1} = \mathbf{v}_j^n - \mu \left(\mathbf{f}\left(\mathbf{v}_{j+1/2}^{n+1/2}\right) - \mathbf{f}\left(\mathbf{v}_{j-1/2}^{n+1/2}\right) \right),$$

where $\mathbf{v}_{j+1/2}^{n+1/2}$ is obtained by a Lax–Friedrichs half-step.

6. The Lax–Friedrichs and Lax–Wendroff schemes, being explicit one-step, can be written as a forward Euler differencing of a semi-discretization. Consider the same setting and answer the same questions as in Exercise 4.10.

(For this exercise it suffices to calculate the eigenvalues using some canned routine for a small instance, say, $J = 10$ or $J = 100$.)

7. (a) Show that the modified PDE for the Lax–Friedrichs scheme for the advection equation is

$$u_t + au_x = \frac{h}{2\mu}\left(1 - \mu^2 a^2\right)u_{xx}. \tag{5.26a}$$

(b) Show that the modified PDE for the Lax–Wendroff scheme for the advection equation is

$$u_t + au_x = -\frac{ah^2}{6}\left(1 - \mu^2 a^2\right)u_{xxx}. \tag{5.26b}$$

8. Show that the box scheme (5.17) can be written as

$$k^{-1}\delta_t \mu_x \mathbf{v}_{j+1/2}^{n+1/2} + h^{-1}\mu_t \delta_x \mathbf{f}_{j+1/2}^{n+1/2} = \mathbf{0}, \tag{5.27}$$

where μ_x is the averaging operator in x, δ_t is the short centered difference in t, etc. (Beware of the clash of the μ's!)

5.4. Exercises

9. Continuing Exercise 1.7 and Example 5.10, show that $\int \bar{\psi}(\psi_{xx} - V'(|\psi|)\psi)dx$ is real.

10. Prove (5.21d). Hint: Use no more than four lines in total.

11. Show that for the choice $\theta = 2/3$,

$$\frac{\theta}{2}D_0(v_j^2) + (1-\theta)v_j D_0 v_j = \frac{v_{j-1} + v_j + v_{j+1}}{3} D_0 v_j.$$

    Compare this with the discretization utilized in Example 5.1.

12. Let M be a real $J \times J$ symmetric positive definite matrix, and consider generalizing (5.23) into

$$M\mathbf{v}_t = L\mathbf{v}$$

    with L skew-symmetric. Also, define the norm

$$\|\mathbf{w}\|_M = (\mathbf{w}^T M \mathbf{w})^{1/2}.$$

    This is sometimes referred to as an "energy norm."

    (a) Show that
    $$\|\mathbf{v}(t)\|_M = \|\mathbf{v}(0)\|_M \quad \forall t \geq 0.$$

    (b) Show that the midpoint method (as well as higher order Gaussian collocation methods) reproduce this "energy conservation":
    $$\|\mathbf{v}^n\|_M = \|\mathbf{v}^0\|_M \quad \forall n \geq 0.$$

    (c) Consider a hyperbolic system with constant coefficients
    $$\mathbf{u}_t + A\mathbf{u}_x = \mathbf{0}$$
    with initial conditions and periodic BCs. Let $T^{-1}AT = \Lambda$, where Λ is diagonal and real, and define
    $$\tilde{M} = T^{-T} T^{-1}.$$
    Show that $\tilde{M}$ is symmetric positive definite and that $\tilde{M}A$ is symmetric.

    (d) For the hyperbolic system (where A is not necessarily symmetric) derive an "energy conservation" rule which the Crank–Nicolson scheme reproduces.

13. Consider the problem
$$u_t = w_x, \quad w_t = u_x, \quad 0 \leq x \leq 2\pi, \ t \geq 0,$$
with periodic BCs and the initial conditions $w_0(x) = 0$, $u_0(x) = e^{-50(x-\pi)^2}$. The exact solution satisfies $u(2l\pi, \cdot) = u_0(\cdot)$ for any integer l.

    Write a program to solve this problem as follows. For the spatial discretization use a centered, five-point, fourth order formula and for the time discretization use RK4. Plot the solution component u at different time steps (in MATLAB use mesh). In a separate figure plot the solution profile $u(t, x)$ at two times, the initial time $t = 0$ and the final time $t = t_f$. If $t_f = 2l\pi$ for some integer l, then calculate least squares and maximum error norms.

Run your program for $t_f = 8\pi$ with spatial steps $h = 2^{-7}\pi$ and $h = 2^{-8}\pi$ and with the time steps $k = h$, $k = 2h$, and $k = 3h$ for each h. What are your observations?

14. Consider the problem
$$u_t = w_x, \quad w_t = u_x + \nu w_{xxt}, \quad 0 \leq x \leq 2\pi, \ t \geq 0,$$
with periodic BC and the initial conditions $w_0(x) = 0$, $u_0(x) = e^{-50(x-\pi)^2}$. The parameter $\nu \geq 0$ is not large; set $\nu = .01$ in your numerical experiments.

(a) Let us write the PDE system as the partial differential algebraic equation (PDAE)
$$u_t = w_x, \tag{5.28a}$$
$$\psi_t = u_x, \tag{5.28b}$$
$$\psi = w - \nu w_{xx}. \tag{5.28c}$$

Show that the Cauchy problem for (5.28) is well-posed and that high wave numbers are not attenuated (as for the case $\nu = 0$).

(b) Write a program to solve this problem as follows. The algebraic condition in time (5.28c) is discretized employing the usual three-point difference formula. The other two equations are discretized as in Exercise 13. Thus, at $t = 0$ we get $\{\psi_j^0\}$ directly from (5.28c). Then, at each step n we obtain u^{n+1} and ψ^{n+1} from the explicit discretization of (5.28a), (5.28b), followed by solving a tridiagonal system based on (5.28c) for w^{n+1}.

Plot the solution component u at different time steps. In a separate figure plot the solution profile $u(t, x)$ at two times, the initial time $t = 0$ and the final time $t = t_f$. Run your program for $t_f = 8\pi$ with spatial steps $h = 2^{-7}\pi$ and $h = 2^{-8}\pi$ and with the time steps $k = h$, $k = 2h$, and $k = 3h$ for each h. What are your observations?

15. Repeat Exercises 13 and 14, replacing RK4 by each of the following alternatives:

(a) The three-step Adams–Moulton method [180]. You can use the three-step Adams–Bashforth as a predictor in a PECE arrangement (recall Section 2.7). Use RK4 for the first few steps to obtain the necessary initial values.

(b) The four-step Adams–Moulton method. You can use the four-step Adams–Bashforth as a predictor in a PECE arrangement.

What are your observations?

Chapter 6
Hamiltonian Systems and Long Time Integration

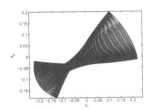

Differential systems often have some qualitative features that are of particular interest. Once such a system is discretized for numerical simulation, the computed solution is expected not only to approximate the differential system but also to approximately reproduce these qualitative features. Depending on the circumstances, it may in fact be worthwhile to attempt to reproduce these features more precisely than the differential solution itself. Methods for doing the latter are generally clustered under the title **geometric integration** (GI). They often give rise to more sophisticated mathematical analyses. Presently, however, most GI enthusiasts hail from the ODE community, where such methods have proved useful. The PDE community remains on the fence, in general, although the need to preserve important solution features and reproduce physical laws has certainly been recognized in the numerical PDE literature for decades. The present chapter, which is devoted to ODEs, and part of the following chapter, where we try to apply some related concepts to nontrivial hyperbolic PDEs in one dimension, should provide a good feeling for the applicability of some such methods.

Perhaps the simplest instances of the above are differential systems with algebraic invariants. These arise frequently in various applications, including but not limited to index reduction in differential algebraic equations (DAEs); see Section 2.8 and [14]. The invariant might represent conservation of energy, momentum, or mass in a physical system [153]. For instance, the ODE in Example 2.8 has the invariant that the energy $(u')^2 + \omega^2 u^2$ is constant for all t. (The initial values determine the constant.) This holds also for other, general autonomous Hamiltonian systems to be considered further below. Attempting to satisfy such invariants in addition to the discretized ODE leads to an overdetermined system, for which only special discretizations, if any at all, would have an exact solution.

Here we concentrate on *Hamiltonian systems* and their numerical integration. This is a very important class of problems in GI to which a lot of attention has been devoted in recent years. There are good recent reference books available on this material, namely, Hairer, Lubich, and Wanner [81] and Leimkuhler and Reich [115]. The lecture notes of Budd and Piggott [41] are useful as well. A classical reference for the mathematical topic is Arnold [3]. The early numerical work of Sanz-Serna and Calvo [150] has also been very influential.

Hamiltonian systems have the important property that the flows they induce are **symplectic**. We explain what this is in Section 6.1. Special methods developed for their numerical integration retain this property at the discrete level and are described in Section 6.2. These methods are claimed to yield superior performance under certain conditions, especially when the time integration is relatively long. Since Hamiltonian systems describe mechanical systems where energy is conserved rather than dissipated, we may well want to integrate them over a long time period.

It must be said that symplectic methods were apparently "sold" rather heavily in the 1990s. Many of the papers describing them and other methods of GI seem to be absorbed with the newness and the mathematical richness of the subject, stopping short of a sober assessment of efficacy. We attempt to do this in a very limited way in Section 6.3 and also in Section 6.4, where we briefly consider highly oscillatory problems.

As in Chapter 2 we denote differentiation by, e.g., $y' = \frac{dy}{dt}$, and use subscripts, e.g., y_n, for the numerical solution.

6.1 Hamiltonian systems

A Hamiltonian system consists of $m = 2l$ differential equations

$$q_i' = \frac{\partial H}{\partial p_i}, \tag{6.1a}$$

$$p_i' = -\frac{\partial H}{\partial q_i} \tag{6.1b}$$

for $i = 1, \ldots, l$, or, in vector notation (with $\nabla_\mathbf{p} H$ denoting the gradient of H with respect to $\mathbf{p}$, etc.),

$$\mathbf{q}' = \nabla_\mathbf{p} H(\mathbf{q}, \mathbf{p}), \quad \mathbf{p}' = -\nabla_\mathbf{q} H(\mathbf{q}, \mathbf{p}).$$

The scalar function $H(\mathbf{q}, \mathbf{p})$, assumed to have continuous second derivatives, is the **Hamiltonian**. A simple instance of a Hamiltonian system is provided in Example 2.8, to which we return in Example 6.1 below. A stiff spring pendulum is another, less simple instance of a Hamiltonian system with

$$H(\mathbf{q}, \mathbf{p}) = \frac{1}{2}\mathbf{p}^T \mathbf{p} + (\phi(\mathbf{q}) - \phi_0)^2 + \varepsilon^{-2}(r(\mathbf{q}) - r_0)^2, \tag{6.2}$$

where $\varepsilon^{-1} \gg 1$ is the stiffness constant. Here $l = 2$.

For a general, autonomous Hamiltonian, differentiating H with respect to time t and substituting (6.1) gives

$$H' = \nabla_\mathbf{p} H^T \mathbf{p}' + \nabla_\mathbf{q} H^T \mathbf{q}' = 0,$$

so $H(\mathbf{q}, \mathbf{p})$ is constant for all t. A typical example to keep in mind is that of a conservative system of particles. Then the components of $\mathbf{q}(t)$ are the generalized positions of the particles, and those of $\mathbf{p}(t)$ are the generalized momenta. The Hamiltonian H in this case is the total energy (the sum of kinetic and potential energies), and the constancy of H is a statement of **conservation of energy**.

Next, consider an autonomous ODE system of order $m = 2$,

$$\mathbf{y}' = \mathbf{f}(\mathbf{y}),$$

6.1. Hamiltonian Systems

with $\mathbf{y}(0) = (y_1(0), y_2(0))^T \in B$, for some set B in the plane. Each initial value $\mathbf{y}(0) = \mathbf{c}$ from B spawns a trajectory $\mathbf{y}(t) = \mathbf{y}(t; \mathbf{c})$, and we can follow the evolution of the set B under this flow,

$$S(t)B = \{\mathbf{y}(t; \mathbf{c}) \,;\, \mathbf{c} \in B\}.$$

We then ask how the area of $S(t)B$ compares to the initial area of B: does it grow or shrink in time? It is easy to see in linear problems that this area shrinks for asymptotically stable problems and grows for unstable problems. Less easy to see, but which can be shown, is the fact that the area of $S(t)B$ remains constant, even for nonlinear problems, if the divergence of $\mathbf{f}$ vanishes:

$$\nabla \cdot \mathbf{f} = \frac{\partial f_1}{\partial y_1} + \frac{\partial f_2}{\partial y_2} = 0.$$

This remains valid for $m > 2$ provided that $\nabla \cdot \mathbf{f} = 0$ with an appropriate extension of the concept of volume in m dimensions.

Now, for a Hamiltonian system with $l = 1$,

$$q' = H_p, \quad p' = -H_q,$$

we have

$$\nabla \cdot \mathbf{f} = \frac{\partial^2 H}{\partial p \partial q} - \frac{\partial^2 H}{\partial q \partial p} = 0;$$

hence the Hamiltonian flow preserves area. In more dimensions, $l > 1$, it turns out that the area of each projection of $S(t)B$ on a $q_i \times p_i$ plane, $i = 1, \ldots, l$, is preserved, and this property is referred to as a **symplectic map**.

Since a Hamiltonian system cannot be asymptotically stable, its stability (*if* it is stable, which is true when H can be considered a norm at each t, e.g., if H is the total energy of a friction free multibody system) is in a sense marginal. The solution trajectories do not simply decay to a rest state, and their long-time behavior is therefore of interest. This leads to some serious numerical challenges.

Example 6.1 The simplest Hamiltonian system is the linear harmonic oscillator. The quadratic Hamiltonian

$$H = \frac{\omega}{2}(p^2 + q^2)$$

yields the linear equations of motion

$$q' = \omega p, \quad p' = -\omega q,$$

or

$$\begin{pmatrix} q \\ p \end{pmatrix}' = \omega J \begin{pmatrix} q \\ p \end{pmatrix}, \quad J = \begin{pmatrix} 0 & 1 \\ -1 & 0 \end{pmatrix}.$$

Here $\omega > 0$ is a known parameter. The general solution is

$$\begin{pmatrix} q(t) \\ p(t) \end{pmatrix} = \begin{pmatrix} \cos \omega t & \sin \omega t \\ -\sin \omega t & \cos \omega t \end{pmatrix} \begin{pmatrix} q(0) \\ p(0) \end{pmatrix}.$$

Hence, $S(t)B$ is just a rotation of the set B at a constant rate depending on ω. Clearly, this keeps the area of B unchanged.

Note that the eigenvalues of J are purely imaginary. Thus, a small "push" (i.e., a perturbation of the system) of these eigenvalues toward the positive half-plane can make the system unstable; see Figure 6.1.

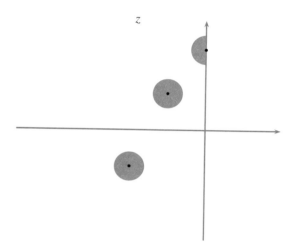

Figure 6.1. *The marginal stability of Hamiltonian systems. If an eigenvalue has a negative real part, then we can draw a circle around it such that perturbations within the circle stay stable. But for an imaginary eigenvalue we cannot draw such a circle. The linear harmonic oscillator matrix ωJ has only imaginary eigenvalues.*

In this simple example the geometric shape of B does not change: it merely gets rotated under the transformation by the solution operator. This much does not hold in general for symplectic maps, though. ∎

In general, we can write the Hamiltonian system (6.1) as

$$\mathbf{y}' = J \nabla H(\mathbf{y}), \tag{6.3a}$$

where

$$\mathbf{y} = \begin{pmatrix} \mathbf{q} \\ \mathbf{p} \end{pmatrix}, \quad J = \begin{pmatrix} 0 & I \\ -I & 0 \end{pmatrix}. \tag{6.3b}$$

The important properties of the matrix J are that it is **skew-symmetric**, i.e., $J^T = -J$ (see the review in Section 1.3.2) and that it is nonsingular and constant in t.

Denote the Jacobian matrix

$$Y(t; \mathbf{c}) = \frac{\partial \mathbf{y}(t; \mathbf{c})}{\partial \mathbf{c}} \tag{6.4}$$

with $\mathbf{y}(0; \mathbf{c}) = \mathbf{c}$ for some arbitrary initial time. The flow is called **symplectic** if the following condition holds:

$$Y^T J^{-1} Y = J^{-1} \quad \forall t. \tag{6.5}$$

For the Hamiltonian system we can differentiate (6.3a) with respect to $\mathbf{c}$ to obtain

$$Y' = J(\nabla^2 H) Y, \quad Y(0) = I, \tag{6.6}$$

where $\nabla^2 H$ stands for the Hessian matrix of H (and not a Laplacian). Differentiating $Y^T J^{-1} Y$ with respect to t and substituting (6.6) then yields

$$\frac{d}{dt}(Y^T J^{-1} Y) = Y^T (\nabla^2 H) J^T J^{-1} Y + Y^T J^{-1} J (\nabla^2 H) Y = 0.$$

Thus, $Y^T J^{-1} Y$ is constant, and since it equals J^{-1} at $t = 0$ the condition (6.5) holds and we have proved that Hamiltonian systems produce symplectic flows.

It can be shown that symplecticity also implies a Hamiltonian flow [81]. These systems possess many beautiful properties which do not usually hold for other, general systems [81, 41, 150]. Hence, it is of great interest to preserve the property (6.5) in numerical discretizations of Hamiltonian systems.

Let us freeze the coefficients in (6.6), as in Section 4.1. Exercise 2 shows that the eigenvalues of $J \nabla^2 H$ are purely imaginary if the symmetric Hessian matrix $\nabla^2 H$ is positive definite. If $\nabla^2 H$ is merely symmetric, however, then eigenvalues with positive and negative real parts may arise. See also Exercise 7 as a warning not to conclude that eigenvalues always tell the whole story.

6.2 Symplectic and other relevant methods

There are different approaches to constructing symplectic numerical methods, of which we consider mainly two: Runge–Kutta, and splitting and composition methods. But before we get to a more systematic development of such methods, let us consider a simple example to motivate the search.

Example 6.2 Returning to the harmonic oscillator of Example 6.1 we now set $\omega = 1$ and consider the evolution of the unit circle under the discrete flow which different numerical discretizations affect. Remember that the exact solution leaves the unit circle invariant at all times t.

The first three methods we consider are the usual:

1. Forward Euler:
$$q_{n+1} = q_n + k p_n, \quad p_{n+1} = p_n - k q_n, \quad n = 0, 1, \ldots, N-1.$$

2. Backward Euler:
$$q_{n+1} = q_n + k p_{n+1}, \quad p_{n+1} = p_n - k q_{n+1}, \quad n = 0, 1, \ldots, N-1.$$

3. Midpoint:
$$q_{n+1} = q_n + k \frac{p_n + p_{n+1}}{2}, \quad p_{n+1} = p_n - k \frac{q_n + q_{n+1}}{2}, \quad n = 0, 1, \ldots, N-1.$$

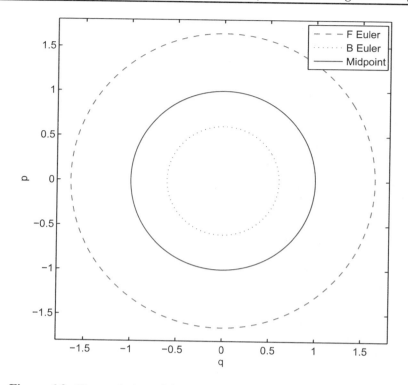

Figure 6.2. *The evolution of the unit circle for the harmonic oscillator, recorded at $t = 100$, using* (i) *forward Euler $k = 0.01$ (dashed),* (ii) *backward Euler $k = 0.01$ (dots), and* (iii) *midpoint $k = 0.5$ (solid). The latter reproduces the initial circle exactly.*

We use these methods to integrate the harmonic oscillator with a constant, relatively coarse step size, starting from initial values on the unit circle. The resulting values of p vs. q at time $t = 100$ are plotted in Figure 6.2.

We see that the forward Euler method increases the original volume significantly, while the backward Euler method shrinks it significantly. The midpoint method reproduces the Hamiltonian, $p^2(t) + q^2(t) = 1$, in this special case.

For the forward and backward Euler methods we have used a relatively small step size of $k = 0.01$. However, with the coarse step size $k = 0.5$ used for the midpoint and other methods below, the forward Euler method comes close to blowing up while the backward Euler method shrinks the unit circle into a tiny circle.

The poor performance of the Euler methods, often cited as the reason to use symplectic methods, is really not that surprising. This constant coefficient Hamiltonian system corresponds to the test equation (2.20) with purely imaginary eigenvalues. The forward Euler stability region intersects with the imaginary axis only at the origin (see Figure 2.6); obviously it is the wrong method to use, much like the unconditionally unstable method of (1.15b). The backward Euler method is just forward Euler in reverse for this symmetric problem. Its excessive damping properties, so useful for highly stiff problems, are a disaster here.

6.2. Symplectic and Other Relevant Methods

> **Note:** Numerical methods for Hamiltonian systems should not dampen solution modes, at least not heavily.

The performance of the midpoint method is "better than life" because here the Hamiltonian is quadratic and the midpoint method reproduces quadratic invariants, as we shall shortly prove. For more general problems this is not possible (nor necessarily desirable; see [16]). Moreover, the midpoint method is implicit. Thus, we turn to other alternatives. In Figure 6.3 we record results corresponding to those of Figure 6.2 for the following methods:

1. The classical four-stage explicit Runge–Kutta of order 4, RK4.

2. Forward Euler for the first ODE, and backward Euler for the second,

$$q_{n+1} = q_n + kp_n, \quad p_{n+1} = p_n - kq_{n+1}, \quad n = 0, 1, \ldots, N-1.$$

This is called **symplectic Euler**. It is an explicit method because the Hamiltonian has a special, *separable* form. (Recall also Example 2.9.)

3. *Staggered midpoint*, as in Example 3.2,

$$q_n = q_{n-1} + kp_{n-1/2}, \quad p_{n+1/2} = p_{n-1/2} - kq_n, \quad n = 1, \ldots, N-1,$$

starting with $p_{1/2} = p_0 - \frac{k}{2}q_0$ and ending with $q_N = q_{N-1} + kp_{N-1/2}$, $p_N = p_{N-1/2} - \frac{k}{2}q_N$. This is called the **Störmer–Verlet** method.

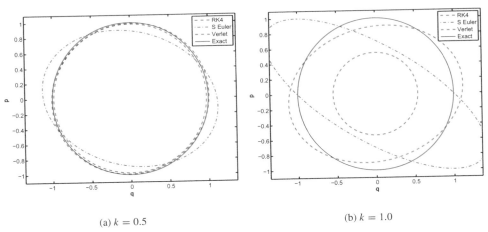

(a) $k = 0.5$ (b) $k = 1.0$

Figure 6.3. *The evolution of the unit circle for the harmonic oscillator, recorded at $t = 100$, using coarse step sizes $k = 0.5$ and $k = 1.0$ and the following methods:* (i) *classical RK4 (dashed),* (ii) *symplectic Euler (dash-dot), and* (iii) *Störmer–Verlet (dashed green). The latter is close to the initial circle, depicted as in Figure 6.2 by solid blue.*

The last three methods are all explicit, unlike implicit midpoint, and none reproduces the Hamiltonian invariant exactly. And yet they all produce decent results with the coarse step size $k = 0.5$. The classical RK4 method is slightly dissipative, as noted in Section 5.3.2, but it is much more suitable for problems with imaginary eigenvalues than forward Euler or any two-stage second order explicit Runge-Kutta method (see Figures 2.6 and 6.11). Its dissipativity is expressed by the fact that the unit circle is slightly smaller in Figure 6.3. For the very coarse step size $k = 1.0$ the shrinkage is significant, though.

The symplectic Euler and the Störmer–Verlet (or just *Verlet*) methods are symplectic, i.e., they reproduce the invariance of $Y^T(t)J^{-1}Y(t)$; see (6.5). This translates, for our simple example with $l = 1$, to keeping the area of the unit circle, although not necessarily its shape, invariant. The symplectic Euler is first order accurate while the other method is second order. Their application here is much simpler than that of RK4. For the coarse step size $k = 1.0$ the difference in accuracy between the two is rather stark, although they both still preserve area. ∎

6.2.1 Symplectic Runge–Kutta methods

In the previous section we saw that the symplectic condition (6.5) is a *quadratic invariant* of the variational differential system (6.6). A first attempt at obtaining a symplectic method could therefore be to look for numerical methods that preserve quadratic invariants. This is also useful in Section 5.3.1 and the next chapter, and it leads to **piecewise polynomial collocation at Gaussian points**; see Sections 2.2 and 5.3.2, as well as [5, 14].

These methods are equivalent to certain implicit Runge–Kutta methods and have order $2s$ when s collocation points are used. Denoting the collocation solution by $\mathbf{y}_\Delta(t)$ (it is a polynomial of degree at most s on each time interval $[t_n, t_{n+1}]$ and it satisfies $\mathbf{y}_\Delta(t_n) = \mathbf{y}_n \; \forall n$) and the collocation points by $t_{nj} = t_n + kc_j$, $j = 1, \ldots, s$, we define $Y_\Delta = \frac{\partial \mathbf{y}_\Delta(t;\mathbf{c})}{\partial \mathbf{c}}$, a matrix whose components are also polynomials of degree at most s on each time interval. The ODE (6.3) is satisfied by $\mathbf{y}_\Delta$ at the collocation points:

$$\mathbf{y}'_\Delta(t_{nj}) = J\nabla H(\mathbf{y}_\Delta(t_{nj})), \quad j = 1, \ldots, s.$$

Therefore, we also have

$$Y'_\Delta(t_{nj}) = J\nabla^2 H(\mathbf{y}_\Delta(t_{nj}))Y_\Delta(t_{nj}), \quad j = 1, \ldots, s,$$

so

$$\begin{aligned}
Y_\Delta^T(t_{n+1})J^{-1}Y_\Delta(t_{n+1}) &- Y_\Delta^T(t_n)J^{-1}Y_\Delta(t_n) \\
&= \int_{t_n}^{t_{n+1}} ((Y_\Delta^T)'J^{-1}Y_\Delta + Y_\Delta^T J^{-1}Y'_\Delta)dt \\
&= k\sum_{j=1}^{s} b_j((Y_\Delta^T)'(t_{nj})J^{-1}Y_\Delta(t_{nj}) + Y_\Delta^T(t_{nj})J^{-1}Y'_\Delta(t_{nj})) \\
&= k\sum_{j=1}^{s} b_j[Y_\Delta^T(\nabla^2 H)J^T J^{-1}Y_\Delta + Y_\Delta^T J^{-1}J(\nabla^2 H)Y_\Delta](t_{nj}) \\
&= 0.
\end{aligned}$$

6.2. Symplectic and Other Relevant Methods

As in Section 5.3.2 the crucial step here is the equality of the integral to the sum (b_j are quadrature weights), and this holds because the integrand is a polynomial of degree at most $2s - 1$ and the points are Gaussian. (See the quick review of quadrature in Section 2.12.4 or a more comprehensive basic text on numerical analysis.) No other quadrature abscissas achieve this precision.

In the Runge–Kutta notation the precision requirement translates into the condition

$$b_i a_{ij} + b_j a_{ji} - b_i b_j = 0, \quad i, j = 1, \ldots, s. \tag{6.7}$$

But there are no other reasonable (nondegenerate) methods that achieve this. The first method of this collocation family is the midpoint method (2.12) which has order $2s = 2$. In Exercise 4 you are asked to show directly that this method is symplectic.

Note that the trapezoidal method (2.8) is not symplectic. However, it is symmetric. More generally, this is referred to as **time reversible**; see, e.g., [84, 81]. Exercise 5 shows that the trapezoidal method is in a sense almost symplectic.

Referring to Example 6.2 we see that the midpoint method both is symplectic and reproduces the Hamiltonian exactly (ignoring roundoff errors). However, in the general case this is not possible [65]. As stated before, attempting to reproduce the Hamiltonian in general does not seem to be a good strategy, whereas devising symplectic methods is.

The big disadvantage of Runge–Kutta methods that are based on Gauss–Legendre collocation is that they are fully implicit ($a_{i,j} \neq 0$, $i, j = 1, \ldots, s$). Thus, they remain implicit even for the popular case where the Hamiltonian is separable, i.e., when it can be written in the form

$$H(\mathbf{q}, \mathbf{p}) = T(\mathbf{p}) + V(\mathbf{q}). \tag{6.8}$$

Typically, T is the kinetic energy and V is the potential energy in (6.8).

The natural partitioning in the Hamiltonian system (6.1) suggests that different Runge–Kutta methods may be applied to (6.1a) and (6.1b). This leads to **partitioned Runge–Kutta** methods. Let $\mathbf{c}$, A, $\mathbf{b}$ be the coefficients of the method for (6.1a) and $\tilde{\mathbf{c}}$, $\tilde{A}$, $\tilde{\mathbf{b}}$ those for (6.1b) (see Section 2.2, specifically (2.10), for notational convention). Then the symplecticity condition is

$$b_i = \tilde{b}_i, \quad b_i \tilde{a}_{ij} + \tilde{b}_j a_{ji} - b_i \tilde{b}_j = 0, \quad i, j = 1, \ldots, s, \tag{6.9}$$

which is to be compared to (6.7). For a separable Hamiltonian the condition $b_i = \tilde{b}_i$ may be dropped.

The *symplectic Euler* method, demonstrated in Example 6.2, applies the forward Euler method to one of the equations in (6.1) and the backward Euler method to the other. To verify that (6.9) holds we calculate

$$1 = 1, \quad 1 \cdot 0 + 1 \cdot 1 - 1 \cdot 1 = 0.$$

The method becomes explicit for a separable Hamiltonian. See [82, 41, 150] for more methods and justification of the above claims.

6.2.2 Splitting and composition methods

Here we follow [41, 115], who in turn follow [128] and [186]. However, the idea of splitting in a more general context is much older than these references, and it occurs repeatedly in the literature on the numerical solution of PDEs in several space variables; see Section 9.3 as well as [159, 76].

The basic idea is to derive the discrete flow over a time step as a composition of simpler, symplectic flows. *This yields a symplectic map!*

Suppose that the Hamiltonian $H(\mathbf{q}, \mathbf{p})$ can be written as a sum of Hamiltonians

$$H = H_1 + H_2 + \cdots + H_s$$

and that the exact flow over a time step $[t_n, t_{n+1}]$ of the ODE system for H_j alone can be obtained: denote it, say, by $\mathbf{y}^j(t; \mathbf{y}^j(t_n))$. (This superscript notation is not a power—just a counter.) Thus

$$(\mathbf{y}^j)' = J \nabla H_j(\mathbf{y}^j).$$

Now, choose

$$\mathbf{y}^1(t_n) = \mathbf{y}_n, \tag{6.10a}$$
$$\mathbf{y}^{j+1}(t_n) = \mathbf{y}^j(t_{n+1}), \quad j = 1, \ldots, s-1, \tag{6.10b}$$

and set the result of the step to be

$$\mathbf{y}_{n+1} = \mathbf{y}^s(t_{n+1}). \tag{6.10c}$$

The obtained method is a composition of flows which are symplectic because they are exact flows of Hamiltonian systems. As such, it is also symplectic, as Exercise 8 indicates.

Example 6.3 For a separable Hamiltonian (6.8) we naturally define the splitting

$$H_1 = T(\mathbf{p}), \quad H_2 = V(\mathbf{q}).$$

Then for $\mathbf{y}^1$ we have

$$(\mathbf{q}^1)' = \nabla_{\mathbf{p}} T(\mathbf{p}^1), \quad (\mathbf{p}^1)' = 0, \quad \mathbf{y}^1(t_n) = \begin{pmatrix} \mathbf{q}_n \\ \mathbf{p}_n \end{pmatrix}.$$

The solution is $\mathbf{p}^1 \equiv \mathbf{p}_n$ constant; hence $\mathbf{q}_{n+1}^1 = \mathbf{q}_n + k \nabla_{\mathbf{p}} T(\mathbf{p}_n)$. Now, for $\mathbf{y}^2$ we have

$$(\mathbf{q}^2)' = \mathbf{0}, \quad (\mathbf{p}^2)' = -\nabla_{\mathbf{q}} V(\mathbf{q}^2), \quad \mathbf{y}^1(t_n) = \begin{pmatrix} \mathbf{q}_{n+1}^1 \\ \mathbf{p}_n \end{pmatrix}.$$

The solution is $\mathbf{q}^2 \equiv \mathbf{q}_{n+1} = \mathbf{q}_{n+1}^1$ constant; hence $\mathbf{p}_{n+1} = \mathbf{p}_n - k \nabla_{\mathbf{q}} V(\mathbf{q}_{n+1})$. This yields once again the symplectic Euler method

$$\begin{aligned} \mathbf{q}_{n+1} &= \mathbf{q}_n + k \nabla_{\mathbf{p}} T(\mathbf{p}_n), \\ \mathbf{p}_{n+1} &= \mathbf{p}_n - k \nabla_{\mathbf{q}} V(\mathbf{q}_{n+1}). \end{aligned} \blacksquare \tag{6.11}$$

Whereas the solution of the simple flows above is exact, their composition is not exact, and this is the source of the error for the resulting method. If we express the solution

6.2. Symplectic and Other Relevant Methods

operators of the two subflows as exponents of their corresponding (Lie derivative) operators L and M, then the local error depends directly on

$$e^{k(L+M)} - e^{kL}e^{kM}.$$

Generally, we have (verify!)

$$e^{kL}e^{kM} - e^{k(L+M)} = \frac{1}{2}k^2(ML - LM) + O(k^3),$$

and an overall first order accuracy in k may result if L and M do not commute.

Fortunately, a simple trick (in hindsight), occasionally referred to as *Strang splitting*, allows restoring an overall second order accuracy. The idea is to replace $e^{kL}e^{kM}$ by

$$e^{k(L+M)} \approx e^{\frac{k}{2}L}e^{kM}e^{\frac{k}{2}L}. \tag{6.12}$$

See Exercise 9.

Example 6.4 Continuing Example 6.3, the formula (6.12) corresponds to taking half a k-step for $\mathbf{p}$, followed by a full k-step for $\mathbf{q}$, followed by another half step for $\mathbf{p}$. Assuming that k is constant, however, we may combine the last half step with the first half step of the next one. This yields the celebrated *Störmer–Verlet* method, which we have derived as a staggered midpoint scheme,

$$\mathbf{q}_{n+1} = \mathbf{q}_n + k\nabla_{\mathbf{p}}T(\mathbf{p}_{n+1/2}),$$
$$\mathbf{p}_{n+1/2} = \mathbf{p}_{n-1/2} - k\nabla_{\mathbf{q}}V(\mathbf{q}_n). \tag{6.13}$$

We start the method by calculating $\mathbf{p}_{1/2} = \mathbf{p}_0 - \frac{k}{2}\nabla_{\mathbf{q}}V(\mathbf{q}_0)$ and end, if necessary, by calculating $\mathbf{p}_N = \mathbf{p}_{N-1/2} - \frac{k}{2}\nabla_{\mathbf{q}}V(\mathbf{q}_N)$.

The scheme (6.13) is the mainstay method for calculations in molecular dynamics. It is explicit and symplectic. Note that its symplecticity follows from its current derivation as a splitting method, and *not* from the derivation in Example 3.2! ■

Example 6.5 The **Henon–Heiles** system has a separable Hamiltonian given by

$$H = \frac{1}{2}\left(p_1^2 + p_2^2\right) + \frac{1}{2}\left(q_1^2 + q_2^2\right) + q_1^2 q_2 - \frac{1}{3}q_2^3; \tag{6.14}$$

see [130]. Please convince yourself that you can derive the corresponding ODE system with your eyes closed; see Exercise 6.

We apply two methods to this system: the RK4 workhorse and the Störmer–Verlet method. For the latter we choose the step size $k = .01$. Since RK4 involves four function evaluations per step we use $k = .04$ although for reasons of overhead the special-purpose symplectic method is more than four times cheaper than RK4.

The initial conditions $(q_1, q_2, p_1, p_2)(0) = (.25, .5, 0, 0)$ yield a chaotic orbit [130, 160]. Indeed at $t = 500$ the two methods yield solutions that are $O(1)$ apart. And yet, the errors in the Hamiltonians satisfy $H(500) - H(0) \approx 2.5 \times 10^{-7}$ for both methods! Both these errors remain small for a long time; see Figure 6.4(a). This shows that the

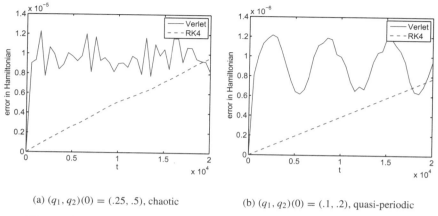

(a) $(q_1, q_2)(0) = (.25, .5)$, chaotic

(b) $(q_1, q_2)(0) = (.1, .2)$, quasi-periodic

Figure 6.4. *Errors in the Hamiltonian over $0 \leq t \leq 20,000$ for RK4 ($k = .04$) and Verlet ($k = .01$). The RK4 error grows linearly but remains very small over this interval. The Verlet error does not clearly grow in time as indeed it should not by theory.*

error in the Hamiltonian, so attractive to calculate due to its availability, can be an extreme underestimator for the pointwise error in the solution.

On the other hand, the initial conditions $(q_1, q_2, p_1, p_2)(0) = (.1, .2, 0, 0)$ yield a quasi-periodic orbit; see [130, 160] and Figure 6.5. Here the two methods yield relatively accurate pointwise solutions for $0 \leq t \leq 2000$, and Figure 6.5 could have been generated by any of them, as far as the eye can tell. The error in the Hamiltonian is depicted in Figure 6.4(b). As it turns out, here the Hamiltonian error is a reasonable error underestimator. Note that while the Hamiltonian error in RK4 grows linearly, that in Störmer–Verlet does not. And yet, one would have to wait a very long time indeed to discount the nonsymplectic method based on the Hamiltonian conservation criterion for this particular example: a mere two million time steps won't do.

The more specialized symplectic method is actually simpler and more accurate for very long times t if the step sizes were to be adjusted to equate the computational costs of the two methods for this small example. Yet, the slightly dissipative RK4 also performs adequately and allows easy time step changes. A more accurate symplectic method of order 4 is proposed in [130].

Even in the quasi-periodic case nasty results may arise when we increase the step size k. McLachlan has pointed out that repeating the calculation leading to the phase portrait of Figure 6.5, but using the standard options of the most popular MATLAB ODE integrator `ode45`, leads to the completely wrong results depicted in Figure 6.6. The problem here is not how to fix the plot: Exercise 6 shows that the same routine `ode45` with more stringent tolerances will easily work well. Rather, the cause for concern is the deceptively good look of the wrong plot and the tendency of many a scholar to accept the results of packaged software as unquestionably true. Here is an important lesson and a chilly reminder to exercise caution with mathematical software!

The average step size that leads to Figure 6.6 is below 0.2. For the RK4 method with constant step size the phase plane plot is still *roughly* correct using $k = .2$ but not with

6.2. Symplectic and Other Relevant Methods

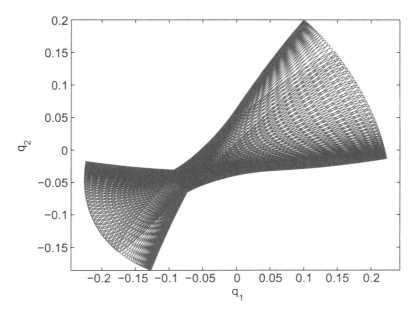

Figure 6.5. *Phase plane plot for $0 \leq t \leq 1000$ of a quasi-periodic orbit for the Henon–Heiles map defined by* (6.14).

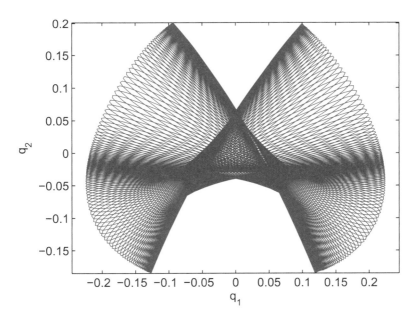

Figure 6.6. *Erroneous phase plane plot produced by MATLAB's solver* `ode45` *with standard options for the quasi-periodic orbit of the Henon–Heiles map. The qualitatively correct plot is in Figure 6.5.*

$k = .4$. For Störmer–Verlet with constant step size the transition to nonsense occurs only at $k > .6$. ∎

Higher order splitting methods are given in [186, 130]. A symplectic splitting method is derived in Section 7.3 for a semi-discretization of the KdV equation of Example 5.1.

6.2.3 Variational methods

We briefly mention, without getting into details, another wide class of integrators. These are based on a **discrete variational principle**. Relevant continuous structure that is available in many ODE applications is thus replicated on the discrete level.

For instance, in the case of basic elasticity the *Lagrangian* is defined as the kinetic energy minus the potential energy,

$$L(\mathbf{q}, \mathbf{q}') = T(\mathbf{q}') - V(\mathbf{q}). \tag{6.15a}$$

Then the *Euler–Lagrange equations*

$$\frac{\partial L}{\partial \mathbf{q}} - \frac{d}{dt}\left(\frac{\partial L}{\partial \mathbf{q}'}\right) = \mathbf{0} \tag{6.15b}$$

express *Hamilton's principle* [3], which states that the correct path of motion of a dynamical system is such that its action has a stationary value. These are the necessary conditions for a critical point of the integral in time of L. Defining the variables $\mathbf{v} = \mathbf{q}'$ and $\mathbf{p} = \frac{\partial L}{\partial \mathbf{v}}$ we can express (6.15b) as

$$\mathbf{q}' = \mathbf{v}, \quad \mathbf{p}' = \frac{\partial L}{\partial \mathbf{q}}, \quad \mathbf{p} = \frac{\partial L}{\partial \mathbf{v}}. \tag{6.15c}$$

Example 6.6 Let

$$T = \frac{1}{2}(\mathbf{q}')^T M \mathbf{q}',$$

where M is a symmetric positive definite mass matrix. Then (please verify) the Euler–Lagrange equations (6.15b) yield Newton's law of motion,

$$M\mathbf{q}'' = -\nabla V(\mathbf{q}).$$

Moreover, in terms of (6.15c) with the momenta $\mathbf{p}$ we can set (with a slight abuse of notation in T)

$$T(\mathbf{p}) = \frac{1}{2}\mathbf{p}^T M^{-1}\mathbf{p}, \quad \mathbf{p} = M\mathbf{v},$$

and obtain the same Newton's law by writing (6.1) for the Hamiltonian,

$$H(\mathbf{q}, \mathbf{p}) = T(\mathbf{p}) + V(\mathbf{q}).$$

Note the sign difference between this Hamiltonian and the Lagrangian in (6.15a). ∎

Next, we can apply to (6.15c) a centered, staggered discretization as before (see Example 6.4 and Section 3.1.2) and obtain the same symplectic method in a natural way.

The idea of a discrete variational principle is not new: it is at the base of early *finite element* developments—see, e.g., Strang and Fix [158]—and is routinely applied in the context of finite volume methods, as in Section 3.1. The approach can be seen as an instance of discretizing before optimizing, a topic to which we return in Section 11.1. Moreover, the utility of the variational approach as a tool for designing useful new symplectic methods that are not naturally derived with other approaches can be challenged. The *advantage* of the discrete variational approach is that it can naturally suggest good discretization methods in a *wider context* that includes more than just Hamiltonian systems, incorporating, for instance, dissipation terms and algebraic constraints (yielding DAEs), and extending to applications such as optimal control [119]. Interesting results have been developed for these methods in the present context, and favorite symplectic methods have been reproduced in this way; see Marsden and West [126] and references therein.

6.3 Properties of symplectic methods

Why is it important to use a symplectic method?

As indicated before, the literature is not always very clear on this. At one end, as we have said earlier, the mathematics here tends to be more involved. At the other end, physicists doing stellar mechanics computations and chemists performing molecular dynamics simulations have used simple symplectic methods such as (6.13) for many years and report good experiences. Indeed, when the competition is forward Euler or backward Euler, or even a two-stage second order explicit Runge–Kutta method, then symplectic methods shine, as is clearly indicated in Example 6.2.

A fairer assessment, comparing symplectic methods to *good* nonsymplectic alternatives such as RK4 or the four-step Adams–Moulton method AM4 given by (2.9), yields the following points:

1. Some favorable error accumulation properties for long times (i.e., many time steps) have been both observed and proved [150]. In particular, the Hamiltonian seems to be well-approximated by such methods without attempting to enforce its constancy. There are impressive results, both theoretical and practical, in this direction.

2. Moreover, symplectic methods often shine, especially when solution phase portraits are considered. See Examples 6.5 and 6.8. This is particularly important when certain statistical properties of the system are of interest [130].

3. However, for any of the methods that we have observed to be symplectic, the step size must be constant! This is a very serious limitation which may overshadow the above point in some practical applications. There are variable-step symplectic and time-reversible methods [81, 115, 83], but they are not as nice and simple as the Störmer–Verlet method. For implicit symplectic methods there is the added disadvantage that complications in the convergence of the nonlinear algebraic equation solver, if they arise, cannot be so simply waved away by cutting down the step size.

4. If the ODE system is large, then an implicit symplectic method necessitates solving a large system of possibly nonlinear algebraic equations at each step. If iterative

methods are used, then the symplectic property could be lost unless the iteration is carried out to a very high accuracy. Attaining such high accuracy may contribute to the expense of the method.

5. The entire development of symplectic methods ignores roundoff errors. Certainly these cannot be expected to be highly structured, as their sources are essentially random. Roundoff errors generally propagate (accumulate) according to similar rules as discretization errors which dominate them. However, linear accumulation of roundoff errors cannot be avoided, and when billions of time steps are applied this may become a factor.

6. Perhaps even more important than the first point above is the following theorem.

Theorem 6.7. *A symplectic method discretizing a Hamiltonian system yields (in infinite precision) a solution which is arbitrarily close to the exact flow of a perturbed Hamiltonian system.*

The error of a symplectic method, then, is not arbitrary but rather is highly structured. There is another, **modified Hamiltonian system** or *ghost Hamiltonian system* that the present numerical solution actually solves (almost) exactly.

This result is proved using **backward error analysis**. See, e.g., Chapter IX of [81] and references therein.

Example 6.8 The modified Kepler problem [150, 84, 81] is a Hamiltonian system with the Hamiltonian

$$H(\mathbf{q}, \mathbf{p}) = \frac{p_1^2 + p_2^2}{2} - \frac{1}{r} - \frac{\alpha}{2r^3},$$

where $r = \sqrt{q_1^2 + q_2^2}$ and we take $\alpha = 0.01$. The ODE system is therefore

$$\mathbf{q}' = \mathbf{p},$$
$$\mathbf{p}' = -\left(\frac{1}{r^3} + \frac{3\alpha}{2r^5}\right)\mathbf{q}.$$

Figure 6.7 depicts phase plain portraits of q_1 vs. q_2 for $0 \leq t \leq 500$ obtained using the relatively coarse step size $k = .1$. We plot (a) the exact solution (or rather, a sufficiently close approximation to it), and solutions by (b) the implicit midpoint method, (c) the classical explicit Runge–Kutta method of order 4 (RK4), and (d) the explicit Störmer–Verlet method. The initial conditions are

$$q_1(0) = 1 - \beta, \quad q_2(0) = 0, \quad p_1(0) = 0, \quad p_2(0) = \sqrt{(1+\beta)/(1-\beta)}$$

with $\beta = .6$.

The exact solution lies on a torus in the phase plane. This feature is reproduced by the two symplectic methods, midpoint and Verlet, but it is not well reproduced by the slightly

6.3. Properties of Symplectic Methods

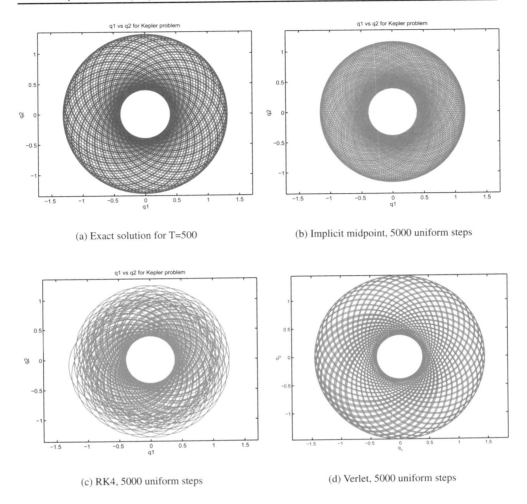

Figure 6.7. *Modified Kepler problem: Approximate solutions in phase plane using step size* $k = .1$.

dissipative, fourth order method RK4.[24] Note that the pattern of the solution and the radius of the torus are off in all these coarse approximations. The symplectic methods look orderly, though, because they accurately approximate another Hamiltonian system!

Is this a good feature to have? This is almost a philosophical question, and the best answer seems to be that it depends on the application and the interpretation. For graphical or virtual reality purposes, a solution which looks realistic, even if wrong, can be better than one that looks wrong. Moreover, the global feature that the solution lies on a torus is depicted by the symplectic computations. The solutions (b) and (d) in Figure 6.7 somehow

---
[24]This step size is particularly nasty to RK4: a smaller $k = .05$ already yields a much better picture, because RK4 becomes less dissipative (see Figure 6.11). But fairness to RK4 is not our purpose here.

look more pleasing than (c) to most people we've surveyed. On the other hand, when we read about simulations of the evolution of the solar system over many millions of years by a symplectic method we may wonder precisely what stellar system the obtained result simulates: would that be the intended physical one?! A picture may lie better than a thousand words. ■

Example 6.9 This example does not relate particularly to Hamiltonian systems or to ODEs, but it does follow an issue raised in Example 6.8, namely, how to measure errors. In introductory classes on numerical analysis (as well as introductory physics and other disciplines) one learns about absolute and relative errors, which are objective means of measuring deviations between exact and approximate quantities. But the tori produced by the symplectic methods in Figure 6.7 have more appeal, compared to the one produced by the nonsymplectic method, than can be quantified by such error measures. Here the *"eye-ball norm"* rules, and not only in the context of computer graphics or image processing.

The latter error measure is harder to qualify, let alone quantify, and here we merely give another example, to make the point rather than attempt to precisely define it. Consider the picture in Figure 6.8. Some of it is not in focus, some overexposed, some in the shade, and there is something ruffled at the edge of the rug. But what we see first and foremost is the center ripple in the rug, which is not the way we expect it to be laid on the floor in the absence of "error." If we view elevation above floor level as error, then the ripple is much more distinguished than a random error of the same absolute value would be because of its clear pattern and the recognition filter that we automatically apply to such a domestic scene. ■

6.4 Pitfalls in highly oscillatory Hamiltonian systems

Consider the Hamiltonian

$$H(\mathbf{q}, \mathbf{p}) = \frac{1}{2}\mathbf{p}^T \mathbf{p} + V(\mathbf{q}) + \frac{1}{2\varepsilon^2}\mathbf{g}(\mathbf{q})^T \mathbf{g}(\mathbf{q}), \tag{6.16}$$

where V and $\mathbf{g}$ are smooth, bounded functions, and the parameter is small, $0 < \varepsilon \ll 1$. Recall the corresponding differential system

$$\begin{aligned} \mathbf{q}' &= \nabla_\mathbf{p} H = \mathbf{p}, \\ \mathbf{p}' &= -\nabla_\mathbf{q} H = -\nabla V(\mathbf{q}) - \varepsilon^{-2} G(\mathbf{q})^T \mathbf{g}(\mathbf{q}), \end{aligned} \tag{6.17}$$

where $G = \frac{\partial \mathbf{g}}{\partial \mathbf{q}}$.

As we saw in Example 2.8, to obtain a pointwise accurate solution we would need to discretize (6.17) with a step size satisfying $k < c\varepsilon$ for some moderate constant c. But this may imply an extremely small step size in applications such as molecular dynamics simulations. Hence, we wish to use a larger step size, giving up pointwise accuracy but still recovering important structural features. Let us remember that here, unlike the usual stiff case considered in Section 2.6, the fast modes do not decay but rather oscillate, and they do affect the exact solution significantly at all times. Thus, unless there is a source

6.4. Pitfalls in Highly Oscillatory Hamiltonian Systems

Figure 6.8. *The "eye-ball norm": The ripple in the rug, interpreted as an error, is much more distinguished than a random error of the same norm would be because of its clear pattern and the recognition filter that we automatically apply to such a domestic scene.*

of damping as in Example 2.9, it is not clear what we can get away with when giving up accurate pointwise approximation.

Let us start with a warning not to be too simpleminded. For instance, the above justifies a quest for a numerical discretization that for any step size k and any ε

- conserves the Hamiltonian reasonably well,
- is stable and efficient, and
- disregards pointwise accuracy of solutions.

But a solution satisfying all these desires is the constant $\mathbf{q}_n = \mathbf{q}(0)$, $\mathbf{p}_n = \mathbf{p}(0)$, $n = 0, 1, \ldots$. Clearly this is not what we have in mind! The question is of interest and relevance when both fast and nontrivial slow solution features are present and the numerical scheme ought to approximate slow solution features well.

Example 6.10 A modification of the notorious **Fermi–Pasta–Ulam problem** (FPU) is presented in the introductory chapter of [81]. It consists of a chain of $\hat{m}$ mass points connected with springs that have alternating characteristics: the odd ones are soft and nonlinear whereas the even ones are stiff and linear.

There are variables $q_1, \ldots, q_{2\hat{m}}$ and $p_1, \ldots, p_{2\hat{m}}$ in which the associated Hamiltonian is written as

$$H(\mathbf{q}, \mathbf{p}) = \frac{1}{4}\left[2\sum_{i=1}^{2\hat{m}} p_i^2 + 2\omega^2 \sum_{i=1}^{\hat{m}} q_{\hat{m}+i}^2 + (q_1 - q_{\hat{m}+1})^4 + (q_{\hat{m}} + q_{2\hat{m}})^4\right.$$
$$\left. + \sum_{i=1}^{\hat{m}-1}(q_{i+1} - q_{\hat{m}+1+i} - q_i - q_{\hat{m}+i})^4 \right]. \qquad (6.18a)$$

The parameter ω relates to the stiff spring constant and is large. This Hamiltonian is conserved as usual by the solution of the corresponding Hamiltonian system. In addition, denote the energy in the ith stiff spring by

$$I_i = \frac{1}{2}(p_{\hat{m}+i}^2 + \omega^2 q_{\hat{m}+i}^2). \qquad (6.18b)$$

Then it turns out that there is an exchange of energies such that the total oscillatory energy

$$I = \sum_{i=1}^{\hat{m}} I_i$$

is an **adiabatic invariant**, satisfying

$$I(\mathbf{q}(t), \mathbf{p}(t)) = I(\mathbf{q}(0), \mathbf{p}(0)) + O(\omega^{-1}) \qquad (6.18c)$$

for exponentially long times t.

As a variation of the example in [81] we choose $\hat{m} = 3$ (yielding an ODE system of size $m = 12$), $\omega = 100$, $\mathbf{q}(0) = (1, 0, 0, \omega^{-1}, 0, 0)^T$, $\mathbf{p}(0) = (1, 0, 0, 1, 0, 0)^T$, and integrate from $t = 0$ to $t = 500$ using RK4 with a constant step size $k = .00025$. The resulting Hamiltonian error is a mere 6.8×10^{-6}, and the oscillatory energies are recorded in Figure 6.9.

The curves depicted in the figure are exact as far as the eye can tell. The "noise" is not a numerical artifact. Rather, small "bubbles" of rapid oscillations occasionally flare up and quickly die away; see [81].

Now, imagine that we fit low-degree polynomials to each of these curves in the least squares sense. Surely I will be fitted with an almost constant straight line, and the long scale trend of the others could also be accommodated by low degree polynomials (of degrees up to 7, say). It is this sort of larger scale behavior that we would ideally like to capture by a method that does not first reproduce all small scale details.

However, satisfying this wish turns out to be a rather nontrivial undertaking. ∎

Example 6.11 Consider, yet again, the harmonic oscillator of Example 6.1, with large ω. We write this in the form (6.16), (6.17) as

$$H(q, p) = \frac{1}{2}[p^2 + \varepsilon^{-2}q^2],$$

6.4. Pitfalls in Highly Oscillatory Hamiltonian Systems

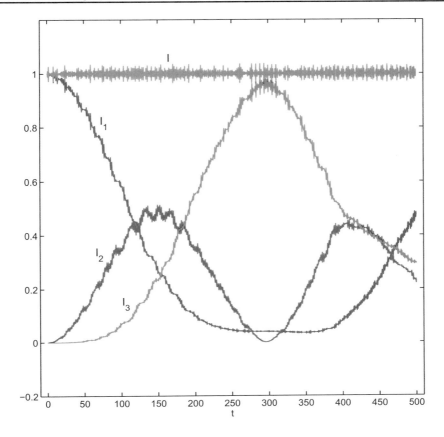

Figure 6.9. *Oscillatory energies for the Fermi–Pasta–Ulam (FPU) problem. Note that the trend of I stays almost constant in time.*

$$q' = p, \quad p' = -\varepsilon^{-2}q.$$

It is not difficult to verify that the Störmer–Verlet method (6.13) is stable only if $k \leq 2\varepsilon$. On the other hand, the midpoint method (2.12) is unconditionally stable. To see this, note that the quadratic Hamitonian is conserved by the implicit midpoint method, so stability is obtained in an energy norm for any k and ε.

Of course, if we use the midpoint method with $k \gg \varepsilon$, then the pointwise numerical solution is wrong. Moreover, it is easy to verify that for a collocation method, in case of a bounded solution, at collocation points (where the ODE is satisfied by the piecewise polynomial approximate function) the numerical solution is $O(\varepsilon/k)\|q\|$. This is *systematically wrong*. Whether it matters depends on the problem in which such a harmonic oscillator is embedded. ∎

Exercise 7 shows that the midpoint method may not always be stable for Hamiltonian systems. Moreover, we must beware of misleading results, as the following example shows.

Example 6.12 This example is from [16]. Consider a linear oscillator with a slowly varying frequency,

$$q' = \omega^2(t)p,$$
$$p' = -\varepsilon^{-2}q,$$

with, e.g., $\omega(t) = 1 + t$.

Since the system is not autonomous, the Hamiltonian is no longer constant in t. But the *adiabatic invariant*

$$J(q, p, t) = H(q, p, t)/\omega(t) = \omega(t)p^2/2 + \varepsilon^{-2}\omega^{-1}(t)q^2/2$$

satisfies for $T = c_1 e^{c_2/\varepsilon}$, where c_i are constants, the "almost conservation"

$$[J(t) - J(0)]/J(0) = O(\varepsilon), \quad 0 \le t \le T;$$

see [3].[25]

Let us see how well this is reproduced by the midpoint method

$$(q_{n+1} - q_n)/k = \omega(t_{n+1/2})^2 (p_{n+1} + p_n)/2,$$
$$(p_{n+1} - p_n)/k = -\varepsilon^{-2}(q_{n+1} + q_n)/2.$$

Specifically, we ask: What ODE does this really approach when $\varepsilon \ll k \to 0$?

Let

$$u_n = (-1)^n \varepsilon^{-1} q_n, \quad v_n = (-1)^{n+1} p_n, \quad \alpha = \frac{k^2}{4\varepsilon}.$$

Then

$$(u_{n+1} + u_n)/2 = -\omega(t_{n+1/2})^2 \alpha (v_{n+1} - v_n)/k,$$
$$(v_{n+1} + v_n)/2 = \alpha (u_{n+1} - u_n)/k.$$

Now, as we let $k \to 0$ for a fixed α we get

$$-\omega^2(t)\alpha v' = u,$$
$$\alpha u' = v.$$

So, the obtained **ghost** ODE is an oscillator with α essentially replacing ε. Hence

$$\left[\hat{J}(t) - \hat{J}(0)\right]/\hat{J}(0) = O(\alpha),$$
$$\hat{J}(u_n, v_n, t_n) = \hat{J}(\epsilon^{-1}q_n, p_n, t_n) = J(q_n, p_n, t_n).$$

---

[25]Note that J of the adiabatic invariant has nothing to do with the skew-symmetric J of (6.3).

6.4. Pitfalls in Highly Oscillatory Hamiltonian Systems

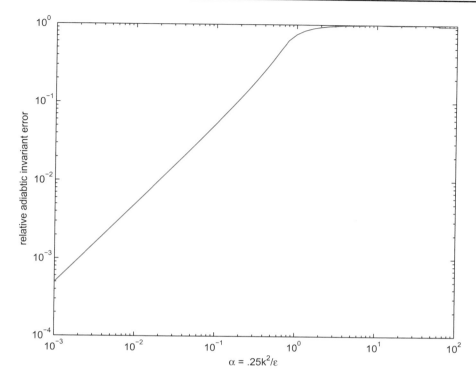

Figure 6.10. *Error in adiabatic invariant vs.* $\alpha = \frac{k^2}{4\epsilon}$ *using the midpoint method. The results for* $\varepsilon = 10^{-5}$, 10^{-6}, *and* 10^{-7} *are virtually indistinguishable when using k proportional to* $\sqrt{\varepsilon}$. *These results are poor unless the constant of proportionality* α *is small, regardless of how small k and* ε *are.*

A wrong adiabatic invariant is consistently approximated!

To verify this numerically we calculate $\Delta J = \max_{0 \le t \le 1} |J(t) - J(0)|/J(0)$ for values of α in the range $[10^{-3}, 10^2]$ as follows: for each of the three values $\varepsilon = 10^{-5}$, 10^{-6}, and 10^{-7} we calculate the corresponding time step $k = \sqrt{4\varepsilon\alpha}$, solve the midpoint equations, and calculate ΔJ. The obtained curves are superimposed in Figure 6.10. They are too close to tell apart.

Thus, unless α is small we may not expect reasonable approximations for the slowly varying adiabatic invariant when using the midpoint method, regardless of how small k is. Similar results are obtained upon using the symplectic, sixth order method of collocation at three Gaussian points [16]. ∎

In general, we may obtain misleading results for highly oscillatory problems when the smooth manifold of the Hamiltonian system

$$\mathbf{q}' = \mathbf{p},$$
$$\mathbf{p}' = -\nabla V(\mathbf{q}) - \varepsilon^{-2} G^T \mathbf{g}(\mathbf{q}),$$

is *not* the solution of the DAE,

$$\mathbf{q}' = \mathbf{p},$$
$$\mathbf{p}' = -\nabla V(\mathbf{q}) - G^T \boldsymbol{\lambda},$$
$$0 = \mathbf{g}(\mathbf{q}),$$

in the limit $\varepsilon \to 0$. We quickly mention two more examples, referring the reader to [15, 16] for additional details.

Example 6.13 Consider a pendulum whose rod is actually a stiff spring, with a spring constant $1/\varepsilon$ for a value of ε satisfying $0 < \varepsilon \ll 1$. With the notation $\mathbf{q} = (q_1, q_2)^T$ we have

$$r = |\mathbf{q}| = \sqrt{q_1^2 + q_2^2},$$
$$\sin \phi = q_1/|\mathbf{q}|;$$

$r(t)$ is now a variable, oscillating rapidly about a constant r_0. In fact, the variables (r, ϕ) form a spherical coordinate system given in terms of the planar coordinate system $\mathbf{q}$. The ODE system describing the **stiff spring pendulum** is

$$\mathbf{q}' = \mathbf{p},$$
$$\mathbf{p}' = -(\phi(\mathbf{q}) - \phi_0)\nabla\phi(\mathbf{q}) - \varepsilon^{-2}(r(\mathbf{q}) - r_0)\nabla r(\mathbf{q}) \quad (6.19)$$

(take $\phi_0 = 0$). The system (6.19) is highly oscillatory. Its numerical simulation leads to challenging questions, unless one is prepared to pay the price for resolving the rapid oscillation in r by taking very small discretization steps.

For the stiff spring pendulum (6.19) the DAE above is the correct limit system. However, we have fast and slow solution modes. Experiments using the implicit midpoint method yield poor results as $k \to 0$ when $\alpha = \frac{k^2}{4\epsilon}$ is large and fixed.

This is because the fast and slow solution modes are strongly coupled! Here r is fast and ϕ is slow: transforming first to an ODE system in r and ϕ, a subsequent midpoint discretization works very well, since the fast and slow modes are decoupled. In general, no method may be expected to work well when $\varepsilon \ll k^2$ if fast and slow modes (the latter to be well reproduced) are strongly coupled. ∎

Example 6.14 Next, consider the **reversed stiff spring pendulum**

$$\mathbf{q}' = \mathbf{p},$$
$$\mathbf{p}' = -\varepsilon^{-2}(\phi(\mathbf{q}) - \phi_0)\nabla\phi(\mathbf{q}) - (r(\mathbf{q}) - r_0)\nabla r(\mathbf{q}), \quad (6.20)$$

so r is slow and ϕ is fast. Such constructs occur within large molecular dynamics problems. Here, unfortunately, the limit DAE as $\varepsilon \to 0$ is different from the obvious one [16].

This example combines the previous two sources of trouble: both coupling of slow and fast modes and poor reconstruction of adiabatic invariants appear. Now, even in the weakly coupled coordinates r and ϕ we must have $\alpha = \frac{k^2}{4\epsilon}$ small; otherwise, a wrong limit ghost DAE is discretized in effect. ∎

The implicit midpoint method, as it turns out, can be particularly difficult to work with in the present circumstances, despite its properties of symplecticity, conservation of quadratic invariants, and unconditional stability for constant coefficient problems. For another unwelcome surprise in a PDE context, see Example 9.6. There are, however, many situations in which effective methods for highly oscillatory problems may be devised. Separate chapters are devoted to such special methods in [115] and [81], and this is currently a fairly hot research area.

However, extreme care must be exercised as no general recipes are available, and caution is the reason we have included the pessimistic but simple examples of this section. In particular, much of the effort in designing methods for highly oscillatory problems has been aimed at molecular dynamics simulations (see again [115]), and yet, despite efforts spanning several decades the method that still reigns in large molecular dynamics codes is Störmer–Verlet (6.13), simply utilized with small steps k and benefitting from a low cost per step. Thus, the most popular family of "stiff highly oscillatory" problems is routinely simulated using a nonstiff, explicit method utilizing a constant time step size! How frustrating.

6.5 Exercises

0. **Review questions**

    (a) What is geometric integration?

    (b) Define a Hamiltonian system and a symplectic map. Does a Hamiltonian system necessarily induce a symplectic map?

    (c) In what sense is the stability of Hamiltonian systems marginal?

    (d) Explain why forward Euler and backward Euler methods generally perform poorly for Hamiltonian systems.

    (e) Define the symplectic Euler and Verlet methods, and describe their properties.

    (f) What is a partitioned Runge–Kutta method? Give an example of one.

    (g) Describe splitting methods in the context of this chapter. Find an example where the Strang splitting clearly produces a more accurate numerical result than the simple splitting it aims at improving.

    (h) What is most perturbing about Figure 6.6?

    (i) Why is Theorem 6.7 important?

    (j) What is the essential difference between highly oscillatory Hamiltonian systems and very stiff ODEs?

1. Write a short program which uses the forward Euler, the backward Euler, and the trapezoidal *or* midpoint methods to integrate a linear, scalar ODE with a known solution, using a fixed step size $k = b/N$, and which finds the maximum error. Apply your program to the ODE

$$\frac{dy}{dt} = (\cos t)y, \qquad 0 \le t \le b,$$

Table 6.1. *Maximum errors for long interval integration of* $y' = (\cos t)y$.

b	N	Forward Euler	Backward Euler	Trapezoidal	Midpoint
1	10	.35e-1	.36e-1	.29e-2	.22e-2
	20	.18e-1	.18e-1	.61e-3	.51e-3
10	100				
	200				
100	1000	2.46	25.90	.42e-2	.26e-2
	2000				
1000	1000				
	10000	2.72	1.79e+11	.42e-2	.26e-2
	20000				
	100000	2.49	29.77	.42e-4	.26e-4

with $y(0) = 1$. The exact solution is

$$y(t) = e^{\sin t}.$$

Verify those entries given in Table 6.1 and complete the missing ones. Make as many (useful) observations as you can about the results in the complete table. Attempt to provide explanations. (Hint: Plotting these solution curves for $b = 20$, $N = 10b$, say, may help.)

2. Let J be a real, skew-symmetric matrix, and let A be a real, symmetric matrix of the same size. Prove the following:

   (a) The nonzero eigenvalues of J are all purely imaginary.

   (b) Let $M = JA$. If λ is an eigenvalue of M, then so is $-\lambda$.

   (c) If A is also positive definite, then the nonzero eigenvalues of M are all purely imaginary. (However, we remark that the latter property does not hold without positive definiteness: if A is merely symmetric, then M will have, in general, eigenvalues with a positive real part.)

3. (a) Show that the midpoint method is symmetric, second-order, and A-stable. How does it relate to the trapezoidal method for a constant coefficient ODE?

   (b) Suppose we allow λ to vary in t, i.e., consider the scalar ODE

   $$y' = \lambda(t)y$$

   in place of the test equation. Show that what corresponds to A-stability holds, namely, using the midpoint method

   $$|y_{n+1}| \leq |y_n| \quad \text{if} \quad \mathcal{R}e(\lambda) \leq 0.$$

6.5. Exercises

(This property is called *AN-stability* [43].) Show that the same cannot be said about the trapezoidal method: the latter is not AN-stable.

4. Show directly (i.e., without casting it as a collocation method and without using (6.7)) that the implicit midpoint method is symplectic.

5. (a) Show that the trapezoidal step (2.8) can be viewed as a half step of forward Euler followed by a half step of backward Euler.

   (b) Show that the midpoint step (2.12) can be viewed as a half step of backward Euler followed by a half step of forward Euler.

   (c) Consider an autonomous system $\mathbf{y}' = \mathbf{f}(\mathbf{y})$ and a fixed step size, $k_n = k$, $n = 1, \ldots, N$. Show that the trapezoidal method applied N times is equivalent to applying first a half step of forward Euler (i.e., forward Euler with step size $k/2$) followed by $N - 1$ midpoint steps and finishing off with a half step of backward Euler.

   Conclude that these two symmetric methods are **dynamically equivalent** [55]; i.e., for k small enough their performance is very similar independently of N, even over a very long time $b = Nk \gg 1$. This is so even though the midpoint method is symplectic whereas the trapezoidal method is not.

   (d) However, if k is not small enough (compared to the problem's small parameter, say, λ^{-1}), then the two methods do not necessarily perform similarly. Construct an example where one of these methods blows up (error $> 10^5$, say) while the other yields an error below 10^{-5}. (Do not program anything: this is a (nontrivial) pen-and-paper question.)

6. (a) Derive the ODE system (6.1) that corresponds to the Hamiltonian defined by (6.14).

   (b) Find sufficiently small absolute and relative error tolerances for ode45 to obtain the qualitatively correct phase plot of Figure 6.5.

7. Consider two linear harmonic oscillators, one fast and one slow, $u_1'' = -\varepsilon^{-2}(u_1 - \bar{u}_1)$ and $u_2'' = -(u_2 - \bar{u}_2)$. The parameter is small: $0 < \varepsilon \ll 1$. We write this as a first order system,

$$\mathbf{u}' = \begin{pmatrix} \varepsilon^{-1} & 0 \\ 0 & 1 \end{pmatrix} \mathbf{v},$$

$$\mathbf{v}' = -\begin{pmatrix} \varepsilon^{-1} & 0 \\ 0 & 1 \end{pmatrix} (\mathbf{u} - \bar{\mathbf{u}}),$$

where $\mathbf{u}(t)$, $\mathbf{v}(t)$, and the given constant vector $\bar{\mathbf{u}}$ each have two components. It is easy to see that $E_F = \frac{1}{2\varepsilon}(v_1^2 + (u_1 - \bar{u}_1)^2)$ and $E_S = \frac{1}{2}(v_2^2 + (u_2 - \bar{u}_2)^2)$ remain constant for all t.

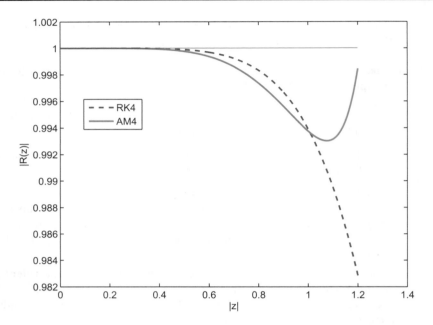

Figure 6.11. *Stability functions of RK4 and AM4 along the imaginary axis. In both cases, $.99 \leq |R(z)| < 1$ for $0 < |z| \leq 1$.*

Next, we apply the time-dependent linear transformation

$$\mathbf{u} = Q\mathbf{x}, \quad \mathbf{v} = Q\mathbf{z},$$

$$Q(t) = \begin{pmatrix} \cos \omega t & \sin \omega t \\ -\sin \omega t & \cos \omega t \end{pmatrix}, \quad K = -Q^T Q' = \begin{pmatrix} 0 & -1 \\ 1 & 0 \end{pmatrix},$$

where $\omega \geq 0$ is another parameter. This yields the coupled system

$$\mathbf{x}' = Q^T \begin{pmatrix} \varepsilon^{-1} & 0 \\ 0 & 1 \end{pmatrix} Q\mathbf{z} + \omega K \mathbf{x}, \tag{6.21a}$$

$$\mathbf{z}' = -Q^T \begin{pmatrix} \varepsilon^{-1} & 0 \\ 0 & 1 \end{pmatrix} Q(\mathbf{x} - \bar{\mathbf{x}}) + \omega K \mathbf{z}, \tag{6.21b}$$

where $\bar{\mathbf{x}} = Q^T \bar{\mathbf{u}}$. We can write the latter system in our usual notation as a system of order 4,

$$\mathbf{y}' = A(t)\mathbf{y} + \mathbf{q}(t).$$

(a) Show that the eigenvalues of the matrix A are all purely imaginary for all ω. (Hint: Show that $A^T = -A$.)

6.5. Exercises

(b) Using the values $\bar{\mathbf{u}} = (1, \pi/4)^T$, $\mathbf{u}(0) = \bar{\mathbf{u}}$, $\mathbf{v}(0)^T = (1, -1)/\sqrt{2}$, and $b = 20$, apply the midpoint method with a constant step size k to the system (6.21) for the following parameter combinations: $\varepsilon = 0.001$, $k = 0.1, 0.05, 0.001$, and $\omega = 0, 1, 10$ (a total of nine runs). Compute the error indicators $\max_t |E_F(t) - E_F(0)|$ and $\max_t |E_S(t) - E_S(0)|$. Discuss your observations.

(c) Show that the midpoint method is unstable for this problem if $k > 2\sqrt{\varepsilon/\omega}$ (see [15]). Conclude that A-stability and AN-stability do not automatically extend to ODE *systems*.

8. Show that the composition of two symplectic maps is also a symplectic map.

9. Show that
$$e^{k(L+M)} = e^{\frac{k}{2}L} e^{kM} e^{\frac{k}{2}L} + O(k^3)$$
for any $s \times s$ matrices L and M.

10. The absolute stability region plots in Section 2.5 for both RK4 and the four-step Adams–Moulton method AM4 suggest that these methods remain stable and only slightly dissipative for a while along the imaginary axis. Check this out by plotting $|R(z)|$ vs. w, where $z = \iota w$ for $0 \le w \le 1.2$. You should get something like Figure 6.11.

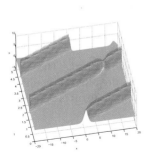

Chapter 7
Dispersion and Dissipation

This chapter is in many ways a continuation of Chapter 5. However, we wish to also extend some of the ideas introduced in Chapter 6 to PDEs, and this is the natural place to do it.

In Section 1.1 we saw that prototype parabolic problems dissipate energy, whereas hyperbolic ones do not. This has very important ramifications on numerical methods: for hyperbolic problems one is torn between the necessity to attenuate numerical solution modes for stability reasons and the fear that too much dissipation may cause the discrete solution to lose sharp qualitative features of the exact one.

We consider exclusively in this chapter hyperbolic-type PDEs that have smooth, wave-like solutions. Methods with little or no dissipativity are then adequate.

Another important wave property is *dispersion*. In Section 7.1 we introduce the concept of numerical dispersion and briefly discuss the wave attributes of dispersion and dissipation in one context. As it turns out, numerical dispersion often arises in high wave-number solution components, even when the PDE is nondispersive, and if unchecked it may give rise to **spurious** or **parasitic** waves. This gives an edge to dissipative methods, because they dampen parasitic dispersion waves. But using dissipative methods may introduce other problems, as we have seen in Chapter 5, particularly Sections 5.1 and 5.2. All the methods considered there suffer from dispersion effects, yet these effects are particularly pronounced when low-order, noncompact methods are applied on coarse meshes. Increasing resolution and order of accuracy, as well as using compact discretizations, generally improve the situation.

In Section 7.2 we briefly discuss numerical methods for the **classical wave equation** and some extensions. These are nondispersive PDEs with a symplectic structure, and we study the performance of the symplectic, compact, and unavoidably dispersive leapfrog method. A staggered mesh (also known as *staggered grid*) discretization results for the corresponding formulation of the wave equation as a first order PDE system.

In Section 7.3 we consider the dispersive, nonlinear *Korteweg–de Vries* (KdV) equation already introduced in Example 5.1. This entire section may be viewed as a case study. Numerical stability as well as artificial dispersion effects arise and we examine the performance of **symplectic**, **compact**, and **multisymplectic** methods in this context.

It is possible to distinguish three classes among mesh-based numerical methods: finite difference/volume, finite element, and *spectral methods*. This book is devoted to the first of these classes, but in Section 3.1.4 we have briefly reviewed the second. In Section 7.4 we likewise review the class of spectral methods. This is a natural place to do it because these methods particularly excel when applied to problems of the types considered in the present chapter.

Finally, in Section 7.5 we briefly mention *Lagrangian* and *particle* methods. These can provide a relief from the dispersion problem, but at a cost of introducing other difficulties.

An excellent, classical reference on waves and the PDEs leading to them is Whitham [182].

7.1 Dispersion

First we introduce the concept of dispersion in PDEs. Then we consider dispersion in PDE discretizations. As usual we utilize a simple example. Consider the advection equation with one Fourier mode as initial data

$$u_t + au_x = 0,$$
$$u(0, x) = e^{-\iota \xi x}. \qquad (7.1)$$

We would like to trace the solution's progress in time, in the exact and approximate cases. Then we would use superposition (i.e., the Fourier transform) for more general initial data.

The exact solution of (7.1) is

$$u(t, x) = e^{\iota \xi (at - x)}.$$

So, this is a wave which propagates with speed a. The speed a is the *phase velocity*.

More generally, let

$$e^{\iota(\omega t - \xi x)}$$

be a solution form for a scalar, constant coefficient PDE, where ξ is the **wave number** and ω is the **frequency**. We define

$\omega = \omega(\xi)$	dispersion relation
$c(\xi) = \frac{\omega(\xi)}{\xi}$	phase velocity
$C(\xi) = \frac{d\omega(\xi)}{d\xi}$	group velocity

For the simple constant-coefficient advection equation (7.1) we have the dispersion relation $\omega = a\xi$, and so the phase and group velocities both simply equal a. Thus, the phase velocity is independent of the wave number and the PDE is *nondispersive*: different Fourier modes travel at the same speed.

More generally, however, the phase velocity $c(\xi)$ may depend on ξ, and this is called *dispersion*. In this case, different Fourier modes travel at different speeds and the evolution of a packet of modes, containing several wave numbers, is more complicated. The KdV equation (5.1) is an example of a dispersive PDE, and so is its linearization, $u_t + \rho u_x + \nu u_{xxx} = 0$. The energy associated with wave number ξ moves asymptotically (as $t \to \infty$) at the group velocity.

7.1. Dispersion

Semi-discretization of the advection equation

Next, on a uniform spatial mesh with step size h consider the semi-discretization

$$v'_j + \frac{a}{2h} D_0 v_j = 0 \qquad \left(v'_j = \frac{dv_j}{dt} \right),$$

which is common to many full discretization schemes where one is interested in retaining features of *smooth* solutions. The solution is clearly

$$v_j(t) = e^{\iota(\omega t - \xi j h)},$$

where the dispersion relation is

$$\omega = \frac{a}{h} \sin(\xi h).$$

The phase velocity is therefore

$$c_1 = a \frac{\sin(\xi h)}{\xi h},$$

and the group velocity is

$$C_1 = a \cos(\xi h).$$

We see that the semi-discretization is **dispersive**, even though the PDE is not! At low wave numbers, $c_1 \approx a \approx C_1$. But at high wave numbers the phase and group velocities differ significantly, and they depend on the wave number. In particular, at $\xi h \approx \pi/2$ we have $C_1 \approx 0$, while at $\xi h \approx \pi$ we have $c_1 \approx 0$. Thus, at high wave numbers the numerical waves are nearly stationary, with the trouble arriving sooner for the group velocity. These are often called *parasitic waves*. The desire to apply as little dissipativity as possible, expressed throughout Chapter 6, now becomes somewhat questionable, if those undamped high wave number modes are going to travel at incorrect speeds (or in other words, if parasitic waves are undamped).

Lax–Wendroff dispersion

For the advection equation (7.1) recall that the Lax–Wendroff scheme reads

$$v_j^{n+1} = \left(1 - \frac{\mu}{2} a D_0 + \frac{\mu^2}{2} a^2 D_+ D_- \right) v_j^n,$$

where k is the step size in time and $\mu = k/h$. Substituting

$$v_j^n = e^{\iota(\omega n k - \xi j h)},$$

we obtain the (implicit) dispersion relation for $\omega(\xi)$

$$\cos(\omega k) + \iota \sin(\omega k) = e^{\iota \omega k} = 1 + \iota \mu a \sin(\xi h) - 2\mu^2 a^2 \sin^2(\xi h/2). \tag{7.2}$$

Thus

$$\tan(\omega k) = \frac{\mu a \sin(\xi h)}{1 - 2\mu^2 a^2 \sin^2(\xi h/2)}.$$

Consider the phase velocity $c_2 = \omega/\xi$. For low wave numbers, $|\xi|h \ll 1$, we have by accuracy
$$c_2 = a(1 + O(h^2)).$$
A more precise expression is given in Exercise 1. For high wave numbers we again obtain
$$c_2 \approx 0, \quad \xi h \approx \pi.$$
For the group velocity we differentiate (7.2) with respect to ξ, obtaining
$$\imath k e^{\imath \omega k} C_2 = \imath h \mu a \cos(\xi h) - h \mu^2 a^2 \sin(\xi h).$$
Again, $C_2 \approx a$ when $|\xi|h \ll 1$, but $\mathcal{R}e\, C_2 \approx 0$ at $\xi h \approx \pi/2$, as $k \to 0$. The Lax–Wendroff scheme supports parasitic waves similarly to the centered semi-discretization.

Leapfrog dispersion

For the leapfrog scheme applied to (7.1),
$$v_j^{n+1} = v_j^{n-1} - \mu a(v_{j+1}^n - v_{j-1}^n),$$
we substitute the solution
$$v_j^n = e^{\imath(\omega n k - \xi j h)}$$
and obtain the implicit dispersion relation (Exercise 2)
$$\sin(\omega k) = \mu a \sin(\xi h). \tag{7.3}$$
Again, for low wave numbers we obtain $\omega \approx a\xi$, hence the phase and group velocities satisfy
$$c_3 \approx a \approx C_3.$$
Following [167] we plot ω vs. ξ in Figure 7.1. From this figure it is clear that again at high wave numbers we can have stagnating parasitic waves,
$$c_3(\pi) = 0, \quad C_3(\pi/2) = 0.$$
In fact, there are even more parasitic waves here than before.

Example 7.1 We repeat the experiments of Example 5.6, but this time with the initial value function
$$u_0(x) = \sin(\eta x) e^{-\eta x^2}, \quad \eta = 50.$$
Thus, we have an oscillatory sinusoidal wave attenuated away from $x = t$. We again set $\mu = 0.5$, use periodic BCs on $[-\pi, \pi]$, and record solutions for the four methods (*Lax–Friedrichs, Lax–Wendroff, leapfrog,* and *leapfrog with dissipativity added*) at $t = 1$. Using a fine step size $h = 0.001\pi$ leads to qualitatively correct approximate solutions, upon employing any of the second order methods. The Lax–Friedrichs scheme performs the

7.1. Dispersion

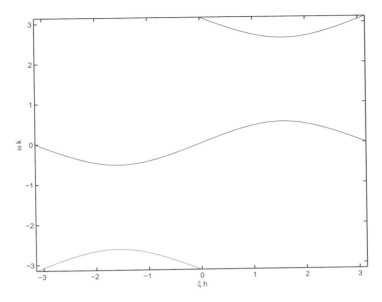

Figure 7.1. *Dispersion relation for the leapfrog scheme applied to $u_t + u_x = 0$, $\mu = 0.5$.*

poorest of them all, producing what looks like an approximation to 0. Coarsening the step size to $h = 0.005\pi$ yields the results recorded in Figure 7.2. The Lax–Wendroff solution is now overly smooth and the leapfrog solution has the oscillations more pronounced, but the pattern is not quite right and a phase error is apparent. Coarsening to $h = 0.01\pi$ further clarifies the observations recorded in Figure 7.3. The Lax–Wendroff solution is much too smooth, while the leapfrog solution shows oscillations which lag behind the exact solution. The reason for the qualitatively different behavior of the two methods is that the Lax–Wendroff scheme is dissipative whereas the leapfrog scheme is not.

One can argue either way as to which solution is worse. The Lax–Wendroff (and certainly the Lax–Friedrichs) solution may lead an unsuspecting scientist to conclude that nothing happens when something does, whereas the leapfrog solution may suggest that something happens, but giving the wrong shape and at the wrong location. Again, a method which takes characteristic directions into account more may be required for this problem when the discretization mesh is so coarse as to render the initial conditions nonsmooth. ■

Box scheme dispersion

The box scheme (5.17) is the only scheme discussed in Chapter 5 that is not based on the long centered difference D_0. This long centered difference is not compact and thus may give rise to spurious modes. Such modes arise also when applying multistep methods for ODEs as in Section 2.1, and methods that appropriately suppress them, such as the Adams methods, are called 0-stable. Here we wonder what happens if short differences are used in both space and time, specifically, the compact box scheme.

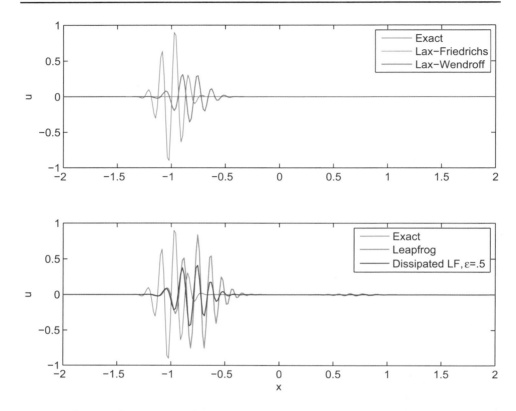

Figure 7.2. *Example 7.1: solving $u_t - u_x = 0$ starting from a tapered sinusoidal wave, using $\mu = 0.5$ and $h = 0.005\pi$. All approximate solutions are visibly wrong at this resolution, with parasitic waves more pronounced for schemes with less dissipativity.*

For the advection equation the scheme reads

$$v_{j+1}^{n+1} + v_j^{n+1} - v_{j+1}^n - v_j^n + \mu a \left(v_{j+1}^{n+1} + v_{j+1}^n - v_j^{n+1} - v_j^n \right) = 0. \tag{7.4}$$

Substituting the Fourier mode for v_j^n as before and centering (7.4) at the middle of the "box," i.e., canceling out $e^{\iota(\omega(n+1/2)k - \xi(j+1/2)h)}$, we obtain

$$\tan(\omega k/2) = \mu a \tan(\xi h/2). \tag{7.5}$$

This dispersion relation is plotted in Figure 7.4. Again $c_3 \approx a \approx C_3$ for low wave numbers, but now we see no parasitic waves! The phase and group velocities do not vanish at high wave numbers, although they are not necessarily correct. This is true for any $\mu > 0$, as befits an implicit scheme. However, it does not mean that there is no dispersion, of course; see Exercise 3. Recall also that the box scheme is conservative and nondissipative.

Example 7.2 We repeat the experiments of Example 7.1 using the box scheme (7.4). The results are displayed in Figure 7.5. The incorrectly located approximate solution profile

7.2. The Wave Equation

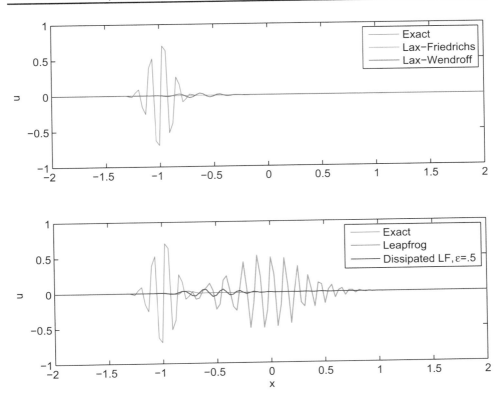

Figure 7.3. *Example 7.1: solving $u_t - u_x = 0$ starting from a tapered sinusoidal wave, using $\mu = 0.5$ and $h = 0.01\pi$. All approximate solutions are visibly wrong at this resolution, with parasitic waves more pronounced for schemes with less dissipativity.*

is unattenuated here, as for the leapfrog scheme in the previous example, although the lag and oscillations are about half as much. The lack of parasitic waves has not made much difference here because these modes are not excited in the previous example either. However, we will see in Section 7.3 an occasion where such spurious modes can become a problem. ∎

7.2 The wave equation

In this section we consider the numerical solution on a uniform mesh of the possibly nonlinear, scalar wave equation

$$\phi_{tt} = c^2 \phi_{xx} - \frac{dV(\phi)}{d\phi}, \quad x_0 \leq x \leq x_{J+1}, \ t > 0, \tag{7.6a}$$

where $c > 0$ is a constant and $V(\phi)$ is a smooth, scalar function. This equation is subject to initial conditions

$$\phi(0, x) = \phi_0(x), \quad \phi_t(0, x) = \phi_1(x), \tag{7.6b}$$

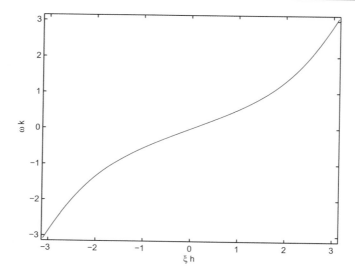

Figure 7.4. *Dispersion relation for the box scheme applied to $u_t + u_x = 0$, $\mu = 0.5$.*

and boundary conditions which may be either periodic,

$$\phi(t, x_0) = \phi(t, x_{J+1}), \qquad (7.6c)$$

or Dirichlet (assume homogeneous for simplicity),

$$\phi(t, x_0) = \phi(t, x_{J+1}) = 0. \qquad (7.6d)$$

As in Section 1.1 the wave equation can be written as a first order system,

$$\mathbf{u}_t + \begin{pmatrix} 0 & c \\ c & 0 \end{pmatrix} \mathbf{u}_x = \mathbf{0}, \qquad \mathbf{u} = \begin{pmatrix} \phi_t \\ -c\phi_x \end{pmatrix}, \qquad (7.7)$$

and it can certainly be discretized this way using the methods introduced in Chapter 5. But handling the initial and boundary conditions becomes less natural and certain conservation laws are not nicely reproduced without staggering the mesh. A more direct treatment results in a very simple method and so we proceed with a direct approach. A more loaded treatment, mathematically speaking, will follow Example 7.3.

Let us denote $V'(\phi) = \frac{dV(\phi)}{d\phi}$. Applying a centered discretization in x yields the semi-discretization

$$\frac{d^2 v_j}{dt^2} = \frac{c^2}{h^2}(v_{j-1} - 2v_j + v_{j+1}) - V'(v_j), \quad j = 1, \ldots, J, \qquad (7.8)$$

with $v_j(0) = \phi_0(x_j)$ and $\frac{dv_j}{dt}(0) = \phi_1(x_j)$ given. Note that unlike D_0 for the first derivative, here the discretization is *compact*. Next, we apply a similar discretization in t as well. This gives the **leapfrog** scheme

$$v_j^{n+1} - 2v_j^n + v_j^{n-1} = c^2 \mu^2 (v_{j-1}^n - 2v_j^n + v_{j+1}^n) - k^2 V'(v_j^n). \qquad (7.9)$$

7.2. The Wave Equation

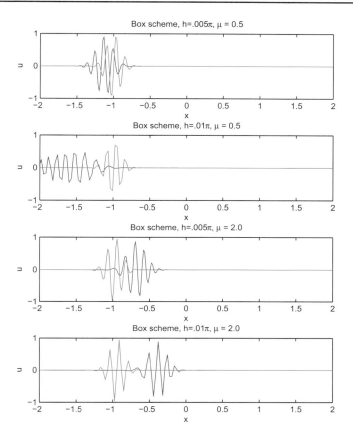

Figure 7.5. *Solving $u_t - u_x = 0$ starting from a tapered sinusoidal wave, using the box scheme with various h and $k = \mu h$. The exact solution at $t = 1$ is plotted in red, and the approximate results are in blue.*

> **Note:** Unlike for the advection equation, the leapfrog scheme is compact and natural for the classical wave equation.

The next step is stability analysis. Let us assume until further notice that $V(\phi)$ is constant, so the low order term $V'(\phi)$ vanishes. The Fourier method yields

$$\hat{v}^{n+1} - 2(1 - 2c^2\mu^2 \sin^2(\zeta/2))\hat{v}^n + \hat{v}^{n-1} = 0.$$

As in Chapter 1 we turn to the characteristic roots of

$$\kappa^2 - 2(1 - 2c^2\mu^2 \sin^2(\zeta/2))\kappa + 1 = 0$$

and find

$$\kappa = 1 - 2c^2\mu^2 \sin^2(\zeta/2) \pm 2c\mu \sin(\zeta/2)\sqrt{c^2\mu^2 \sin^2(\zeta/2) - 1}.$$

If the expression under the square root sign is positive, then there is a root with magnitude greater than 1 and the method is unstable. Otherwise we get two roots of modulus 1: the method is **stable** and **nondissipative** under the CFL condition

$$c\mu \le 1. \tag{7.10}$$

(Actually, our analysis holds only for $c\mu < 1$. But here it can be argued that $c\mu = 1$ is okay as well [159].)

Note that the exact solution of the pure initial value problem is given by[26]

$$\phi(t, x) = \frac{1}{2}[\phi_0(x - ct) + \phi_0(x + ct)] + \frac{1}{2c}\int_{x-ct}^{x+ct} \phi_1(\xi)d\xi. \tag{7.11}$$

Recall that the characteristic curves are given by

$$\frac{dx}{dt} = \pm c.$$

We see that the solution at a point (t, x) is influenced only by the data in the interval $[x - ct, x + ct]$. This forms the famous domain of dependence typical of hyperbolic problems, depicted in Figure 7.6. Recall also Figures 1.7, 1.8, and 1.9 for the advection equation. The CFL condition (7.10) implies that the domain of dependence of the numerical

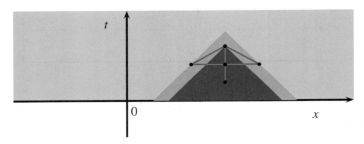

Figure 7.6. *Domain of dependence for* $\phi_{tt} = \phi_{xx}$. *Note that the CFL stability condition implies that the domain of dependence of the PDE is contained in that of the difference scheme.*

scheme must contain that of the PDE. This is an obvious requirement on physical grounds. If it is not met, then arbitrary changes in the data within the PDE domain of dependence, but outside the numerical one, would affect the exact solution arbitrarily without affecting the numerical solution at the top point in Figure 7.6.

For a general first order hyperbolic problem, handling Dirichlet BC requires care and is discussed in Chapter 8. But here it is simple: setting $v_0^n = v_{J+1}^n = 0 \, \forall n \ge 1$, the solution of (7.9), $j = 1, 2, \ldots, J$, is uniquely defined for given initial conditions, regardless of how v_{-l} or v_{J+1+l} are defined for $l \ge 1$. Thus, we can extend the problem into a Cauchy problem by padding with zeros.

---

[26]Again we are interested in investigating properties of a numerical method for a simple model problem to get at the main issues, not to actually obtain the solution for that particular model.

7.2. The Wave Equation

For initial conditions we have

$$v_j^0 = \phi_0(x_j), \qquad (7.12a)$$

but we also need v_j^1 accurate to second order. Using Taylor's expansion gives

$$\phi(k, x) = \phi(0, x) + k\phi_t(0, x) + \frac{k^2}{2}\phi_{tt}(0, x) + O(k^3).$$

Then we apply the semi-discretization (7.8) which defines

$$v_j^1 = \phi_0(x_j) + k\phi_1(x_j) + \frac{\mu^2 c^2}{2}\left(\phi_0(x_{j-1}) - 2\phi_0(x_j) + \phi_0(x_{j+1})\right) \qquad (7.12b)$$

for $j = 1, \ldots, J$. The condition (7.12b) is modified in an obvious way when the full equation (7.6a), with V not constant, is solved.

Example 7.3 Consider the wave equation on the interval $-10 \leq x \leq 10$ with $c = 1$ and the initial conditions

$$\phi(0, x) = e^{-\alpha x^2}, \quad \phi_t(0, x) = 0,$$

where $\alpha > 0$ is a parameter. As evident from (7.11), the exact solution has a particularly simple form for these initial conditions. The initial shape splits into two pulses of a lower amplitude moving toward the boundaries at wave speeds ± 1. With periodic boundary conditions, the waves then get reflected back so $\phi(20l, x) = \phi(0, x)$ for all nonnegative integers l. With homogeneous Dirichlet BC there is also a reflection at the boundaries but the sign is reversed, so the period in time is doubled: $\phi(40l, x) = \phi(0, x)$.

Figure 7.7 displays the exact and approximate solutions for $\alpha = 1$ at $t = 400$. After 20,000 time steps the resulting solution is still very good in that the pulse shape and its location are qualitatively correct. There is no meaningful difference between the errors upon using the BC (7.6c) or (7.6d).

Next consider the sine-Gordon equation, for which we do not know the exact solution, where in (7.6a) we take

$$\frac{dV(\phi)}{d\phi} = \sin(\phi).$$

We use periodic boundary conditions and record the solution at $t = 400$ in Figure 7.8. The solution profile is qualitatively correct based on accuracy checks done by halving h and k and repeating the calculation.

It is tempting to conclude that the leapfrog scheme always performs satisfactorily for the classical wave equation. Alas, upon sharpening the initial profile to $\alpha = 10$, we obtain familiar spurious wiggles in the approximate solution, like those encountered when dispersion mingles with low mesh resolution in Section 7.1; see Figure 7.9. These artificial wiggles spread toward the boundaries as t increases further, and although no solution blowup is observed, the solution profile becomes qualitatively wrong. Computing a solution for the sine-Gordon equation with the sharper profile $\alpha = 10$ would require k and h to be substantially smaller than in Figure 7.8 in order to obtain a meaningful result. ∎

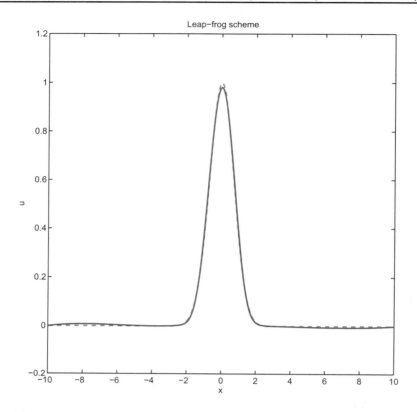

Figure 7.7. *Example 7.3: approximate (solid blue) and exact (dashed red) solutions using leapfrog for the wave equation: $\alpha = 1$, $t = 400$, $k = 0.02$, $h = 0.04$. There is no sign of instability and the maximum error is .019.*

Let us next connect the leapfrog scheme to Hamiltonian ODEs and Chapter 6. Note that the semi-discretization (7.8) can be written as a system

$$\frac{d^2\mathbf{v}}{dt^2} = -B\mathbf{v} - V'(\mathbf{v}). \tag{7.13}$$

(This involves a slight, harmless abuse of notation concerning $V(\mathbf{v})$.) The matrix B is symmetric positive definite, and with the BC (7.6d) it is also tridiagonal; see Exercise 5. Clearly the ODE (7.13) is *Hamiltonian*, i.e., it is in the form (6.1) with

$$H(\mathbf{v}, \mathbf{w}) = \frac{1}{2}\mathbf{w}^T\mathbf{w} + \frac{1}{2}\mathbf{v}^T B\mathbf{v} + V(\mathbf{v}). \tag{7.14}$$

The Hamiltonian H is conserved by the flow (namely, the solution of (7.13)) and defines a symplectic map. The *leapfrog* scheme (7.9) is, in fact, none other than the *Störmer–Verlet* method for (7.13) and hence is a symplectic discretization for it.

7.2. The Wave Equation

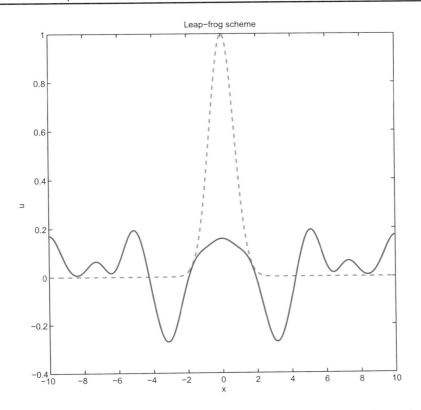

Figure 7.8. *Example 7.3: approximate solution (solid blue) and initial profile (dashed red) using leapfrog for the sine-Gordon wave equation: $\alpha = 1$, $t = 400$, $k = 0.005$, $h = 0.01$.*

Indeed, the semi-discretization (7.8) can be written as a first order ODE system,

$$\frac{dv_j}{dt} = w_j,$$
$$\frac{dw_j}{dt} = \frac{c^2}{h^2}\left(v_{j-1} - 2v_j + v_{j+1}\right) - V'(v_j).$$

A subsequent, *staggered* discretization, as in Section 6.2 and in Example 3.2 but in time, gives

$$\frac{v_j^{n+1} - v_j^n}{k} = w_j^{n+1/2}, \quad j = 1, \ldots, J,$$
$$\frac{w_j^{n+1/2} - w_j^{n-1/2}}{k} = \frac{c^2}{h^2}\left(v_{j-1}^n - 2v_j^n + v_{j+1}^n\right) - V'(v_j^n).$$

This allows elimination of the w-quantities, which yields the leapfrog scheme (7.9).

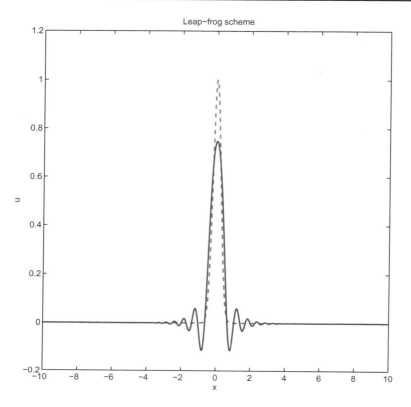

Figure 7.9. *Example 7.3: approximate (solid blue) and exact (dashed red) solutions using leapfrog for the wave equation $\phi_{tt} = \phi_{xx}$: $\alpha = 10$, $t = 400$, $k = 0.01$, $h = 0.02$. The maximum error is .25 and artificial wiggles (or waves) appear.*

Example 7.4 The same approach can be applied to **Maxwell's equations**—see Example 3.6. Now the derivation is less obvious, and it leads to *Yee*'s celebrated 1966 scheme; see [185, 163]. Thus, we have a staggered mesh in space as well as in time. The discretization, which for smooth coefficients has accuracy order (2, 2), reads

$$\mu \left(H1^{n+1/2}_{i,j+\frac{1}{2},k+\frac{1}{2}} - H1^{n-1/2}_{i,j+\frac{1}{2},k+\frac{1}{2}} \right)$$
$$= \frac{k}{h} \left[E2^n_{i,j+\frac{1}{2},k+1} - E2^n_{i,j+\frac{1}{2},k} - E3^n_{i,j+1,k+\frac{1}{2}} - E3^n_{i,j,k+\frac{1}{2}} \right],$$
$$\epsilon \left(E1^{n+1}_{i+\frac{1}{2},j,k} - E1^n_{i+\frac{1}{2},j,k} \right)$$
$$= \frac{k}{h} \left[-H2^{n+1/2}_{i+\frac{1}{2},j,k+1/2} + H2^{n+1/2}_{i+\frac{1}{2},j,k-\frac{1}{2}} + H3^{n+1/2}_{i+\frac{1}{2},j+\frac{1}{2},k} - H3^{n+1/2}_{i+\frac{1}{2},j-\frac{1}{2},k} \right]$$
$$- k\sigma_{i+\frac{1}{2},j,k} E1^{n+1/2}_{i+\frac{1}{2},j,k},$$

7.2. The Wave Equation

etc. Note that here μ is *not* a mesh ratio; rather, it is a parameter (permeability) which appears in the problem statement. The parameter function σ (conductivity) is independent of time. If σ is defined to be a positive constant $\sigma_{i,j,k}$ in each cell as depicted in Figure 3.5, then the value between cells can be defined by the harmonic average

$$\sigma_{i+\frac{1}{2},j,k} = \frac{2}{\sigma_{i,j,k}^{-1} + \sigma_{i+1,j,k}^{-1}}.$$

Also define $E1^{n+1/2} = (E1^n + E1^{n+1})/2$.

Although this looks more complicated than the simple wave equation, it is possible to eliminate the **H**-quantities in a way similar to the elimination of w_j in Example 7.3. We simply write the equations for $E1$, say, at time levels $n+1$ and n and subtract, obtaining at the right-hand-side expressions involving $H2$ and $H3$. The result can be substituted for in terms of **E**-components. We leave this to Exercise 8.

Whether or not **H** is eliminated, we expect a method with order, stability, and long-time characteristics that are similar to that of the leapfrog scheme for the wave equation. ∎

The wave equation is also a *Hamiltonian PDE* [41, 38, 137] and can be seen to satisfy the **local conservation law**,

$$\left[\frac{1}{2}u_1^2 + \frac{1}{2}u_2^2 + V(\phi)\right]_t + [cu_1 u_2]_x = 0, \quad (7.15a)$$

$$\text{where} \quad u_1 = \phi_t, \quad u_2 = -c\phi_x \quad (7.15b)$$

(recall (7.7) and do Exercise 7). This is referred to as a **multisymplectic structure**, but so are other similar concepts; see Exercise 9.

If we are to discretize (7.15), recovering nodal mesh quantities $v_j^n \approx \phi(t_n, x_j)$, then again it makes sense to use a staggered mesh with

$$(u_1)_j^{n+1/2} = \frac{v_j^{n+1} - v_j^n}{k}, \quad (u_2)_{j+1/2}^n = -c\frac{v_{j+1}^n - v_j^n}{h}.$$

Using short differences in (7.15a) as well yields the conclusion that a discrete version of the conservation law is obeyed by the discretization. See also Example 7.5. Further, the intermediate quantities u_1 and u_2 can subsequently be eliminated, and the same leapfrog scheme (7.9) is obtained yet again (Exercise 12).

We conclude that the leapfrog scheme with uniform step sizes k and h satisfies several conservation laws and reproduces certain PDE solution structure. At the same time, the mesh resolution must be sufficiently fine to resolve the real waves, lest artificial ones be introduced due to numerical dispersion.

Example 7.5 (Speech synthesis) This example is a case study based on [174]. It concerns acoustics simulation for articulatory speech synthesis. A widely used physical model for sound propagation in ducts such as occurring in the vocal tract and in wind instruments is a one-dimensional tube described by an area function. Excitations are placed in the tube and their propagation is approximated by a digital ladder (or waveguide) filter. The resulting numerical task is the real-time simulation of wave propagation in a one-dimensional acoustical tube, with a cross-section area that depends on both time and space, and in the presence of damping terms of varying strength.

The mathematical model

Let us then consider a tube of length L and area function $A(t, x)$ with $0 \leq x \leq L$ along the axis of the tube. We assume that the physical quantities of pressure $\hat{p}$, air density $\hat{\rho}$, and air velocity $\hat{u}$ depend only on x and t, and introduce the scaled variables $p(t, x)$ (dimensionless pressure or density deviation) and $u(t, x)$ (volume velocity with dimension of area), by $p = \hat{\rho}/\rho_0 - 1$, $u = A\hat{u}/c$, where ρ_0 is the mass density of air and c is the speed of sound. The linear lossless acoustics of the tube is governed by the equations

$$(u/A)_t + cp_x = 0, \qquad (7.16a)$$
$$(Ap)_t + cu_x = -A_t. \qquad (7.16b)$$

While (7.16) is on solid theoretical footing, realistic models must also account for energy losses. The latter are much harder to describe within a one-dimensional model as essentially higher-dimensional phenomena such as boundary layer effects play an important role here. The damping term we consider is of the form

$$-d(A)u + D(A)u_{xx}.$$

See [174] for further justification and experimentation. We are therefore led to consider the PDE

$$(u/A)_t + cp_x = -d(A)u + D(A)u_{xx}, \qquad (7.17a)$$
$$(Ap)_t + cu_x = -A_t. \qquad (7.17b)$$

Note that near constrictions in the tract the damping coefficients d and D can become large, whereas in other areas they may be small. This gives the model different properties in different spatial regions. For ease of presentation we consider the boundary conditions

$$u(0, t) = u_g(t), \quad p(L, t) = 0, \qquad (7.17c)$$

where u_g is a prescribed volume velocity source.

Note that $A(t, x)$ could be described in practice by a dynamical model, resulting in a dynamical wall model; however, we assume here that the function A is given.

Discretization

The PDE (7.16) corresponds, in case that $A \equiv 1$, to the first order system (7.7) for the classical wave equation (7.6a) with $V \equiv 0$ there. Since A is not constant, and since we have additional terms in (7.17), we want to discretize the latter directly. However, since the damping terms may be very small in some sections of the tube we want to retrieve the leapfrog scheme for (7.6a) in the appropriate limit case.

Thus we use a staggered mesh, both in space and in time. Still assuming $A \equiv 1$, consider for simplicity a uniform mesh in space, $x_0 < x_1 < \cdots < x_J < x_{J+1}$, with

7.2. The Wave Equation

$h = x_{j+1} - x_j$, and likewise a mesh in time $0 = t_0 < t_1 < \cdots$, with $k = t_{n+1} - t_n$. Let also $x_{j+1/2} = (x_j + x_{j+1})/2$ and $t_{n+1/2} = (t_{n+1} + t_n)/2$. We can integrate (7.16a) on a mesh volume in a manner similar to deriving the box scheme in Section 5.2.4. This yields

$$0 = \frac{1}{kh} \int_{t_{n-1/2}}^{t_{n+1/2}} \int_{x_{j-1/2}}^{x_{j+1/2}} (u_t + cp_x) dt dx = \frac{u_j^{n+1/2} - u_j^{n-1/2}}{k} + c \frac{p_{j+1/2}^n - p_{j-1/2}^n}{h},$$

which is exact by the Gauss divergence theorem if we interpret $u_j^{n+1/2}$, etc., as line integrals, e.g.,

$$u_j^{n+1/2} = \frac{1}{h} \int_{x_{j-1/2}}^{x_{j+1/2}} u(t_{n+1/2}, x) dx.$$

This is a second order approximation for the value at edge midpoint $(t_{n+1/2}, x_j)$.

Likewise, we integrate (7.16b) on *another*, nearby mesh volume,

$$0 = \frac{1}{kh} \int_{t_n}^{t_{n+1}} \int_{x_j}^{x_{j+1}} (p_t + cu_x) dt dx = \frac{p_{j+1/2}^{n+1} - p_{j+1/2}^n}{k} + c \frac{u_{j+1}^{n+1/2} - u_j^{n+1/2}}{h};$$

see Figure 7.10.

Again, second order accurate pointwise values can be obtained upon interpretation of these quantities as midpoint values at the edges of the control volume. What results is clearly a staggered mesh scheme, as the volume velocity variables u live at the mesh nodes in space and at half-mesh nodes in time, whereas the pressure variables p live between spatial mesh nodes and at the temporal mesh nodes.

Now, substituting the centered approximations

$$u_j^{n+1/2} = -\frac{\phi_j^{n+1} - \phi_j^n}{ck}, \qquad p_{j+1/2}^n = \frac{\phi_{j+1}^n - \phi_j^n}{h} - 1,$$

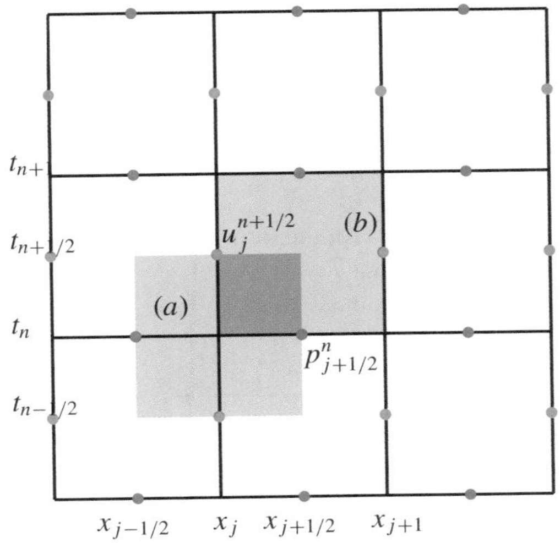

Figure 7.10. *The staggered mesh, control volumes, and unknowns for Example 7.5.*

we can obtain the leapfrog scheme approximating (7.6a) with $V = 0$ there. On our uniform mesh this leapfrog scheme is the usual (7.9). The staggered scheme therefore inherits various conservation properties from the differential system, including symplecticity and multisymplecticity.

Next we use the above as the basis for developing a discretization for the acoustic equation (7.17). The variables u and p continue to live on the same separate space–time mesh locations and are considered constant in the cells around the nodes. The area function A, on the other hand, is given externally and is assumed to be defined everywhere.

Integrating (7.17a) as above and approximating $\frac{1}{A}$ by a piecewise linear function we get

$$\frac{u_j^{n+1/2}/\tilde{A}_j^{n+1/2} - u_j^{n-1/2}/\tilde{A}_j^{n-1/2}}{k} + c\,\frac{p_{j+1/2}^n - p_{j-1/2}^n}{h}$$
$$= \frac{1}{kh}\int_{t_{n-1/2}}^{t_{n+1/2}}\int_{x_{j-1/2}}^{x_{j+1/2}} [-d(A)u(x,t) + D(A)u_{xx}(x,t)]\,dx\,dt$$

with

$$\frac{1}{\tilde{A}_j^{n+1/2}} = \left(\frac{1}{A_{j-1/2}^{n+1/2}} + \frac{2}{A_j^{n+1/2}} + \frac{1}{A_{j+1/2}^{n+1/2}}\right)/4.$$

In order to discretize the volume integral of $-d(A)u$ above we consider u at $t_{n+1/2}$ rather than at t_n, thus obtaining an implicit scheme that remains stable even for very large values of d without requiring very small time steps. This touch of backward Euler turns out to be practically as accurate for the present mathematical model as a more moderate attenuation of the hitherto centered time discretization. We then integrate over x, which yields $-\widehat{d}_j^{n+1/2} u_j^{n+1/2}$, where

$$\widehat{d}_j^{n+1/2} = \left(d\left(A_{j-1/2}^{n+1/2}\right) + 2d\left(A_j^{n+1/2}\right) + d\left(A_{j+1/2}^{n+1/2}\right)\right)/4.$$

For the term $D(A)u_{xx}$ above we first replace u_{xx} with

$$u_j'''^{n+1/2} = \frac{1}{h^2} D_+ D_- u_j^{n+1/2} = \frac{u_{j+1}^{n+1/2} - 2u_j^{n+1/2} + u_{j-1}^{n+1/2}}{h^2}$$

for $j = 1, \ldots, J$. Then we proceed as for the term $d(A)u$, leading to $\widehat{D}_j^{n+1/2} u_j'''^{n+1/2}$, where we used a similar notation for $\widehat{D}$ as for $\widehat{d}$.

We finally obtain

$$\frac{u_j^{n+1/2}/\tilde{A}_j^{n+1/2} - u_j^{n-1/2}/\tilde{A}_j^{n-1/2}}{k} + c\,\frac{p_{j+1/2}^n - p_{j-1/2}^n}{h}$$
$$+ \widehat{d}_j^{n+1/2} u_j^{n+1/2} - \widehat{D}_j^{n+1/2} u_j'''^{n+1/2} = 0. \qquad (7.18a)$$

Likewise, integrating (7.17b) over the cell labeled (b) in Figure 7.10 yields

$$\frac{\widehat{A}_{j+1/2}^{n+1} p_{j+1/2}^{n+1} - \widehat{A}_{j+1/2}^n p_{j+1/2}^n}{k} + c\,\frac{u_{j+1}^{n+1/2} - u_j^{n+1/2}}{h} + \frac{\widehat{A}_{j+1/2}^{n+1} - \widehat{A}_{j+1/2}^n}{k} = 0 \qquad (7.18b)$$

7.2. The Wave Equation

with
$$\widehat{A}_{j+1/2}^n = \left(A_j^n + 2A_{j+1/2}^n + A_{j+1}^n\right)/4.$$

To correspond to the BCs (7.17c) we set $x_0 = 0$, $x_{J+1/2} = L$. At a given time t_n, $n \geq 0$, assume we already know $\{p_{j+1/2}^n\}_{j=0}^J$ and $\{u_j^{n-1/2}\}_{j=0}^J$. For $n = 0$ we obtain this from the initial data. Advancing by one time step then involves the following:

1. Use (7.18a) to obtain $\{u_j^{n+1/2}\}_{j=1}^J$. This involves the inversion of a tridiagonal matrix which can be done rather efficiently. If $D = 0$, then no matrix inversion is needed at all. Note that the tridiagonal system involves only the u nodes.

2. Set $u_0^{n+1/2} = u_g(t_{n+1/2})$ by the left BC.

3. Use (7.18b) to obtain $\{p_{j+1/2}^{n+1}\}_{j=0}^{J-1}$. This can be done explicitly.

4. Set $p_{J+1/2}^{n+1} = 0$ by the right BC.

For stability, the CFL condition must generally hold, i.e.,
$$\frac{ck}{h} < 1.$$

In more general situations we find that stability depends also on the area function $A(t, x)$, and instability may occur sometimes already for ck close to but still below h. Setting $k = .5h/c$, say, is safe. The damping term does not make matters worse here because it is discretized implicitly.

We have assumed above that the length of the spatial interval L is constant. In general, however, this length varies with time, $L = L(t)$. To take care of this we can dynamically change the step size $h = L/(J + 1)$ to correspond to changes in $L(t)$. This is a simple example of a **moving mesh**. We have

$$x_j(t) = \frac{x_j(0)}{L(0)} L(t)$$

and similarly for $x_{j+1/2}(t)$, $j = 1, \ldots, J$. For a corresponding mesh quantity $w_j(t)$ we now have $w_j = w(t, x_j(t))$. Hence

$$\frac{d}{dt} w_j(t) = (w_t)_j + (w_x)_j \frac{d}{dt} x_j(t) = (w_t)_j + (w_x)_j \frac{x_j(0)}{L(0)} \frac{d}{dt} L(t).$$

Generally, an approximation to the rightmost term above must therefore be added in both equations (7.18) reflecting, for example, the fact that $u_j^{n+1/2}$ and $u_j^{n-1/2}$ no longer relate to the same location in space. In the present application, however, fortunately the length $L(t)$ of the tract changes slowly, so this adjustment to the difference equations proves unnecessary in practice.

In [174] there are descriptions of several applications of the scheme developed above for sound synthesis. An online demo can be found on van den Doel's website,

`http://www.cs.ubc.ca/~kvdoel/` ∎

7.3 The KdV equation

This entire section may be viewed as one *case study*. It is based on [12], which in turn extends [11].

The design and development of symplectic methods for Hamiltonian ODEs, to which Chapter 6 is devoted, yields very powerful numerical schemes with beautiful geometric properties. Symplectic and other symmetric methods have been noted for their superior performance, especially for long time integration. Recall that Hamiltonian systems describe, for instance, the motion of frictionless, energy-conserving mechanical systems. Thus, they possess **marginal stability** (see Figure 6.1) which corresponding symplectic numerical schemes mimic. This "living at the edge of stability" is feasible, at least for sufficiently small (and possibly many!) time steps, because of the implied geometric structure that such discrete schemes conserve.

On the other hand it has long been known that conservative discretization schemes for nonlinear, nondissipative PDEs governing wave phenomena tend to become numerically unstable, and dissipation has subsequently been routinely introduced into such numerical schemes, as described in detail in Chapter 5. In particular, conservative difference schemes are known to occasionally yield numerical solutions which at first look fine but at a later time may suddenly explode, as in Example 5.1. See also Example 9.6. Consequently, non-dissipative schemes have generally been discouraged in important sectors of the scientific computing community, especially for long time integration. Typical work on pseudospectra (see, e.g., [168]) when applied to stability studies of ODEs, must also assume that eigenvalues are placed off the imaginary axis and into the left half-plane, so that sufficiently small circles of stability can be drawn around them: in the context of Hamiltonian systems this corresponds to using a slightly dissipative discretization scheme.

Thus, the common beliefs of two established communities seem headed to a clash upon considering symplectic time differencing for Hamiltonian semi-discretizations and multisymplectic discretizations of certain Hamiltonian PDEs. The purpose of this study is to examine this situation numerically, by designing methods for the Korteweg–de Vries (KdV) equation (5.1) reproduced here,

$$u_t + (\alpha u^2 + \nu u_{xx})_x = 0. \tag{7.19}$$

Specifically, we attempt to see whether carefully designed, conservative finite difference and finite volume discretizations can remain stable and deliver sharp solution profiles for a long time. If yes, then to what extent is symplectic structure essential in such methods?

There are several good reasons to use the KdV equation as a prototype for our comparative study:

- It is a model nonlinear hyperbolic equation with smooth solutions for all times.
- It is **nondissipative**, although, unlike the advection and the classical wave equations, it is **dispersive**. To see this, consider the linearization

$$u_t + \rho u_x + \nu u_{xxx} = 0, \qquad -\infty < x < \infty, \; t \geq 0, \tag{7.20}$$

where $\rho = 2\alpha \bar{u}$, with $\bar{u}$ some constant estimate for the size of u. Using the notation of Section 1.1 we write the linear PDE as $u_t = P(\partial_x)u$. The symbol is

$$P(\iota \xi) = \iota(\nu \xi^3 - \rho \xi).$$

7.3. The KdV Equation

This shows that the Cauchy problem is stable but conservative, i.e., as for the advection equation there is no attenuation of high wave number modes. The dispersion relation is given by

$$\omega = \xi(\rho - \nu\xi^2).$$

Dividing by ξ we also see that for $\nu \neq 0$ the phase velocity depends on the wave number (recall Section 7.1); hence this PDE is dispersive.

Thus, the KdV equation is a natural test bed for comparing conservative vs. dissipative discretizations.

- The KdV equation (7.19) is notorious: a lot of previous attention has been devoted to it. Moreover, the unexpected, "nonlinear" instability displayed in Example 5.1 is tantalizing.

- The KdV equation is well known to satisfy a lot of conservation laws. Let us highlight, in particular, three properties:

    - Let $\|u(t)\|^2 = \int u^2(x,t)dx$. Then it is easy to see **norm preservation**

    $$\|u(t)\| = \|u_0\| \quad \forall t.$$

    - It is **a Hamiltonian PDE** (e.g. [129, 38]); i.e., it can be written as

    $$u_t = \frac{\partial}{\partial x}\frac{\delta H}{\delta u}, \quad H = \int \left(\frac{\alpha}{3}u^3 - \frac{\nu}{2}(u_x)^2\right)dx.$$

    - It has a **multisymplectic structure**; see [36, 37, 188, 11] and Exercise 9. This arises from writing it as

    $$L\mathbf{z}_t + K\mathbf{z}_x = \nabla_{\mathbf{z}}S(\mathbf{z}), \tag{7.21}$$

    where K and L are skew-symmetric, constant matrices, and

    $$\mathbf{z} = (\phi, u, v, w)^T,$$
    $$\phi_x = u, \quad v = -\nu u_x, \quad w = v_x + \frac{1}{2}\phi_t - \alpha u^2.$$

> **Note:** Rather than getting technical, let us simply say that *multisymplectic* = *symplectic in both space and time*.

These properties may be used to design different discretizations for (7.19), and the question then is, Which stability and qualitative accuracy properties are obtained?

- When $|\nu| \ll 1$ the PDE (7.19) is somehow close to the Burgers equation, its dispersion limit. But the Burgers equation often develops solutions with shock discontinuities, discussed at length in Chapter 10, whereas the solution of (7.19) remains always smooth. For the Burgers equation a nondissipative scheme cannot avoid developing wiggles similar to those observed in Section 5.2 as soon as a shock forms [97, 173]. Thus, interesting numerical phenomena crop up when approximating (7.19) with small ν using a large spatial step h.

- When $|\nu|$ is not small the term νu_{xxx} leads to stiffness: fast waves can destabilize schemes. Often these fast waves are not resolved in time. Indeed it is easy to see, as in Example 5.1, that an explicit scheme must satisfy the very restrictive condition

$$k = O(|\nu|^{-1}h^3). \qquad (7.22)$$

This is unacceptable in most practical situations.

- Many recent schemes that have appeared in the literature are fully implicit. But this can be cumbersome! On the other hand, explicit schemes are too limited. So, we consider semi-explicit splitting schemes below, where the term νu_{xxx} is discretized implicitly, whereas the rest is discretized explicitly.

- Symplectic discretizations may not be compact, whereas multisymplectic discretizations must be. There is no such difference for the classical wave equation of the previous section, where the same discretization is compact, symplectic, and multi-symplectic. Here we can therefore ask, Is the extra structure preservation worthwhile?

There are also reasons that make the choice of the KdV equation as a prototype for our study less ideal. In particular it is completely integrable, which is uncharacteristic of nonlinear hyperbolic PDEs, and it models only low wave number phenomena in one dimension. But despite the negatives there are sufficiently many positives to make the present study of interest: indeed, it employs much of the accumulated wisdom of the previous chapters as well as the present one.

7.3.1 Schemes based on a classical semi-discretization

The usual semi-discretization in space reads

$$\frac{dv_j}{dt} + \frac{\alpha}{2h}\left(\theta D_0 v_j^2 + 2(1-\theta)v_j D_0 v_j\right) + \frac{\nu}{2h^3} D_0 D_+ D_- v_j = 0, \qquad (7.23)$$

where θ is a parameter.

The choice $\theta = 1$ yields a Hamiltonian system arising from approximating

$$H_\Delta = h \sum_j \left(\frac{\alpha}{3}v_j^3 - \frac{\nu}{2h^2}(v_{j+1} - v_j)^2\right), \qquad (7.24)$$

whereas the choice $\theta = 2/3$ yields discrete norm preservation

$$\|v(t)\|^2 = h \sum_j v_j^2 = \|v(0)\|^2$$

(recall Section 5.3). Thus, one must make a choice regarding which of these two properties to preserve. However, it turns out that usually there is no major qualitative difference in

7.3. The KdV Equation

practice between choosing one of these alternatives or the other. In particular, replacing $\theta = 2/3$ by $\theta = 1$ in the explicit leapfrog scheme of Example 5.1 does not have a significant impact on that scheme's instability.

Full discretization: The midpoint rule

We next consider applying time discretization to the large ODE system (7.23). An explicit scheme is usually too limited as we have seen, and we abandon any such possibility. On the other hand, the implicit *midpoint* rule preserves the properties of the semi-discretization: with $\theta = 1$ it yields a symplectic scheme (because the ODE (7.23) is Hamiltonian and midpoint is a symplectic method), whereas with $\theta = 2/3$ it yields a norm-preserving scheme

$$\|v^n\|^2 = \|v^0\|^2 \quad \forall n$$

(because midpoint reproduces quadratic invariants), and hence stability in the l_2-sense is guaranteed. The scheme reads

$$v_j^{n+1} = v_j^n - \frac{\mu}{2} D_0 \left[\theta \alpha (v_j^{n+1/2})^2 + \frac{\nu}{h^2} D_+ D_- v_j^{n+1/2} \right]$$
$$- (1-\theta) \mu \alpha v_j^{n+1/2} D_0 v_j^{n+1/2}, \quad v_j^{n+1/2} = (v_j^n + v_j^{n+1})/2. \quad (7.25)$$

As before, $\mu = k/h$.

To advance in time using the fully implicit midpoint scheme (7.25) requires the solution of a nonlinear system of algebraic equations for v^{n+1}. Newton's method amounts to (fully) linearizing the nonlinear terms in the equation. This is also called *quasi-linearization*. Thus, with $\hat{v}$ known near v, we write

$$v^2 \approx \hat{v}^2 + 2\hat{v}(v - \hat{v}) = 2\hat{v}v - \hat{v}^2.$$

Also

$$vv_x \approx \hat{v}\hat{v}_x + \hat{v}(v_x - \hat{v}_x) + (v - \hat{v})\hat{v}_x = -\hat{v}\hat{v}_x + \hat{v}v_x + v\hat{v}_x,$$

but to simplify notation we consider only $\theta = 1$. The Newton iteration therefore reads

$$v_j^{n+1,s} = v_j^n - \frac{\mu}{4} D_0 \left(\alpha \left[\left(v_j^n + v_j^{n+1,s-1} \right) \left(v_j^n + v_j^{n+1,s} \right) - \frac{1}{2} \left(v_j^n + v_j^{n+1,s-1} \right)^2 \right] \right.$$
$$\left. + \frac{\nu}{h^2} D_+ D_- \left(v_j^n + v_j^{n+1,s-1} \right) \right) \quad (7.26a)$$

for $s = 1, 2, \ldots$. For the initial iterate we can choose

$$v_j^{n+1,0} = v_j^n. \quad (7.26b)$$

The equations (7.26a) form a linear system for $v^{n+1,s}$. Newton's method is well known to converge (locally) quadratically, and since the initial iterate (7.26b) is $O(k)$ away from the solution v^{n+1}, the first Newton iteration is already $O(k^2)$ away from the solution at step $n + 1$. This may not always be sufficiently accurate, but two Newton iterations are normally more than sufficient. There is a modified Newton technique, widely used in initial value ODE software, which typically necessitates only one Jacobian matrix assembly and decomposition and costs less than two Newton iterations per time step.

In numerical experiments the implicit midpoint scheme stays stable over a wide range of parameters for a long time. For a given discretization mesh it is often one of the more accurate among the schemes considered here. However, it does develop tiny spatial wiggles for relatively large h and k, and the symplectic version can even become unstable and blows up in some extreme cases. The l_2-stability for $\theta = 2/3$ does prevent blowups but it does not guarantee stability for u_x, and we get meaningless sawtooths in corresponding extreme cases. Moreover, when a sharp-sawtooth approximate solution tries to move fast, the nonlinear solver eventually falls apart.

A semi-explicit, symplectic method

Fully implicit schemes are cumbersome and expensive to use. So, consider a symplectic, semi-explicit scheme, where the nonlinear part of the KdV equation is discretized explicitly and the rest is discretized implicitly. This is a **splitting** method (see Section 6.2.2 and also 9.3). Starting from the Hamiltonian semi-discretization $v_t = (2h)^{-1} D_0 \nabla H_\Delta(v)$, we split the Hamiltonian into three parts,

$$H_\Delta = h \sum_j \left(\frac{\alpha}{3}(v_j)^3 - \frac{\nu}{2h^2}(v_{j+1} - v_j)^2 \right)$$

$$= h \left(\sum_j \frac{\alpha}{3}(v_{2j})^3 + \sum_j \frac{\alpha}{3}(v_{2j+1})^3 - \sum_j \frac{\nu}{2h^2}(v_{j+1} - v_j)^2 \right)$$

$$\equiv H_\Delta^1 + H_\Delta^2 + H_\Delta^3.$$

The ODE $v_t = (2h)^{-1} D_0 \nabla H_\Delta^3(v)$ has constant coefficients, so it is discretized by the unconditionally stable implicit midpoint method. This requires setting up and decomposing the matrix just once! For half a time step, let us denote the inverse of this matrix by M. To the rest we apply odd–even splitting; e.g., the ODE $v_t = (2h)^{-1} D_0 \nabla H_\Delta^1(v)$ at even spatial mesh points

$$(v_{2j})_t = \frac{\alpha}{2}((v_{2j+1})^2 - (v_{2j-1})^2), \quad (v_{2j+1})_t = 0,$$

is integrated exactly from t_n to t_{n+1}. Then the same mechanism is applied to the odd part of the mesh.

Our semi-explicit, symplectic scheme is now obtained by a Strang-type (also, Verlet) splitting,

$$U = M v^n;$$
$$U_{odd} = U_{odd} - \frac{\alpha \mu}{4}((U_{odd+1})^2 - (U_{odd-1})^2);$$
$$U_{even} = U_{even} - \frac{\alpha \mu}{2}((U_{even+1})^2 - (U_{even-1})^2);$$
$$U_{odd} = U_{odd} - \frac{\alpha \mu}{4}((U_{odd+1})^2 - (U_{odd-1})^2);$$
$$v^{n+1} = MU. \tag{7.27}$$

7.3. The KdV Equation

A constant coefficient Fourier stability analysis demands restricting the time step to

$$2|\alpha u_{\max}|\mu < 1.$$

As in Chapter 6, note that the first M-partial step at level n follows the last M-partial step at level $n - 1$ and so they can be combined into one partial step with M^2. The operator splitting is thus staggered, which allows retaining the second order accuracy (see Exercise 6.9).

The selling feature of this semi-explicit scheme is that the work per time step is roughly an order of magnitude less than for a fully implicit method. Just how much faster the semi-explicit scheme is depends of course on various factors, including the implementation details of the nonlinear solver for a fully implicit scheme.

To carry out the step defined at each time level in (7.26a) or in a semi-explicit scheme requires in principle solving a 5-diagonal linear system of equations. For the semi-explicit scheme on an infinite spatial mesh we have

$$A\mathbf{v}^{n+1} = \mathbf{b},$$

where

$$A = \begin{pmatrix} \ddots & \ddots & \ddots & & & \\ & \ddots & \ddots & \ddots & \ddots & \\ & & \ddots & \ddots & \ddots & \ddots \\ & & -\eta & 2\eta & 1 & -2\eta & \eta \\ & & & \ddots & \ddots & \ddots & \ddots \\ & & & & \ddots & \ddots & \ddots \\ & & & & & \ddots & \ddots & \ddots \end{pmatrix}, \quad \eta = \frac{\mu\nu}{2h^2}.$$

With periodic boundary conditions, however, we have a matrix of finite size with nonzeros at the corners,

$$A = \begin{pmatrix} 1 & -2\eta & \eta & & & & & & -\eta & 2\eta \\ 2\eta & 1 & -2\eta & \eta & & & & & & -\eta \\ -\eta & 2\eta & 1 & -2\eta & \eta & & & & & \\ & -\eta & 2\eta & 1 & -2\eta & \eta & & & & \\ & & \cdot & \cdot & \cdot & \cdot & \cdot & & & \cdot \\ & & & \cdot & \cdot & \cdot & \cdot & \cdot & & \\ & & & & -\eta & 2\eta & 1 & -2\eta & \eta \\ & & & & & -\eta & 2\eta & 1 & -2\eta \\ \eta & & & & & & -\eta & 2\eta & 1 \\ -2\eta & \eta & & & & & & -\eta & 2\eta & 1 \end{pmatrix}$$

Rapid techniques exist for the solution of such systems of linear equations (a matrix of this type is called **circulant**) using the **fast Fourier transform** (FFT); see [100, 178] and the review in Section 2.12.6.

7.3.2 Box schemes

All the schemes we have seen so far use a noncompact semi-discretization of the form (7.23), so in particular they cannot be multisymplectic.

Hence we wish to apply a compact discretization in both x and t centered at a *cell (box)*, or a finite volume, with the corners

$$(x_j, t_n), \ (x_{j+1}, t_n), \ (x_j, t_{n+1}), \ (x_{j+1}, t_{n+1}).$$

Like the implicit midpoint rule the schemes obtained here are fully implicit, but they are also compact, so we hope to avoid those extra wiggles that may appear when using the noncompact semi-discretization (7.23).

We can start developing a box scheme as in Section 5.2.4, arriving at (5.16) and (5.17), where now

$$f(u) = \alpha u^2 + \nu u_{xx}.$$

But this must be followed with a spatial discretization of u_{xx}. Employing the usual compact three-point discretization for the latter in (5.16b) yields

$$f_j^{n+1/2} = \int_{t_n}^{t_{n+1}} \left[\alpha v_j(t)^2 + h^{-2} \nu D_+ D_- v_j(t) \right] dt.$$

Further, using the trapezoidal rule and plugging the result in (5.16a) gives the discretized term νu_{xxx} as

$$h^{-3} \nu \mu_t \delta_x D_+ D_- v_{j+1/2}^{n+1/2}$$
$$= \frac{\nu}{2h^3} \left[v_{j+2}^{n+1} - 3v_{j+1}^{n+1} + 3v_j^{n+1} - v_{j-1}^{n+1} + v_{j+2}^n - 3v_{j+1}^n + 3v_j^n - v_{j-1}^n \right].$$

The corresponding computational molecule (or stencil) is depicted in Figure 7.11. The control volume is thus expanded to include three mesh cells.

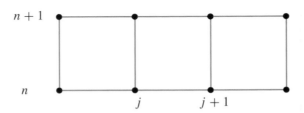

Figure 7.11. *Computational molecule for the KdV box schemes.*

A direct extension of (7.4) now reads

$$k^{-1} \delta_t \mu_x v_{j+1/2}^{n+1/2} + h^{-1} \mu_t \delta_x \left[\alpha (v_{j+1/2}^{n+1/2})^2 + h^{-2} \nu D_+ D_- v_{j+1/2}^{n+1/2} \right] = 0,$$

7.3. The KdV Equation

which we call the **narrow box** scheme.

But now, given that the u_{xxx} term has taken us to the wider stencil depicted in Figure 7.11, we need not restrict ourselves to the narrow cell we had before when discretizing first derivatives. For instance, we can obtain a fourth order approximation for $u_{j+1/2}^n$ in (5.16b) by integrating the cubic interpolant of $u(t_n, x)$ at x_i, $i = j-1, j, j+1, j+2$. For $(u^2)_x$ we can consider the wider stencil, too. Define extrapolating averages as

$$v_{R,j+1/2} = (1-3b)v_{j+1} + bv_{j+2}, \quad v_{L,j+1/2} = (1-3b)v_j + bv_{j-1},$$

where $0 \leq b \leq 1/3$ is a parameter, and use $(v_{R,j+1/2})^2 - (v_{L,j+1/2})^2$ for $u^2(t, x_{j+1/2})_x$. With c another parameter, our general KdV box scheme (in all its gory glory) is then written as

$$\frac{1}{k}\left[2c(v_{j+1}^{n+1} + v_j^{n+1} - v_{j+1}^n - v_j^n) + (1-2c)(v_{j+2}^{n+1} + v_{j-1}^{n+1} - v_{j+2}^n - v_{j-1}^n)\right]$$
$$+ \frac{\alpha}{h}\left[bv_{j+2}^{n+1} + (1-3b)v_{j+1}^{n+1} - (1-3b)v_j^{n+1} - bv_{j-1}^{n+1}\right.$$
$$\left. + bv_{j+2}^n + (1-3b)v_{j+1}^n - (1-3b)v_j^n - bv_{j-1}^n\right] \quad (7.28)$$
$$+ \frac{\nu}{h^3}\left[v_{j+2}^{n+1} - 3v_{j+1}^{n+1} + 3v_j^{n+1} - v_{j-1}^{n+1} + v_{j+2}^n - 3v_{j+1}^n + 3v_j^n - v_{j-1}^n\right] = 0.$$

These are compact, centered, implicit, conservative schemes of order (2, 2). If we choose $c = \frac{13}{24}$, corresponding to the cubic interpolant mentioned earlier, then the accuracy order improves to (2, 4). The choice $b = 0$, $c = 1/2$ yields the narrow box scheme. The choice $b = 1/4$ (hence also $1 - 3b = 1/4$), $c = 3/8$, corresponds to a **multisymplectic box** scheme, derived from the discrete conservation of (7.21); see [188, 11].

With $\rho = 2|\alpha u_{\max}|$ it is possible to show (Exercise 14) that for large values of

$$s = \rho \nu^{-1} h^2$$

and very large μ the scheme becomes unstable unless $b \geq 1/4$.

Having introduced several potentially interesting discretizations we next test them numerically on two examples.

Example 7.6 Consider the same choice of parameters used to demonstrate instability in Example 5.1. Noting that $2|\alpha u_{\max}| \approx 2.6$, we take this to be the Courant number for the semi-explicit method. Thus, we safely set

$$\mu = .2.$$

Here we select $h = .005$, $k = .001$, and record relative errors at $t = 5$ and at $t = 10$ in the discrete Hamiltonian (7.24) (denoted *Error-Ham*) and in the l_2-norm (denoted *Error-norm*). Note that these are often *underestimators* for the maximal pointwise error; see, e.g., Example 6.5. The results are displayed in Table 7.1.

The main observation concerning these results is that the multisymplectic box is the least accurate, and by a significant margin. On the other hand, the semi-explicit, symplectic method shines, given its much smaller computational cost per time step. The narrow box scheme is the most accurate on the given mesh.

Table 7.1. *Error indicators in Hamiltonian and norm for Example 7.6 using* $h = .005$, $k = .001$.

Method	t	Error-Ham	Error-norm
Multisymplectic Box	5	5.0e-1	2.1e-3
Narrow Box	5	6.0e-5	4.9e-4
Semi-explicit Symplectic	5	3.6e-3	8.9e-4
Implicit midpoint	5	8.4e-5	8.9e-4
Multisymplectic Box	10	4.8e-1	2.0e-3
Narrow Box	10	8.6e-5	4.8e-4
Semi-explicit Symplectic	10	3.7e-3	8.7e-4
Implicit midpoint	10	1.2e-4	8.7e-4

Next we apply our discretization schemes using rough resolutions. By this we mean that the discretization step h is too large to enable reconstruction of all the solution details visible at $t = 10$ in Figure 5.1. We keep $\mu = .2$.

At first we set $h = .02$, thus $k = .004$. The solutions are graphed in Figure 7.12. Note the extra wiggles that appear in the solution curves of those schemes that are based on the semi-discretization (7.23). The box schemes do not show such wiggles. The narrow box scheme produces a curve which is less smooth but closer to the exact solution, whereas the multisymplectic box scheme produces a very stable, wiggle-free, not very accurate solution. Exercise 15 attempts to explain all this.

Next we make things really rough by setting $h = 1/30$; hence $k = 1/150$. Now the semi-explicit, symplectic scheme blows up before reaching $t = 10$, and so do the narrow box and the symplectic midpoint schemes. The multisymplectic box scheme is the only one that continues to produce smooth, wavy solution profiles which, although having little to do with the exact solution, remain very stable and could look like the exact solution of another problem of the same type; see Figure 7.13(a) and recall Theorem 6.7.

As mentioned earlier, setting $\theta = 2/3$ rather than $\theta = 1$ in (7.23) and applying implicit midpoint in time produces a scheme which is unconditionally stable in the l_2-sense. For the present example the solution indeed does not blow up in this case, unlike its symplectic comrade. However, it does not look like much more than noise either, even though the l_2-norm is in relative error of only 3.9e-8 (which comes from the accuracy tolerance of the nonlinear solver); see Figure 7.13(b). ∎

Example 7.7 The following setting is also taken from [188]. We choose

$$\nu = -1, \ \alpha = -3, \ \rho = 0,$$
$$u^0(x) = 6\text{sech}^2(x), \ u(-20, t) = u(20, t).$$

After a short while the solution has two solitons, a tall and narrow one (maximum height = 8), and a shorter, wider one. These pulses both move with time to the right,

7.3. The KdV Equation

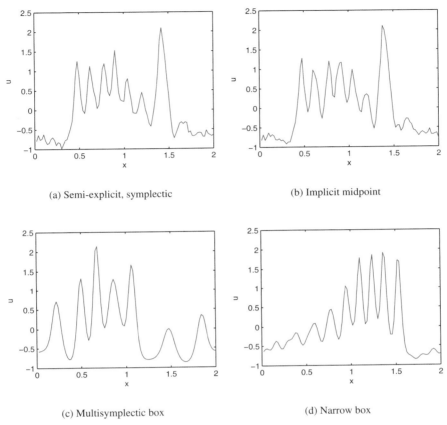

Figure 7.12. *Solutions for Example 7.6 for a rough resolution $h = .02$ at time $t = 10$. Note the wiggles in the noncompact schemes (a) and (b).*

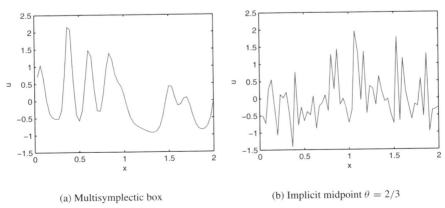

Figure 7.13. *Solutions for Example 7.6 for a very rough resolution $h = 1/30$ at time $t = 10$. These profiles have little to do with the exact solution, yet note the smooth variation in scheme (a) in contrast to the very large spatial derivative in (b).*

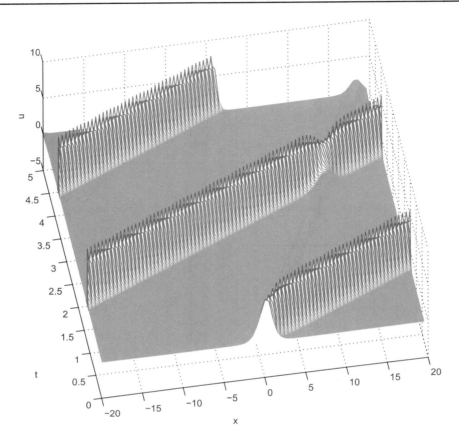

Figure 7.14. *KdV solution, a waterfall plot for Example 7.7. The semi-explicit scheme with $h = .05$, $k = .001$ was used for the calculations, and each 10th mesh value in both time and space (i.e., a sparser mesh by a factor of 100) was utilized for the plot.*

wrapping around as is usual for periodic boundary conditions. But the taller soliton moves faster, hence the two occasionally merge and then split apart again. See Figure 7.14 for a "waterfall" figure depicting the progress of the solitons in time. Figure 7.15 shows full resolution snapshots at times $t = 4$ and $t = 100$.

Turning to the numerical discretization, note that $2|\alpha u_{\max}| \approx 48$. For the semi-explicit scheme we take $h = .05$, $k = .001$.

But the remaining three fully implicit schemes do not share the stability restriction of the semi-explicit scheme. For these we take $h = .05$, $k = .005$. This yields roughly equal CPU time per time step for all schemes. Results are recorded in Table 7.2.

The most striking effect in the results of Table 7.2 is that these error underestimators hardly deteriorate for a long time. Note that with $k = .001$ it takes 100,000 time steps to arrive at $t = 100$. These error underestimators thus appear to be too good to be true! (A similar phenomenon occurs in Example 6.5.) Indeed, the pointwise error does deteriorate: the shape of the solitons is preserved by all schemes other than midpoint with $k = .005$, but their location is not. This well-known numerical dispersion effect is evident in Figure 7.15.

7.3. The KdV Equation

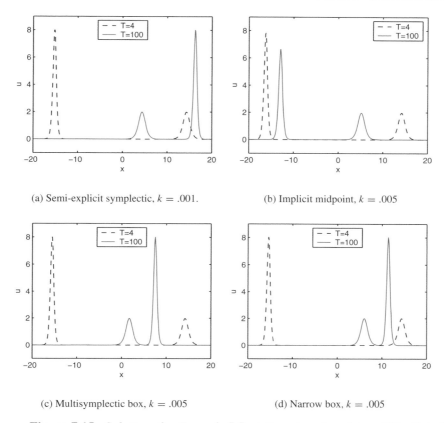

Figure 7.15. *Solutions for Example 7.7 at times $t = 4$ and $t = 100$. Note how crisply all of the conservative schemes except midpoint capture the solitons after a long time integration. However, notice also the significant phase error.*

Further experimentation with finer space–time meshes indicates that the phase shift is significantly the smallest in the semi-explicit plot Figure 7.15(a), where the computation was indeed performed with a smaller time step.

Note how crisply the solitons are captured in Figure 7.15. The correct amplitude is preserved even at $t = 100$ by the semi-explicit, symplectic method (7.27). Exercise 16, and even more so Exercise 17, should convince you that this is no mean feat. ∎

Running the above examples with $c = 13/24$ yields qualitatively similar results to those previously displayed. Yet, choosing $c = 13/24$ appears to improve the time accuracy, especially in terms of conserving the l_2-norm, without compromising stability at all. Here is free lunch.

Thus we end this long exploration with the conclusion that conservative schemes for conservative problems should not be discounted too quickly, although they do require more care in handling. Another conclusion, one that surprised us, is that the winners—the semi-explicit scheme (7.27) and the box schemes (7.28) with $c = 13/24$ and $b = 0$ or $b = 1/4$—are not among "the usual suspects" and did not appear in the literature before

Table 7.2. *Error indicators in Hamiltonian and norm for Example 7.7 using* $h = .05$.

Method	t	k	Error-Ham	Error-norm
Multisymplectic box	4	.005	5.8e-3	8.8e-4
Narrow box	4	.005	1.5e-3	2.3e-4
Semi-explicit symplectic	4	.001	3.5e-4	4.5e-4
Implicit midpoint	4	.005	6.6e-2	2.0e-2
Implicit midpoint	4	.001	1.2e-4	5.0e-4
Multisymplectic box	100	.005	5.8e-3	8.9e-4
Narrow box	100	.005	1.5e-3	2.3e-4
Semi-explicit symplectic	100	.001	3.5e-4	4.5e-4
Implicit midpoint	100	.005	3.7e-1	1.2e-1
Implicit midpoint	100	.001	1.1e-4	5.0e-4

the relatively recent [11, 12]. Finally, whereas these box schemes have been shown to have favorable properties in preserving structure, it seems that their essential ingredient (in addition to being conservative and symmetric) is compactness, i.e., the use of short differencing avoiding D_0 in both space and time.

7.4 Spectral methods

Throughout most of this book we consider low order methods. More precisely, we use low order methods to expose concepts and numerical issues, due to their simplicity and flexibility. But this of course does not mean that the preferred methods to use for a given application are necessarily of low order. The choice of a suitable method depends on the special challenges and accuracy requirements of the application at hand. For many problems accumulated experience suggests that the ideal order of accuracy is not far from four.

In several places of this text we do provide and demonstrate fourth order methods and higher, in both time and space. These methods are all based on local polynomial approximations; see, in particular, Sections 2.1, 2.2, and 3.1. For both multistep methods in Chapter 2 and spatial derivative discretizations in Chapter 3 we have seen that if we insist on a linear difference method, then more mesh points are required for higher order approximations. If we continue to increase the order by including more and more mesh points then eventually one high-order polynomial approximation is obtained. Our approximate solution, say, in one space variable, may then be written for each t as

$$v(t, x) = \sum_{i=1}^{J} \alpha_i(t) \phi_i(x), \qquad (7.29)$$

7.4. Spectral Methods

where $\phi_i(x)$ are *global* basis functions, as in Review Section 2.12.3. Note that t is just a parameter in (7.29). In favorable circumstances this approach can lead to approximations that are more accurate than any finite difference order l, i.e., the approximation error is $O(J^{-l})$ *for every positive finite l*. This is called **spectral accuracy**.

The next observation is that these global basis functions may be chosen suitably for different instances, not necessarily to form a polynomial approximation: whereas locally anything sufficiently smooth looks like a polynomial according to Taylor's theorem, the same does not hold globally. This opens the door to a large variety of methods called *spectral methods*. The present section offers a taste of these methods, their triumphs and limitations. There are several texts and monographs available, including the pioneering Gottlieb and Orszag [71], the comprehensive Canuto et. al. [45], the useful Fornberg [63], and the delightful Trefethen [169].

There are *Galerkin* spectral methods [45], but most people these days concentrate on **collocation** spectral methods. Moreover, for time-dependent PDE problems, typically an ODE discretization method of the sort explored in Chapters 2 and 6 is coupled with a spectral discretization of the spatial derivatives. This is along the lines of the representation (7.29), where the time variable is separated from space variables.

Let us then concentrate on the spatial variable x. We choose global basis functions $\phi_i(x)$ and a discretization mesh $\{x_j\}_{j=1}^J$ and consider forming a $J \times J$ **differentiation matrix** as in Section 2.12.6. (Recall also the implicit difference methods of (3.8) and (3.9).) This operator yields values $v_x(t, x_j)$, $j = 1, \ldots, J$, given values $v_j = v(t, x_j)$, $j = 1, \ldots, J$, according to (7.29).

If the given PDE is subject to periodic BC, as is the case in most examples considered in this chapter (and unlike many other examples that arise in practice), then it is best to use trigonometric polynomial basis functions. The effect of the differentiation matrix is then achieved by an FFT, as described in Section 2.12.6. This is where the spectral accuracy of the resulting spatial discretization over a uniform mesh may imply that we may use fewer mesh points and/or obtain less numerical dispersion!

Example 7.8 Let us return to the classical wave equation and the specifications of Example 7.3. We retain the leapfrog method in time but replace the spatial discretization by $\mathcal{F}^{-1}\left(-\xi^2 \mathcal{F}(v^n)\right)$, where $\mathcal{F}$ denotes the discrete Fourier transform in space. This has the effect of applying the square of the differentiation matrix to the mesh function $v^n = \{v_j^n\}_{j=1}^J$ at the current time level. The scheme in time and space is then

$$v^{n+1} = v^{n-1} + 2v^n + k^2 \mathcal{F}^{-1}\left(-\xi^2 \mathcal{F}(v^n)\right). \tag{7.30}$$

Numerical results with time step sizes corresponding to those in Example 7.3 are recorded in Table 7.3. For $\alpha = 1$ the recorded error is due only to the time discretization, and for $J \geq 128$ there is no difference in the plots of exact and approximate solutions as far as the eye can tell. This also holds for much smaller J, although then care must be taken to plot the results using a more judicious interpolation than the standard piecewise linear (i.e., "broken line") method. For $J = 1024$ ($h \approx .02$) the scheme blows up because a stability restriction on the time step is violated.

For the nonlinear sine-Gordon equation we subtract $k^2 \sin(v^n)$ from the right-hand side of (7.30). A plot that is qualitatively similar to Figure 7.8 is obtained with $k = .005$ and $J = 128$. The corresponding spatial step $h \approx .16$ is far coarser than in Figure 7.8.

Table 7.3. *Maximum errors at $t = 400$ using the spectral method (7.30) with time step size k.*

	$J = 64$	$J = 128$	$J = 256$	$J = 512$
$\alpha = 1, k = .02$	3.4e-3	3.4e-3	3.4e-3	3.4e-3
$\alpha = 10, k = .01$	8.9e-1	9.6e-2	9.6e-2	9.6e-2

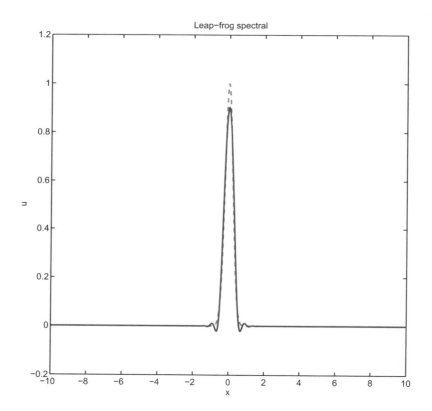

Figure 7.16. *Example 7.8: approximate (solid blue) and exact (dashed red) solutions using the spectral leapfrog method for the wave equation $\phi_{tt} = \phi_{xx}$: $\alpha = 10$, $t = 400$, $k = 0.01$, $h = 20/256$. The maximum error is .096 and artificial wiggles (or waves) appear.*

For $\alpha = 10$ at $J \geq 128$ the solution plot depicted in Figure 7.16 has visible dispersion, but it is less pronounced than in Figure 7.9. This dispersion arises mainly due to the time differencing. For $J = 64$ the results are disastrous as the spurious waves spread throughout the domain; the corresponding finite difference method of Example 7.3 deteriorates more gracefully as h is increased.

7.4. Spectral Methods

In summary, the spectral method does very well in the setup of the present example. Of course, the cost of a double FFT is higher for the same number of spatial mesh points than the cost of calculating a finite difference array, but when J is 10 times smaller a more efficient and capacious method results. ∎

If the boundary conditions are not periodic, then the trigonometric polynomial approximation is not suitable. In such a case we resort to a high degree polynomial approximation. But this requires care, both in terms of choosing the basis functions describing the space of polynomials of degree $< J$ and in terms of selecting the collocation points. If we choose them equi-spaced as in the Fourier case, then rather large interpolation errors may result using a very large degree polynomial. Recalling Sections 2.12.2 and 2.12.3, it is much preferable to use the **Chebyshev points**. On the interval $[-1, 1]$ and including the boundary points, these are given by

$$x_j = \cos\left(\frac{\pi j}{J}\right), \quad j = 0, 1, \ldots, J.$$

Now we can construct $(J+1) \times (J+1)$ differentiation matrices based on polynomial interpolation of degree J at these points, with the usual scaling and translation applying for any interval $[a, b]$ other than $[-1, 1]$.

Moreover, the effect of the Chebyshev differentiation matrix can be obtained using FFTs as well; see, e.g., Chapter 8 of [169]. Note also that upon obtaining a solution on this mesh, a polynomial evaluation is required to plot the solution reasonably on some possibly uniform, denser plotting mesh.

Depending on the BC, then, we may use Chebyshev spectral collocation or Fourier spectral collocation to replace the spatial derivatives of our time-dependent PDE. This can be easily done also in more than one space dimension on a rectangular (or box in three dimensions) domain. The resulting ODE system in time is usually discretized as above by finite differences. The eigenvalues of the resulting differentiation matrix imply a restriction on the time step yielding absolute stability; see Section 2.5. This *necessary* condition for stability is often close enough to being practically sufficient as well. In Example 7.8 we saw that k must be less than $h = 20/J$ times some factor smaller than 1 for the explicit leapfrog. For Chebyshev collocation, unfortunately, often the time step size restriction is much more severe than for trigonometric polynomial collocation. However, spectral method aficionados shrug this off because due to accuracy considerations the time step size k is usually taken much smaller than $(b-a)/J$ anyway. One could even think of an explicit method for the KdV equation with ν not small [63, 169].

The basic disadvantages of spectral methods are clear as well. As explained in basic texts on numerical methods, there is a reason why standard robust general-purpose interpolation software never uses high degree polynomials! And this reason is relevant here. Thus, spectral methods work well if the solution and the coefficient functions used to define the PDE are very smooth. For the speech synthesis Example 7.5, for instance, this would *not* be the case. Similarly, complex geometries and detailed boundaries in more than one space variable would cause spectral methods to collapse. However, for many models considered by applied mathematicians and for many other mathematical models that arise in the physical and engineering sciences, conditions are right for spectral methods to shine.

7.5 Lagrangian methods

This chapter and Chapter 5 highlight not only strengths but also some limitations in finite difference (and finite volume and finite element) discretizations for PDEs of the hyperbolic type: if we want to mimic the conservative nature of such problems, even in the presence of discontinuities or of steep solution features that are not well resolved by the mesh, then unpleasant surprises may arise. If instead we introduce lots of artificial dissipativity or viscosity, then the problem of oversmoothing arises, as we have seen when using the Lax–Friedrichs scheme in Figures 5.4, 5.5, 7.2, and 7.3. Section 5.2.3 and Chapter 10 offer techniques based on upwinding that do improve upon the Lax–Friedrichs performance. Moreover, more accurate simulations using higher order methods are less susceptible to significant dispersion; see, e.g., Section 7.4. Still, the essential dilemma remains: we really do not know how to faithfully integrate complex systems of hyperbolic type in several space dimensions over funny geometries for a long time at a reasonable price.

Some of the trouble appears to be inherent in methods that utilize a mesh. This is an **Eulerian approach**: one sits at a fixed mesh point and observes the flow that passes by. Instead, a **Lagrangian approach** would have the observer watch the world while sitting on a particle that is part of the flow. This leads to **particle methods**. But unlike Example 2.9 that deals with cloth simulation, here there may not be an underlying mesh at all. Such methods are therefore also referred to as **meshless**.

In Lagrangian methods the material derivative $\frac{D\mathbf{u}}{Dt}$ defined in Example 5.4 becomes an ordinary derivative: $\frac{du_i}{dt}$ for the ith particle. Specifying the way these particles interact with neighboring particles and with the boundary then yields a set of ODEs to be solved in time.

However, the "law of preservation of difficulty" is at work here, too. Maintaining everything on a prespecified mesh as in the Eulerian approach makes life much easier in many aspects, including programming, evaluating, and displaying the resulting solution and assessing the quality of the obtained solution in rigorous terms. Using a Lagrangian method instead, we are left without any of these advantages.

Specifying just precisely how these particles interact yields different methods. One class of such techniques especially popular in the astrophysics community is called **smooth particle hydrodynamics** (SPH). We shall not dwell on this much here; see Monaghan [136] for a recent, long review of these methods (which originated in the 1970s) and [64] for a large state-of-the-art hydrodynamic code implementing it for astrophysical purposes. Another method of this sort is called **material point**; see [162] and references therein. We do note, in general, that some significant tweaking may be required to make such methods work, and theory for them is at a generally unsatisfactory state; yet they can occasionally be made to work to an impressive effect.

7.6 Exercises

0. **Review questions**

    (a) Define dispersion, phase velocity, and group velocity for a linear PDE.

    (b) Is the advection equation dispersive? Is the wave equation? Is the KdV equation?

7.6. Exercises

(c) Which of the following schemes for the advection equation is dispersive: Lax–Wendroff, leapfrog, box?

(d) What is the effect of dissipativity on dispersive numerical schemes?

(e) Show that the centered three-point semi-discretization of the classical wave equation is Hamiltonian and that the subsequent application of the Verlet method yields the leapfrog scheme.

(f) What is the sine-Gordon equation? Answer for it the previous question.

(g) Name as many properties of the KdV equation as you can that the wave equation does not have.

(h) Why is the semi-explicit method derived for the KdV equation symplectic but not multisymplectic?

(i) Write down specifically the narrow box and the multisymplectic box schemes for the KdV equation (7.19).

(j) Give an example for which use of a spectral method is recommended and an example for which it is not recommended.

1. Use the dispersion relation (7.2) to show that for the Lax–Wendroff scheme applied to (7.1) the phase velocity satisfies

$$c_2 \approx a \left[1 - \frac{1}{6}(\xi h)^2 (1 - \mu^2) \right]$$

when $|\xi|h \ll 1$. (See [133].)

2. Prove the dispersion relation (7.3) for the leapfrog scheme.

3. For the box scheme (7.4):

   (a) Prove the dispersion relation (7.5).

   (b) Setting $a = 1$, calculate (approximately) phase and group velocities at $\pi - h$ and at $\pi/2$, for $\mu = 0.5$ and for $\mu = 2.0$. What do you observe? Try to explain Figure 7.5.

4. The advection equation with variable coefficient

$$u_t + a(x)u_x = 0, \quad a(x) = .2 + \sin^2(x - 1),$$

for $0 \leq x \leq 2\pi$, $t \geq 0$, with periodic BC and $u_0(x) = e^{-100(x-1)^2}$, was used to create the cover page of [169] and shows the initial pulse propagating with variable speed. The corresponding program (called p6) is available at

http://www.comlab.ox.ac.uk/oucl/work/nick.trefethen

A spectral discretization with $J = 128$ points was used in space and the leapfrog scheme of Exercise 2.4 (see also (7.30)) was applied in time with $k = h/4$.

Write a program that solves this problem using the same time discretization and the following space discretizations: (i) the centered second order discretization of u_x that creates the usual leapfrog scheme, and (ii) the centered five-point, fourth order scheme derived in Section 3.1.1. Add also a variant of order (4, 4) that combines the five-point discretization in space with RK4 in time: for the latter you may set $k = h$.

Try your program for $J = 128$, $J = 192$, and $J = 256$ using $h = 2\pi/J$. Explain your observations.

5. Consider the semi-discretization (7.8) for the wave equation (7.6a) with V constant and Dirichlet BC (7.6d).

   (a) Write this as an ODE system
   $$\frac{d^2\mathbf{v}}{dt^2} = -B\mathbf{v}$$
   with B a tridiagonal, symmetric positive definite matrix.

   (b) Find a nonsingular matrix C such that
   $$B = C^T C.$$

   (c) Define $\mathbf{w}$ by
   $$C^T \mathbf{w} = \frac{d\mathbf{v}}{dt}$$
   and show that the semi-discretization can be written as
   $$\begin{pmatrix} \mathbf{v} \\ \mathbf{w} \end{pmatrix}_t = L \begin{pmatrix} \mathbf{v} \\ \mathbf{w} \end{pmatrix}$$
   with L skew-symmetric. Conclude that
   $$\|\mathbf{v}(t)\|^2 + \|\mathbf{w}(t)\|^2 = \|\mathbf{v}(0)\|^2 + \|\mathbf{w}(0)\|^2 \quad \forall t \geq 0.$$

   (d) Repeat the numerical experiments reported in Example 7.3 with V constant, using the leapfrog scheme and measuring
   $$E_n = [\|\mathbf{v}^n\|^2 + \|\mathbf{w}^n\|^2]^{1/2} - [\|\mathbf{v}^0\|^2 + \|\mathbf{w}^0\|^2]^{1/2}.$$
   You will need to define $\mathbf{w}^n$ sensibly and locally for this, based on the computed mesh values of v in time and space. Plot E_n vs. t_n for $\alpha = 1$ and $\alpha = 10$. What are your conclusions?

6. For the system of conservation laws (5.4) we can define the two-step, staggered mesh scheme
   $$\mathbf{v}^{n+1/2}_{j+1/2} = \mathbf{v}^{n-1/2}_{j+1/2} - \mu\left(\mathbf{f}(\mathbf{v}^n_{j+1}) - \mathbf{f}(\mathbf{v}^n_j)\right),$$
   $$\mathbf{v}^{n+1}_j = \mathbf{v}^n_j - \mu\left(\mathbf{f}(\mathbf{v}^{n+1/2}_{j+1/2}) - \mathbf{f}(\mathbf{v}^{n+1/2}_{j-1/2})\right),$$
   where $\mu = k/h$.

7.6. Exercises

(a) Show that this scheme is second order accurate in both k and h.

(b) Apply a Fourier analysis to the frozen-coefficient problem. Is the scheme dissipative? Is it conservative?

(c) Show that when applied to the simple advection equation

$$u_t = u_x,$$

this scheme can be considered as equivalent to the leapfrog scheme applied to the classical wave equation $u_{tt} = u_{xx}$.

(d) Apply the scheme to the simple advection equation with the initial data functions of Examples 5.6 and 7.1. Use $\mu = 0.5$ and try $h = 0.01\pi$ and $h = 0.005\pi$. Plot your results and compare them to Figures 5.4, 7.2, and 7.3. What are your conclusions?

7. Show that the wave equation (7.6a) satisfies the local conservation law (7.15).

8. (a) Write down the full set of algebraic equations discretizing Maxwell's equations as in Example 7.4 (and Example 3.6).

(b) Eliminate the **H**-unknowns, obtaining equations in **E** alone. What do these approximate?

9. A *multisymplectic PDE* has been defined by Bridges [36, 37] to be one that can be written in the form (7.21), repeated here,

$$L\mathbf{z}_t + K\mathbf{z}_x = \nabla_\mathbf{z} S(\mathbf{z}),$$

where K and L are constant, skew-symmetric matrices.

(a) Show that the wave equation (7.6a) can be written in this form.

(b) Show that the KdV equation (7.19) can be written in this form. (See the lines following (7.21) for a choice of **z**.)

10. The problem (5.28) of Exercise 5.14 reduces for $\nu = 0$ to the classical wave equation written in first order form. The methods and considerations of Section 7.2 are therefore applicable.

(a) Design a compact, centered, second order discretization on a staggered mesh for the PDE system (5.28).

(b) Repeat the experiments of Exercise 5.14 with the new scheme. What are your observations regarding stability and dispersion?

(You should see some improvements with the current scheme despite its lower accuracy order.)

11. This "exercise" describes a potential term project.

One of the most widely used variants in the class of *Boussinesq systems* was first proposed by Boussinesq himself (see the wave bible Whitham [182, p. 462]). It is given by

$$w_t + (uw)_x = 0, \tag{7.31a}$$

$$u_t + \left(w + \frac{1}{2}u^2\right)_x = -\nu w_{xtt}. \tag{7.31b}$$

Here (w, u) are scalar-valued functions of (t, x) with $w(t, x)$ required to be positive for all (t, x). For shallow water flow in a one-dimensional channel aligned in the x-direction, $w(t, x)$ is the height of the water surface above the channel bottom and $u(t, x)$ is the average velocity in the x-direction.

Instead of w often the elevation $\eta(t, x)$ of the free surface from the water column at rest w_0 is employed. Writing $w = w_0 + \eta$, and assuming for simplicity that $w_0 \equiv 1$ and $|\eta| \ll 1$, we obtain

$$\eta_t + u_x + (u\eta)_x = 0, \tag{7.32a}$$

$$u_t + \left(\eta + \frac{1}{2}u^2\right)_x = -\nu \eta_{xtt}. \tag{7.32b}$$

Moreover, from (7.32a) $\eta_{ttx} \approx -u_{txx}$, so another nearby Boussinesq system is [180, 29, 138]

$$\eta_t + u_x + (u\eta)_x = 0, \tag{7.33a}$$

$$u_t + \left(\eta + \frac{1}{2}u^2\right)_x = \nu u_{xxt}. \tag{7.33b}$$

(a) Show that the PDE system (7.31) is multisymplectic (as per the notation in Exercise 9) for

$$\mathbf{z} = (w, q, \psi, \phi, r, p)^T,$$

$$u = \phi_x, \quad q = uw, \quad \psi = w_t, \quad w = r_x, \quad p = w + \phi_t + \nu \psi_t + \frac{1}{2}u^2.$$

Hint: Construct K and L with

$$S(\mathbf{z}) = pw - \frac{1}{2}w^2 + \frac{1}{2}q^2/w - \frac{\nu}{2}\psi^2.$$

(b) Apply the box scheme to the obtained PDE form. This yields a multisymplectic discretization. Then proceed to eliminate all auxiliary unknowns, resulting in the full discretization

$$\delta_t M_x w + \delta_x M_t (uw) = 0, \tag{7.34a}$$

$$\delta_t M_t M_x u + \delta_x M_t^2 \left(w + \frac{1}{2}u^2\right) = -\nu \delta_x \delta_t^2 w. \tag{7.34b}$$

7.6. Exercises

Note that the first of these is centered at $(j+1/2, n+1/2)$, whereas the second is centered at $(j+1/2, n)$. This formula is certainly compact, spanning one spatial mesh subinterval and two time steps. Show that it is implicit with a 4-diagonal Jacobian matrix.

(c) Investigate the numerical properties of the scheme (7.34). Design a numerical experiment and compare its performance to other schemes that you may either dream up or look up in the literature.

(d) Design a numerical scheme for the system (7.33). Discuss computational issues and compare its performance with the scheme for (7.31) and (7.32).

Hint: Recalling Exercises 10 and 5.14 may help.

12. Carry out the staggered mesh formulation outlined following (7.15) and obtain the leapfrog method for the wave equation.

13. Suppose that the wave equation with $V' = 0$ is defined in a spatial domain $[-\pi, \pi]$ and that at $x = 0$ there is a (virtual) "wall" such that the spatial step, h_-, at $x < 0$ is half that of the spatial step, h_+, at $x \geq 0$.

    (a) Recalling the staggered mesh underlying the leapfrog scheme, carefully extend the method for this case of a nonuniform mesh.

    (b) Design a box scheme for the wave equation and show that it is not affected at all by a nonuniform spatial mesh.

14. Let k/h increase unboundedly in (7.28) when applied to the linearized KdV (7.20), and consider the scheme corresponding to a steady state $u_t = 0$.

    (a) Show that the obtained linear difference scheme has the characteristic polynomial
    $$p(\zeta) = (\zeta - 1)[(sb + 1)(\zeta^2 + 1) + (s(1 - 2b) - 2)\zeta],$$
    where $s = \rho v^{-1} h^2$.

    (b) Show that all roots of the characteristic polynomial are in the unit circle, hence that the scheme is stable in this limit, if
    $$\begin{cases} s < \frac{4}{1-4b}, & b < 1/4, \\ \text{unconditionally}, & b \geq 1/4. \end{cases}$$

15. Perform a dispersion analysis for the various KdV schemes used in Section 7.3. Show that schemes based on the semi-explicit discretization (7.23) all support spurious low-frequency modes corresponding to high wave numbers, whereas the box schemes do not support such modes.

16. The semi-explicit splitting framework can be used to construct other schemes, too, and not necessarily conservative ones, either. Thus, (7.27) is replaced by applying $U = Mv^n$, followed by solving the Burgers equation (i.e., (7.19) with $\nu = 0$) for one

time step, followed by another application of $v^{n+1} = MU$, where M is the inverse of the solution operator for half a step of (7.19) with $\alpha = 0$.

Consider applying the classical RK4 to the semi-discretization of Burgers's equation (i.e., (7.23) with $\nu = 0$) in this context. Repeat the experiments of Examples 7.6 and 7.7. Can you recover the soliton shapes as well as the conservative schemes do in Figure 7.15, even for $t = 100$?

17. Repeat Exercise 16 where the Burgers equation is now approximated by a one-sided scheme,
$$U_j = U_j - \mu\alpha \begin{cases} (U_{j+1})^2 - (U_j)^2, & U_j < 0, \\ (U_j)^2 - (U_{j-1})^2, & U_j \geq 0. \end{cases}$$
This is an instance of a **semi-Lagrangian** method.

18. This "exercise" describes a potential term project.

    The *Kuramoto–Sivashinsky* (KS) equation
    $$u_t = -u_{xxxx} - (u_{xx} + uu_x) \tag{7.35}$$
    has attracted much attention for several decades among researchers interested in dynamical systems and reaction-diffusion problems [112, 99]. It admits chaotic solutions. The nonlinear term transfers energy from low wave numbers to high wave numbers, and in the latter regime the fourth order term acts to stabilize while the second order term acts to destabilize.

    In [105] there is an impressive picture of a solution for the case of periodic BCs on $[0, 32\pi]$ and initial data
    $$u_0(x) = \cos(x/16)(1 + \sin(x/16)),$$
    together with the program used to generate that figure. The method used couples elements briefly described in Sections 7.4 and 9.3.3.

    Your task is to investigate the behavior of this problem further, using the above references and others, and to devise numerical methods for its solution that are different from the one given in [105], along the lines of the methods developed in this chapter and in Section 9.3. Notice that here, as in the KdV equation, the leading terms of the PDE are pleasantly constant coefficient ones, so some form of splitting comes to mind.

    Test your methods on the above problem and also using the initial conditions
    $$u_0(x) = e^{-(x-16\pi)^2}.$$

    Discuss your findings.

Chapter 8

More on Handling Boundary Conditions

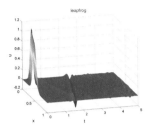

In previous chapters we have steered clear of potentially problematic boundary conditions (BC). We have dealt mostly with pure initial value problems, but we also discussed **periodic** boundary conditions for hyperbolic-type equations, particularly in Example 1.2 and Section 7.3.1. Furthermore, in Section 3.1.3 we handled **Neumann** and **Dirichlet** BC in simple situations. **Ghost unknowns** are introduced for both periodic and Neumann BC; see, e.g., Figure 3.6.

This, of course, does not account for all situations related to BC that may arise in practice. In this chapter we discuss handling other such issues, mainly how to prescribe the boundary conditions in some contexts. We also consider the effect on stability. We still work essentially in one space dimension, although several ideas do extend to multidimensional problems.

As usual, life is simpler for parabolic problems, and we discuss these in Section 8.1. Here we concentrate in particular on handling inconsistencies that often arise at the domain corners where initial and boundary data meet.

Hyperbolic problems may yield more difficulties, or perhaps unexpected challenges, and we discuss these in Section 8.2, which is by far the longest section in the present chapter. Essentially, the problem is to prescribe BC in a manner that avoids unwanted phenomena such as spurious error waves reflecting off the boundary back into the domain of integration. Other phenomena such as order reduction and stability may also arise from the meeting of the different operators that apply inside the domain and on the boundary, on both continuous and discrete levels.

Finally, in Section 8.3 we briefly consider the approximation of infinite domains by finite ones, which involves prescribing appropriate BC.

8.1 Parabolic problems

The basics of handling BC for parabolic problems are covered already in Section 3.1.3. We can generalize and abstract this as follows. For the general difference scheme (3.27c) reproduced here,

$$\sum_{i=-l}^{r} \gamma_i v_{j+i}^{n+1} = \sum_{i=-l}^{r} \beta_i v_{j+i}^n,$$

253

the range of j where the scheme is applied depends on the type of BC, on the scheme used to approximate them, and on the extent of the computational molecule l, r. If we use some fixed linear extrapolation,

$$v_j^n = \sum_{i=1}^{q} b_{ji} v_i^n + f_j, \quad j = 0, \ldots, -r, \tag{8.1}$$

to approximate the BC at the left boundary, and similarly at the right boundary, then the scheme can be written in the matrix form (3.27b) or (3.27d), i.e.,

$$Q_1 v^{n+1} = Q_0 v^n, \quad Q = Q_1^{-1} Q_0,$$

where Q is a large but finite matrix incorporating the BC. In the explicit case, $Q = Q_0$ is a banded matrix with the β_i on its diagonals ($i = -l, \ldots, r$), and with two blocks involving additional rows corresponding to the BC added at the top left and at the bottom right.

The stability condition is still

$$\|Q\| \leq 1,$$

and a necessary condition for this to hold is

$$\rho(Q) \leq 1.$$

The general analysis of stability is complicated and can be found in Varah [176]. The overall conclusion is simple and comforting, though, saying basically that barring weird situations there are no instability traps to beware of.

> **Note:** For parabolic PDEs, sensible consistent BC approximations turn out to be stable.

It often occurs in practice that initial and boundary values do not agree at their points of meeting, say, $(t, x) = (0, 0)$ and $(0, 1)$. It makes sense then to interpret the boundary conditions as holding only for $t > 0$, to avoid ambiguity. The jumps at the corners are smoothed by the diffusion operator as soon as we step into the interior of the domain of definition.

Example 8.1 (Adsorption in a finite bath) This chemical engineering model for multi-component adsorption from a finite bath is due to Liapis and Rippin [120]. We use it as a case study, although we stop short of calculating any numerical solutions. Rather, we emphasize the use of the knowledge and experience accumulated hitherto in order to design promising numerical discretizations. Exercise 1 then completes the computational task. Alternatively, see Viera and Biscaia [177].

The mathematical model consists of a system of parabolic PDEs,

$$\frac{\partial y_i}{\partial t} + \frac{\partial z_i}{\partial t} = \frac{\lambda_i}{r^2} \frac{\partial}{\partial r} \left(r^2 \frac{\partial y_i}{\partial r} \right), \tag{8.2a}$$

$$z_i = k_i(\mathbf{y}) y_i, \quad t > 0, \ 0 \leq r \leq 1, \tag{8.2b}$$

$$\frac{du_i}{dt} = -3\alpha \lambda_i \frac{\partial y_i}{\partial r}\bigg|_{r=1}, \quad i = 1, \ldots, N_e. \tag{8.2c}$$

8.1. Parabolic Problems

The z_i are known functions of $\mathbf{y} = (y_1, \ldots, y_{N_e})$. The constants λ_i and α are known. The unknowns $u_i(t)$ basically "live" on the spatial boundary $r = 1$.

There are initial conditions,

$$y_i(0, r) = z_i(0, r) = 0, \quad u_i(0) = u_{i0} > 0, \tag{8.3a}$$

boundary conditions at $r = 0$ arising from symmetry,

$$\frac{\partial y_i}{\partial r}(t, 0) = 0, \tag{8.3b}$$

and boundary conditions at $r = 1$,

$$\frac{\partial y_i}{\partial r}(t, 1) = \beta_i [u_i(t) - y_i(t, 1)]. \tag{8.3c}$$

For actual values of the various parameters and functions mentioned above, see Exercise 1. This model features relatively large time derivatives near the initial time, corresponding to higher mass transfer rates for the separation of mixture components occurring at the beginning of the separation process [177].

Let

$$\bar{y}_i(t) = 3 \int_0^1 r^2 y_i(r, t) dr,$$

$$\bar{z}_i(t) = 3 \int_0^1 r^2 z_i(r, t) dr.$$

Then multiplying (8.2a) by r^2 and integrating with respect to r yields

$$\frac{d\bar{y}_i}{dt} + \frac{d\bar{z}_i}{dt} = 3\lambda_i \frac{\partial y_i}{\partial r}\bigg|_{r=1} = -\alpha^{-1} \frac{du_i}{dt}.$$

Thus, the global mass balance

$$u_i(t) + \alpha(\bar{y}_i(t) + \bar{z}_i(t)) = u_{i0} \tag{8.4}$$

is an (integral) invariant of the system.

An inconsistency occurs at the corner $(0, 1)$. Crawling in r along the initial line $t = 0$ we have from (8.3a), $\frac{\partial y_i(0, r)}{\partial r}|_{r=1} = 0$, as well as $y(0, 1) = 0$; and yet $u_i(0) \neq 0$, so (8.3c) cannot hold at the corner. But the entire problem is nice and parabolic, so any boundary jump is immediately smoothed inside the PDE domain of definition.

It seems best to insert (8.3c) into (8.2c), obtaining

$$\frac{du_i}{dt} = -3\alpha \lambda_i \beta_i [u_i - y_i(t, 1)]. \tag{8.5}$$

Then integrate (8.2a), (8.2b), (8.5) for $t > 0$ with the initial conditions (8.3a) holding at $t = 0$ and the boundary conditions (8.3b) and (8.3c) holding for $t > 0$.

Next, consider a semi-discretization for this problem. We can use the same method as for the diffusion equation in Example 3.1. Thus, let $r_j = jh$, where $h = (J+1)^{-1}$ is the discretization step in r, and $r_{j+1/2} = r_j + h/2$. With $y_{i,j}(t)$ approximating $y_i(t, r_j)$, etc., we obtain

$$\frac{dy_{i,j}}{dt} + \frac{dz_{i,j}}{dt} = \frac{\lambda_i}{r_j^2 h^2} \left(r_{j+1/2}^2 (y_{i,j+1} - y_{i,j}) - r_{j-1/2}^2 (y_{i,j} - y_{i,j-1}) \right),$$
$$j = 1, \ldots, J+1, \ i = 1, \ldots, N_e, \quad (8.6a)$$
$$z_{i,j} = k_i(\mathbf{y}_j) y_{i,j}, \quad j = 0, \ldots, J+1, \ i = 1, \ldots, N_e, \quad (8.6b)$$
$$\frac{du_i}{dt} = -3\alpha \lambda_i \beta_i (u_i - y_{i,J+1}), \quad i = 1, \ldots, N_e. \quad (8.6c)$$

At $r = 0$ we get a division by 0 if we apply (8.6a). Writing instead

$$\frac{\lambda_i}{r^2} \frac{\partial}{\partial r} \left(r^2 \frac{\partial y_i}{\partial r} \right) = \lambda_i \left[\frac{\partial^2 y_i}{\partial r^2} + \frac{2}{r} \frac{\partial y_i}{\partial r} \right]$$

and applying l'Hôpital's rule to obtain

$$\frac{2}{r} \frac{\partial y_i}{\partial r} \bigg|_{r=0} = 2 \frac{\partial^2 y_i}{\partial r^2}$$

gives

$$\frac{dy_{i,0}}{dt} + \frac{dz_{i,0}}{dt} = \frac{3\lambda_i}{h^2} (y_{i,1} - 2y_{i,0} + y_{i,-1}). \quad (8.6d)$$

Because the boundary conditions are Neumann or mixed, we have extended the spatial discretization stencil to include the boundaries. The boundary conditions yield

$$y_{i,-1} = y_{i,1}, \quad (8.6e)$$
$$y_{i,J+2} = y_{i,J} + 2h\beta_i(u_i - y_{i,J+1}). \quad (8.6f)$$

The two side conditions (8.6e)–(8.6f) can be used to eliminate the ghost unknowns $y_{i,-1}$ and $y_{i,J+2}$. Substituting this and a once-differentiated (8.6b) into (8.6a) yields (including also (8.6c)) an ODE system of size $(J + 3)N_e$. The initial conditions are

$$y_{i,j}(0) = 0, \quad j = 0, \ldots, J+1, \ i = 1, \ldots, N_e, \quad (8.7a)$$
$$u_i(0) = u_{i0}, \quad i = 1, \ldots, N_e. \quad (8.7b)$$

The semi-discrete system (8.6)–(8.7) may best be further discretized in time by an implicit method which has stiff decay, such as BDF or Radau collocation (see Section 2.6 and [14]). These methods are implemented in general-purpose codes such as DASSL [34] or RADAU5 [85]. The system is not evaluated at the initial line, so the inconsistency problem is avoided altogether, except perhaps in the automatic choice of the first step by a general-purpose time integration code. If the latter is a problem, then simply turn off the automatic step selection mechanism for the first time step and choose that first step size yourself. (This should be easy to do in RADAU5, where no variable order mechanism is utilized.) After all, the method-of-lines ODE or DAE (8.6) already contains the errors from the spatial discretization where the discretization mesh is typically not chosen automatically.

8.2. Hyperbolic Problems

Without the last set of manipulations leading to an ODE, the system (8.6) is an index-1 DAE of the form (2.29) (see Section 2.8), which is of a type that can be solved using `DASSL` or `RADAU5`. Solving this DAE, and not the ODE, may well be necessary if differentiation of (8.6b) must be avoided (e.g., because of the appearance of square root terms of near-0 values in k_i). Note that the system (8.6) can be reduced to a semi-explicit index-1 DAE by a simple change of variables,

$$\zeta_{i,j} = y_{i,j} + z_{i,j}, \quad \eta_{i,j} = y_{i,j} - z_{i,j}.$$

As an example, we write down the application of the backward Euler method (2.7) to the system (8.6), with a time step k and $\mu = \frac{k}{h^2}$:

$$y_{i,j}^{n+1} + z_{i,j}^{n+1} - \mu \frac{\lambda_i}{r_j^2} \left(r_{j+1/2}^2 \left(y_{i,j+1}^{n+1} - y_{i,j}^{n+1} \right) - r_{j-1/2}^2 \left(y_{i,j}^{n+1} - y_{i,j-1}^{n+1} \right) \right)$$
$$= y_{i,j}^n + z_{i,j}^n, \quad j = 1, \ldots, J+1, \ i = 1, \ldots, N_e,$$
$$z_{i,j}^{n+1} = k_i \left(\mathbf{y}_j^{n+1} \right) y_{i,j}^{n+1}, \quad j = 0, \ldots, J+1, \ i = 1, \ldots, N_e,$$
$$u_i^{n+1} + k3\alpha \lambda_i \beta_i \left(u_i^{n+1} - y_{i,J+1}^{n+1} \right) = u_i^n, \quad i = 1, \ldots, N_e,$$
$$y_{i,0}^{n+1} + z_{i,0}^{n+1} - \mu 3\lambda_i \left(y_{i,1}^{n+1} - 2y_{i,0}^{n+1} + y_{i,-1}^{n+1} \right) = y_{i,0}^n + z_{i,0}^n,$$
$$y_{i,-1}^{n+1} = y_{i,1}^{n+1},$$
$$y_{i,J+2}^{n+1} = y_{i,J}^{n+1} + 2h\beta_i \left(u_i^{n+1} - y_{i,J+1}^{n+1} \right).$$

The implicit discretization of (8.6) does involve solving a nonlinear system of algebraic equations with a block structure (blocks of size $O(N_e)$ because of the dependence of z_i on $\mathbf{y}$) at each time step. If we want to avoid this and consider an explicit method (with a small, $O(h^2)$ time step, though) for the ODE formulation, then the ODE does get evaluated at the initial line $t = 0$. However, the inconsistency translates into a nonzero (but only $O(h)$) value for the ghost unknown $y_{i,N+1}(0)$. If the time horizon is not large, then a simple, constant step, forward Euler method in time may work well for a rough accuracy that could be sufficient for some engineering purposes. ■

8.2 Hyperbolic problems

In general, an initial-boundary value hyperbolic problem is well-posed only if some types of BC are imposed selectively. We must understand this first, before moving on to numerical discretizations.

8.2.1 Boundary conditions for hyperbolic problems

Example 8.2 Consider once again the simple advection equation, this time on a strip,

$$u_t - u_x = 0, \quad 0 \leq x \leq 1, \ t > 0,$$
$$u(0, x) = u_0(x).$$

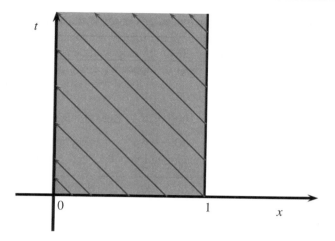

Figure 8.1. *Characteristics for the advection equation $u_t - u_x = 0$ in case of an initial-boundary value problem.*

For the pure Cauchy problem we know that the solution is $u(t, x) = u_0(t+x)$. The solution consists of the initial data propagated leftward at speed $\frac{dx}{dt} = -1$.

The solution propagation on a strip is depicted in Figure 8.1. We see that the values of u at the left boundary $(t, x = 0)$ are determined by the initial values and by the boundary values at the right boundary $(t, x = 1)$. Thus, only solution values at the right boundary may be specified.

For the same advection equation on the quarter-space,

$$x \geq 0, \quad t \geq 0,$$

we need no BC.

On the other hand, for the advection equation

$$u_t + u_x = 0$$

the characteristics propagate rightward. Hence, on the same positive quarter-space we do need BC imposed at the boundary $x = 0$, as well as initial data. ∎

Let us generalize the observations made in Example 8.2. For a hyperbolic system

$$\mathbf{u}_t + A\mathbf{u}_x = \mathbf{0}, \tag{8.8a}$$

the matrix A is diagonalizable and has real eigenvalues,

$$T^{-1}AT = \Lambda = \begin{pmatrix} \lambda_1 & & \\ & \ddots & \\ & & \lambda_s \end{pmatrix}. \tag{8.8b}$$

8.2. Hyperbolic Problems

Let us arrange the eigenvalues so that the negative ones—say, there are m of those—appear first on the diagonal of Λ,

$$\Lambda = \begin{pmatrix} \Lambda^I & \\ & \Lambda^{II} \end{pmatrix}, \tag{8.8c}$$

and define the **characteristic variables**

$$\begin{pmatrix} \mathbf{w}^I \\ \mathbf{w}^{II} \end{pmatrix} \equiv \mathbf{w} = T^{-1}\mathbf{u} \tag{8.8d}$$

with $\mathbf{w}^I$ denoting the first m components of $\mathbf{w}$. Then, in the case that A is constant we have

$$\mathbf{w}_t + \Lambda \mathbf{w}_x = \mathbf{0}, \tag{8.8e}$$

so the PDEs for $\mathbf{w}$ decouple and can be examined individually. For $\mathbf{w}^I$, as in Example 8.2, we only need BC at the right end of the interval in x, if that interval end is finite. Similarly, for $\mathbf{w}^{II}$ we only need BC at the left end of the interval in x, if it is finite.

So, we may specify

$$\mathbf{w}^{II}(t, 0) = S^I(\mathbf{w}^I(t, 0) + \mathbf{g}^I(t)), \quad s - m \text{ conditions} \tag{8.9a}$$

$$\mathbf{w}^I(t, 1) = S^{II}(\mathbf{w}^{II}(t, 1) + \mathbf{g}^{II}(t)), \quad m \text{ conditions,} \tag{8.9b}$$

where S^I is $(s - m) \times m$ and S^{II} is $m \times (s - m)$. The BC describe how waves are reflected at the boundaries.

In general, then, the allowable BC must satisfy that $s - m$ conditions are specified at $x = 0$ and m conditions are specified at $x = 1$ with the BC matrices multiplying $\mathbf{w}^{II}(t, 0)$ and $\mathbf{w}^I(t, 1)$ being nonsingular. (This assumes that no eigenvalue vanishes. For $\lambda_l = 0$ it does not matter at which end the BC on w_l is imposed.) Note that the characteristic curves are again straight lines given by

$$\frac{dx}{dt} = \lambda_l, \quad l = 1, \ldots, s.$$

Example 8.3 For the classical *wave equation*

$$\phi_{tt} - c^2 \phi_{xx} = 0$$

considered in Section 7.2, specifying ϕ at both ends works well; see, e.g., numerical results in Example 7.3. To examine it in light of our recently acquired knowledge, let us write the linear wave equation in first order form for $u_1 = \phi_t$, $u_2 = -c\phi_x$. Then we have, as in Chapter 1, a system of the form (8.8a) with $\mathbf{u} = (u_1, u_2)^T$ and

$$A = \begin{pmatrix} 0 & c \\ c & 0 \end{pmatrix}.$$

The eigenvector and eigenvalue matrices for A are

$$T^{-1} = \begin{pmatrix} 1 & -1 \\ 1 & 1 \end{pmatrix}, \quad \Lambda = \begin{pmatrix} -c & 0 \\ 0 & c \end{pmatrix}.$$

So, the characteristic variables are $w^I = u_1 - u_2$ and $w^{II} = u_1 + u_2$. We should specify one boundary condition at each end. Any $\phi_t + \alpha c \phi_x$ would do at $x = 0$, except $\alpha = 1$, and any such condition would do at $x = 1$, except $\alpha = -1$. ∎

If $A = A(t, x)$ is not constant, then the results above still hold, except that the characteristics are not straight lines in general, and $\mathbf{w}$ satisfies a PDE of the form

$$\mathbf{w}_t + \Lambda \mathbf{w}_x + B \mathbf{w} = \mathbf{0}.$$

This can be solved by the iteration

$$\mathbf{w}_t^{\nu+1} + \Lambda \mathbf{w}_x^{\nu+1} + B \mathbf{w}^\nu = \mathbf{0},$$

so we obtain the previous form (8.8e) with a harmless inhomogeneity added. The results regarding the type and location of the BC remain unchanged.

If the problem is nonlinear, then, in general, additional surprises can materialize. A special situation occurs for nonlinear conservation laws (5.4). Consider one conservation law,

$$u_t + f(u)_x = 0, \quad a(u) = \frac{\partial f}{\partial u}.$$

The characteristic direction is

$$\frac{dx}{dt} = a(u).$$

Since the solution remains constant along characteristics, $a(u)$ remains constant: *the characteristics of conservation laws (5.4) are straight lines.* See, for instance, Figure 10.1. However, unlike the constant coefficient case, the value of $a(u)$, even its sign, depends on the data! For nonlinear problems, the locations of the boundary conditions are not independent of the values of the initial and boundary data.

Example 8.4 For the inviscid Burgers equation

$$u_t + u u_x = 0$$

defined on the positive quarter-space $x, t \geq 0$, the initial data $u_0(x) \equiv -1$ produce the solution $u(t, x) \equiv -1$, requiring no boundary data to be specified at $x = 0$. But the initial data $u_0(x) \equiv 1$ must be supplemented by boundary data at $(t, 0)$ and produce a solution which is not necessarily constant everywhere. *Please verify that for this scalar example $(s = 1)$ the first u_0 yields $m = 1$ whereas the second yields $m = 0$, and then check conditions* (8.9). ∎

8.2. Hyperbolic Problems

In summary, the PDE problem is well-posed if BC are prescribed on part of the boundary. These BC are appropriately called **inflow boundary conditions** because they must be prescribed where the flow direction is into the domain. This is the general rule of thumb also for problems in more space variables and on spatial domains which are not necessarily rectangles or cubes.

8.2.2 Boundary conditions for discretized hyperbolic problems

We now turn to the numerical solution of initial-boundary value problems for hyperbolic systems. While the PDE problem requires inflow BC, a numerical method may well also require **outflow boundary conditions**. For example, the Lax–Wendroff, Lax–Friedrichs, and leapfrog schemes for the advection equation all expect a neighbor both left and right of any spatial location x_j.

Two questions arise, in general:

- If the numerical scheme requires boundary data where the PDE problem does not, how should this additional information be imposed? In short, how should outflow BC be specified?

- What are the implications of boundary conditions on the stability of the resulting scheme?

These questions are, of course, interrelated. For both, the quest is not to improve upon, but rather to avoid deterioration of, results for the pure initial value problem.

Example 8.5 For the simple advection equation on a strip Ω as in Example 8.2,

$$u_t - u_x = 0, \quad 0 \le x \le 1, \, t > 0,$$
$$u(0, x) = u_0(x),$$
$$u(t, 1) = g_1(t),$$

consider two different discretization schemes:

1. The upwind scheme (1.15a) reads

$$v_j^{n+1} = v_j^n + \mu(v_{j+1}^n - v_j^n), \qquad j = 0, 1, \ldots, J.$$

To complete description of the scheme we need to specify v_j^0 and v_{J+1}^n for all relevant j and n. (NB $(J+1)h = 1$.) Using the given initial and boundary conditions we set

$$v_j^0 = u_0(jh), \quad j = 0, \ldots, J+1,$$
$$v_{J+1}^n = g_1(nk), \quad n = 1, 2, \ldots,$$

and there is nothing more to it.

2. The leapfrog scheme (1.15c) reads

$$v_j^{n+1} = v_j^{n-1} + \mu(v_{j+1}^n - v_{j-1}^n), \qquad j = 1, \ldots, J.$$

Recall that this scheme is stable in the absence of BC if the CFL condition holds, which here requires $\mu \leq 1$.

To complete description of the scheme we must specify v_j^0, v_j^1, v_0^n, and v_{J+1}^n for all relevant j and n. Obviously, we specify

$$v_j^0 = u_0(jh), \quad j = 0, \ldots, J+1.$$

We also saw before in Section 1.2 how to specify v_j^1, $j = 1, \ldots, J$, so assume this has been done. For the boundary condition at the right end we specify

$$v_{J+1}^n = g_1(nk), \quad n = 1, 2, \ldots.$$

This leaves the left boundary v_0^n, for which no values are provided by the PDE problem specification.

One might carelessly specify a value arbitrarily, say,

$$v_0^n = 0.$$

(There seems to be an implicit human need to minimize what we don't know, and since we don't know v_0^n, why not make it 0?) However, such an arbitrary assignment is inconsistent with the value of the exact solution which is propagated from within the domain Ω. Combined with the nondissipative leapfrog scheme, this inconsistency generates an error wave which propagates undamped back into the interior of Ω. For instance, with

$$u_0(x) \equiv 1, \quad g_1(t) \equiv 1,$$

the exact solution equals 1 everywhere, but the approximate solution is

$$v_j^n = \begin{cases} 1, & x_j > t_n, \\ 1 - (-1)^j, & x_j \leq t_n \leq 1. \end{cases}$$

A better idea for specifying v_0^n for the advection equation with $a > 0$ is **simple extrapolation** from interior to boundary solution values, i.e.,

$$v_0^n = v_1^n. \quad \blacksquare$$

A more careful approach for specifying outflow BC is to extrapolate along the outgoing characteristic curves from within the domain Ω. This is consistent with the behavior of the PDE. For $u_t - u_x = 0$, the characteristic curve hitting $(t, x) = ((n+1)k, 0)$ crosses time-level n at (nk, k). Thus, the value v_0^{n+1} should be the same as $v^n(x = k)$, except that the latter is not defined. We specify it by a linear interpolation in x,

$$v^n(x = k) = h^{-1}(kv_1^n + (h-k)v_0^n);$$

see Figure 8.2. Note that $h - k \geq 0$ because the CFL condition is obeyed.

We rewrite the obtained **absorbing boundary condition**[27] as

$$v_0^{n+1} = v_0^n + \mu(v_1^n - v_0^n).$$

---
[27] Absorbing BC also go by other names, depending on circumstances and taste, e.g., **nonreflecting BC**.

8.2. Hyperbolic Problems

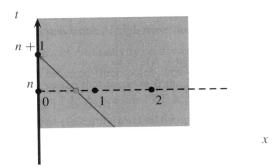

Figure 8.2. *Extrapolation along an outgoing characteristic to obtain an absorbing boundary condition.*

and observe that the formula is identical to the one-sided scheme (1.15a) considered in Chapter 1 (see also Exercise 1.6).

Let us generalize the principle demonstrated in Example 8.5 for a hyperbolic problem (8.8) with boundary conditions (8.9).

> **Note:** *A general recipe for absorbing BC reads as follows:* extrapolate along the m outgoing characteristic variables $\mathbf{w}^I(t, x)$ for x near 0, and use the obtained values together with the given BC (8.9a) to specify $\mathbf{v}_0^{n+1}$. Similarly, near the right end of the interval in x extrapolate along $\mathbf{w}^{II}(t, x)$ and use this together with (8.9b) for $\mathbf{v}_{J+1}^{n+1}$.

Absorbing boundary conditions work very well in practice when they can be applied. However, specifying them may require some fancy manipulation. The simple order-0 extrapolation mentioned at the end of Example 8.5 often also works well in practice, especially if the PDE is discretized by a dissipative scheme, and it is simpler to apply. (However, see Example 8.6 and Table 8.1.)

Stability

Turning to the question of general stability of the obtained scheme for an initial-boundary value problem, we consider again the general form (3.27b) or (3.27d),

$$v^{n+1} = Qv^n,$$

where now the matrix Q includes the effects of a scheme for boundary conditions, as described above. The question is, given that the scheme for a pure initial value problem is stable, does it remain stable after incorporating the effects of the boundary conditions? In other words, have the boundary conditions introduced any new eigenvalue into Q which has modulus > 1? A necessary condition for stability is that there exist no eigensolution (i.e., a mesh function in x) $\phi \in l_2$ for the equation[28]

$$Q\phi = z\phi \qquad (8.10)$$

such that $|z| > 1$. This is called the **Godunov–Ryabenkii condition**.

For any eigensolution pair (z, ϕ) of Q we may consider ϕ as an initial value mesh function v^0, provided $\phi \in l_2$. Then

$$v^{n+1} = Qv^n = \cdots = Q^{n+1}v^0 = Q^{n+1}\phi = z^{n+1}\phi.$$

(The superscript on v is the time level, while on z it is a power.) Obviously, if $|z| > 1$ we obtain a blowup as $n \to \infty$, i.e., an instability. The blowup is extreme if we choose the eigenvector corresponding to the offending eigenvalue as an initial value, but other initial value functions will generally contain a component in the direction of ϕ as well.

To be more concrete, let us consider the leapfrog scheme for the advection equation, as in Example 8.5, but on a quarter-space $x \geq 0$. We have

$$z^2 \phi_j = \phi_j + z\mu D_0 \phi_j, \quad j = 1, 2, \ldots, \qquad (8.11a)$$

and the question is, What should ϕ_0 be? For a given z the equation (8.11a) is a difference equation with constant coefficients for ϕ. We therefore write

$$\phi_j = \sum_i \alpha_i (\tau_i)^j,$$

where α_i are arbitrary coefficients and τ_i are the roots of the characteristic equation

$$z^2 = 1 + \mu(\tau - \tau^{-1})z. \qquad (8.11b)$$

Now, we may only consider $\phi \in l_2$, so the general form of ϕ_j cannot contain components with $|\tau_i| > 1$. Thus

$$\phi_j = \sum_{|\tau_i|<1} \alpha_i (\tau_i)^j. \qquad (8.11c)$$

Rewriting (8.11b), we are thus looking at roots τ of the quadratic equation

$$\tau^2 - \frac{z^2 - 1}{\mu z}\tau - 1 = 0 \qquad (8.11d)$$

which satisfy $|\tau| < 1$ when $|z| > 1$. This would give an eigensolution which destroys stability.

It is easy to check that for $|z| > 1$ there is at most one solution of the quadratic equation (8.11d) with $|\tau| < 1$. So, $\phi_j = \alpha \tau^j$. Now, consider a general boundary condition form

$$\phi_0 = \sum_{j=1}^p b_j \phi_j. \qquad (8.11e)$$

---

[28] By $\phi \in l_2$ we mean that $\|\phi\|$ remains bounded as $h \to 0$.

8.2. Hyperbolic Problems

We would like to design BC of this form which lead to an overall stable scheme. Note that z is an eigenvalue corresponding to an eigenvector

$$\{\phi_j = \alpha \tau^j\}_{j=0}^{\infty}$$

provided

$$1 = \sum_{j=1}^{p} b_j \tau^j. \tag{8.11f}$$

So, if the equation (8.11f) has solutions with $|\tau| < 1$, then the choice of (8.11e) leads to instability!

As it turns out, reasonably conceived BC approximations tend not to yield $|\tau| < 1$. *The necessary stability condition usually holds, given that the pure IVP scheme is stable.*

However, this condition is only necessary. In particular, we must worry about cases where

$$z = \pm 1.$$

In such cases, a small perturbation to $z(\varepsilon) = \pm(1 + \varepsilon)$ yields $|z| > 1$ when $0 < \varepsilon \ll 1$. Note from (8.11d) that $|\tau| = 1$ when $z = \pm 1$, i.e., τ is a borderline case as well. For the perturbed values of z we respectively obtain

$$\tau \approx \pm \left(-1 + \frac{\varepsilon}{2\mu}\right),$$

as well as a root satisfying $|\tau| > 1$. A dangerous situation may therefore arise when $z = 1 + \varepsilon$ and $\tau = -1 + c\varepsilon$, or when $z = -1 - \varepsilon$ and $\tau = 1 - c\varepsilon$, where $c > 0$ is some constant.

1. If we use an extrapolation of the form

   $$D_+^q \phi_0 = 0$$

   for (8.11e), with q a natural number, then we obtain $(\tau - 1)^q = 1$, and hence $\tau = 1$ and $z = \pm 1$ are possible. The combination of $\tau = 1$ and $z = -1$ allows for unstable perturbations for the leapfrog scheme.

2. If we extrapolate along the outgoing characteristic, as in Example 8.5, then

   $$z = 1 + \mu(\tau - 1).$$

   Here, only $z = 1$ and $\tau = 1$ are supported, and so no unstable perturbations result.

The unstable perturbation described above creates in effect a wave that grows linearly in t/k. This is a parasitic wave going rightward into the domain Ω, i.e., in the wrong direction. The growth is only linear in n, so it has a limited effect in practice. However, when solving the problem on a strip as in Example 8.5, and not on a quarter-space, the wave reflects back into the domain from $x = 1$, and then, when it reaches $x = 0$, it grows once again. The overall effect is an intolerable growth proportional to $(t/k)^t$.

Thus, for the leapfrog scheme the BC $D_+\phi_0 = 0$ is not stable, whereas the extrapolation along the characteristic is. On the other hand, if a dissipative scheme such as Lax–Wendroff is used for the discretization of the PDE, then there is really no similar difficulty of instability, because the small perturbations generated at the boundary get damped by the scheme as they travel in the interior of Ω.

Example 8.6 Continuing Example 8.5, we now solve it numerically using

1. the Lax–Wendroff scheme, and
2. the leapfrog scheme.

At $x = 1$ we impose the BC
$$u(t, 1) = 0.$$
We fix $h = 0.1$ and impose the initial condition that
$$v_j^0 = \begin{cases} 0, & j \neq 1, \\ 1, & j = 1. \end{cases}$$

It may be argued that these initial conditions are mesh-dependent (we are envisioning $u_0(x)$ with support that is contained in $(0, 2h)$, like a roof basis function in finite elements), but our purpose here is merely to check BC effects on error propagation. The exact solution for $t > 2h$ (say) vanishes identically.

We use three boundary conditions at $x = 0$:

1. $D_+ v_0^{n+1} = 0$,
2. $D_+^2 v_0^{n+1} = 0$,
3. absorbing BC.

The results are recorded in Table 8.1, where maximum errors are measured at $t = 10$ for $k = .05$, and any error below 10^{-13} is rounded to 0. The errors in the Lax–Wendroff scheme are of course unrealistically good—they simply mean that no instability is caused by any of these BCs. The performance of the leapfrog scheme is more interesting and alarming. The results agree with the preceding analysis. Additional calculations also confirm that the errors can be improved significantly by adding dissipativity to the leapfrog scheme as discussed in Section 5.2.

Table 8.1. *Errors at $t = 10$ for the Lax–Wendroff and leapfrog difference schemes applied to the equation $u_t - u_x = 0$ with different choices of BCs at $x = 0$. The side conditions are all homogeneous except $v_1^0 = 1$; step-sizes are $h = 0.1$, $k = 0.05$.*

Scheme	$v_0^{n+1} = v_1^{n+1}$	$v_0^{n+1} = 2v_1^{n+1} - v_2^{n+1}$	$v_0^{n+1} = v_0^n + \mu(v_1^n - v_0^n)$
Lax–Wendroff	0	0	0
leapfrog	3.3e+4	1.1e+9	3.0e-2

8.2. Hyperbolic Problems

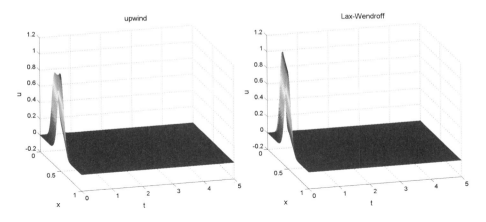

(a) Upwind: no artificial wave but inaccurate

(b) Lax–Wendroff: a little dissipativity helps

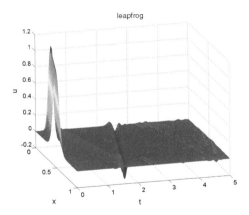

(c) Leapfrog: artificial waves reflect off boundaries

Figure 8.3. *Artificial wave reflection for the leapfrog scheme, Example 8.6. The upwind scheme does not need an outflow BC. For Lax–Wendroff and leapfrog the simple extrapolation $D_+ v_0^{n+1} = 0$ is used at the outflow boundary $x = 0$. Here $u_0(x) = e^{-100(x-.5)^2}$, $h = .005\pi$, and $k = .5h$.*

A more realistic initial value function is depicted in Figure 8.3. Here $u_0(x) = e^{-100(x-.5)^2}$ is a pulse. The rest of the details, including PDE, domain, and inflow BC remain as before. This pulse propagates leftward unchanged for our advection equation, and should disappear for $t \geq 1$.

We apply the upwind, Lax–Wendroff, and leapfrog schemes and display results for $0 \leq t \leq 5$ in Figure 8.3. Observe that the pulse does disappear, of course, when using the upwind scheme (for which no outflow BCs are necessary), but the amplitude diminishes noticeably. For the second order schemes we use the simple extrapolation $D_+ v_0^{n+1} = 0$. The slightly dissipative Lax–Wendroff scheme performs best here. The leapfrog scheme performs poorly. Note the artificial waves repeatedly reflecting off both boundaries for the leapfrog scheme, in agreement with what has been explained earlier. In a more realistic context, where small waves may be expected as part of the true solution, encountering such artificial waves is worse than their mere maximum error value would indicate, because it may be hard to tell them apart from the real ones. ∎

> **Note:** If applying absorbing BCs is simply too fancy, then try a simple extrapolation together with a dissipative scheme.

The perturbation argument we've made prior to the last example appeals to intuition. A general theory covering the distance between necessary and sufficient conditions for stability in the present context was given by Gustaffson, Kreiss, and Sundstrom [77]. The conditions they came up with are referred to as the *GKS theory*. Their paper is hard to read, though. A nice interpretation of these results in terms of *group velocity* (see Section 7.1) was given by Trefethen [167].

8.2.3 Order reduction for Runge–Kutta methods

> **Note:** This entire subsection constitutes one case study in which we put into use several techniques encountered in the present chapter and in previous ones.

Consider the conservation law

$$u_t + f(u)_x = 0, \quad x \geq 0, \; t \geq 0, \tag{8.12a}$$
$$u(t, 0) = g(t), \quad u(0, x) = u_0(x). \tag{8.12b}$$

Thus, we are assuming that f and the initial and boundary conditions are such that the characteristic curves propagate to the right, unlike in Examples 8.5 and 8.6. We first semi-discretize in space, obtaining ODEs in time for the unknowns $v_j(t) \approx u(t, x_j)$. Suppose the discretization in space is sufficiently accurate so we can concentrate on the error in time. Recall that the classical RK4 (2.10), (2.11) is very popular for the ensuing time-discretization. The main reason is that it has a reasonable absolute stability region that contains a sizable segment of the *imaginary* axis (recall Section 5.3.2). Another important reason is that it is fourth order accurate, which is sufficient and effective for many practical purposes. But do we actually realize this fourth order accuracy in calculations?

Let us examine the question of accuracy order. If the unknowns are denoted $\mathbf{v}(t) = (v_1(t), \ldots, v_J(t))^T$, then we can write the semi-discretization with the obvious use of the

8.2. Hyperbolic Problems

boundary condition at $x = 0$ as a DAE,

$$\mathbf{v}' = F(t, h, v_0, \mathbf{v}), \tag{8.13a}$$

$$v_0(t) = g(t), \tag{8.13b}$$

where h is the spatial discretization step size. But this DAE yields the ODE

$$\mathbf{v}' = F(t, h, g, \mathbf{v}). \tag{8.14}$$

In the ODE literature there are known cases where the order of an RK method is reduced due to stiffness. This is essentially because the intermediate stages in (2.10) generally have a lower accuracy order, and the higher order at the end of each time step is achieved because of cancellation of error terms due to the integration: if there is no integration (as in the extreme stiffness limit or for the algebraic part of a DAE), then the higher accuracy order is lost. Here, however, the ODE (8.14) is not stiff, and discretizing it by the classical fourth order Runge–Kutta method RK4 should yield fourth order accuracy globally in time. No ODE order reduction should be experienced.

And yet, as it turns out, this is not the entire story. Let us proceed with a simple numerical test.

Example 8.7 Following Carpenter et al. [46], consider the simplest advection equation

$$u_t + u_x = 0, \quad x, t \geq 0,$$

with the chosen solution $u(t, x) = \sin(t - x)$, and thus

$$u(0, x) = u_0(x) = -\sin(x),$$
$$u(t, 0) = g(t) = \sin(t).$$

Since RK4 is an explicit method we must keep $\mu = k/h = O(1)$ as k is refined, and in fact we always keep μ fixed below. Therefore, to realize fourth order accuracy a fourth order discretization is used in space as well.

Because of the Dirichlet BC, use of the wider stencil $D_0(I - \frac{1}{6}D_+D_-)$ is awkward. We employ the implicit scheme (3.8) instead. Also, at the outflow, set at $x = 1$, we use the higher order absorbing BC,

$$v_{J+1} = 4v_J - 6v_{J-1} + 4v_{J-2} - v_{J-3}. \tag{8.15a}$$

At the left end the given boundary condition is applied

$$v_0 = g(t). \tag{8.15b}$$

In the present context this means that $\sin(t_n)$, $\sin(t_n + k/2)$, $\sin(t_n + k/2)$ again, and $\sin(t_n + k)$ are used for the intermediate stages of the current time step.

The results are listed in Table 8.2 under err_g. These are maximum absolute errors at $t = 1$ for a sequence of step sizes each being half its predecessor. We also list the rate of convergence $rate_g$ defined by

$$rate = \log_2 \frac{err(2k)}{err(k)}.$$

Table 8.2. *Numerical errors in maximum norm at $t = 1$ for the linear advection equation $u_t + u_x = 0$ using RK4 and $\mu = 1$. Note error order deterioration, especially in err_g.*

$h = k$	err_g	$rate_g$	$err_{g'}$	$rate_{g'}$
$.1 \times 2^0$	2.83e-05		2.67e-05	
$.1 \times 2^{-1}$	5.15e-06	2.5	2.20e-06	3.6
$.1 \times 2^{-2}$	1.39e-06	1.9	1.63e-07	3.8
$.1 \times 2^{-3}$	3.54e-07	2.0	1.08e-08	3.9
$.1 \times 2^{-4}$	8.88e-08	2.0	7.11e-10	3.9
$.1 \times 2^{-5}$	2.22e-08	2.0	7.54e-11	3.2
$.1 \times 2^{-6}$	5.56e-09	2.0	8.18e-12	3.2

Clearly from Table 8.2 the error is only $O(k^2)$ and not $O(k^4)$ as anticipated. ∎

In fact, the unexpected error order reduction observed in Example 8.7 has two sources. But first, let us note something we *don't* see. When discussing order reduction in ODE methods we assume that the ODE system is fixed as the step size k is reduced, whereas here we also reduce h, so the ODE system is actually different for different k. Hence the above example *does not* violate ODE theory.

The first source of order reduction may be observed already at points (t_n, x_j) which are common to fine and coarse meshes. Assuming a spatial discretization of order 4, one would be tempted to naively write for the solution of the fully discretized problem

$$v_j^n - u(t_n, x_j) = [v_j^n - v_j(t_n)] + [v_j(t_n) - u(t_n, x_j)] = O(k^4) + O(h^4).$$

However, this does not take into account the fact that the size of the ODE system (8.14) increases with refinement. The "constant" of the $O(k^4)$ error term may depend on this size and through it, on h^{-1}, which may decrease the order in k for a fixed mesh ratio μ. Such an effect may be expected for hyperbolic PDEs, although not for parabolic ones; see Section 1.1.2.

The second effect may occur if we compare errors at all points for different meshes, including fine mesh points which are not coarse mesh points. In particular, the first spatial interior mesh point x_1 gets closer and closer to the boundary as h gets smaller. The error in v_1^n may be larger then, because there is no integration of $g(t)$ on the boundary! In such a case we could get an *error boundary layer*.

A simple remedy for the second error effect is to replace (8.15b) at each time step by its derivative: at each time level n, replace (8.15b) by the ODE

$$v_0' = g'(t), \quad v_0(t_n) = g(t_n), \tag{8.16}$$

and apply the same RK4 discretization to the entire ODE system so obtained. This alleviates the problem of error boundary layer, independent of the linearity of the PDE and the actual use of RK4: it should work for nonlinear conservation laws and for other, higher order Runge–Kutta schemes equally well! Also, no derivatives of g other than the first are used. However, the first effect of error order reduction may persist.

8.3. Infinite or Large Domains

Example 8.8 Continuing the previous example, results using (8.16) are listed in Table 8.2 under the heading $err_{g'}$. These errors are certainly satisfactory, although the rate insists on remaining less than the ideal 4.

For the very simple advection equation it is possible to recover the full fourth order accuracy [46]. However, the resulting method variant does not produce appreciably better results than (8.16), and worse, it is not generalizable. ∎

Example 8.9 Finally, consider the nonlinear conservation law problem

$$u_t + uu_x = 0, \quad x, t \geq 0,$$

with

$$u(0, x) = u_0(x) \equiv 1,$$
$$u(t, 0) = g(t) = e^{-t}.$$

This is the famous Burgers equation which we discuss more thoroughly in Chapter 10. Here let us say only that with these initial-boundary conditions, the characteristic curves $\frac{dx}{dt} = u$ all migrate upper-right, and they do not cross, so no shocks arise.

We now find the exact solution at a fixed t. For $x \geq t$ it is obvious that $u(t, x) \equiv 1$. For each $0 < x < t$ there exists a τ, $x < \tau < t$, such that the straight-line characteristic emanating from $(\tau, 0)$ hits (t, x), and so $u(t, x) = e^{-\tau}$. We have

$$\frac{t - \tau}{x - 0} = e^{\tau},$$

which defines a transcendental equation for $\tau = \tau(t, x)$. Solving this to a high accuracy (by Newton's method for each x, starting from a good initial guess) yields the exact solution $u(t, x) = e^{-\tau}$ at this point in space–time.

In Figure 8.4(a) we display the exact and approximate solutions, using $k = h = .01$ and (8.16). The solution has a jump in first derivative at $x = t = 1$, and even though the approximate solution looks very similar to the exact one, the error plot reveals that this mild form of discontinuity does have an effect. Certainly no fourth order accuracy may be anticipated here! ∎

8.3 Infinite or large domains

Some issues regarding handling boundary conditions also arise in problems that are independent of time, which is hardly surprising. One very important such issue is the handling of infinite spatial domains. In many mathematical models of physical problems the spatial domain is assumed infinite, and we generally know that the solution and its derivatives should decay or become constant "far enough" from where the action takes place. In practice we must replace this infinite domain with a finite one, and the question becomes which boundary conditions to impose at the boundary of this finite domain, and how large to make it, in order to approximate the solution to the original problem sufficiently well.

Indeed, the original domain does not need to be infinite for such a question to arise: for a given, large spatial domain, one is often particularly interested in what happens in some smaller subdomain. So, if we try to restrict the problem to the subdomain of interest, a similar question regarding BC arises.

272 Chapter 8. More on Handling Boundary Conditions

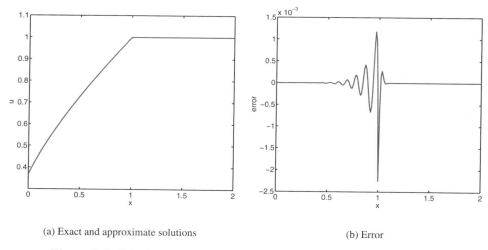

(a) Exact and approximate solutions   (b) Error

Figure 8.4. *Solutions and error for the Burgers equation in a case without jump discontinuities. Here $t = 1$, $h = k = .01$. The maximum error is $\approx 10^{-3}$.*

For boundary value ODEs, which is what remains from a PDE in one spatial variable at steady state, as x increases there is typically a subspace of exponentially growing modes (i.e., fundamental solution components) and its complement of nongrowing modes. The BC then must be chosen so as to kill off the growing modes, because these could rapidly blow up. See, e.g., [10, 108] and Exercise 5. The resulting type of BC is often referred to as **radiating boundary conditions**.

When the problem depends on time as well there is an added difficulty, especially in the hyperbolic case. The issue is clear and simple: we do not want artificial waves reflecting off our introduced boundary into the new spatial domain! The problem then is not dissimilar to that of specifying outflow BC. **Absorbing boundary conditions** and **nonreflecting boundary conditions** usually yield the desirable behavior. When working with these proves to be too fancy in practice, try using a simple extrapolation, which often works well.

But more delicate situations also arise in applications, especially in more than one space variable. A lot of research has been done on this. The topic in its entirety is outside the scope of this text, and we refer to Hagstrom [80], Taflove [163], and LeVeque [116] for more.

8.4 Exercises

0. **Review questions**

    (a) Define periodic, Dirichlet, and Neumann BC.

    (b) Why can't the situation described in Example 8.2 and Figure 8.1 occur for parabolic problems?

8.4. Exercises

(c) Define characteristic variables and explain their importance for determining BC for hyperbolic problems.

(d) Is the initial-boundary value problem for the classical wave equation well defined with the BC
$$u_1(0) = 0, \quad u_2(1) = 0?$$
What about the BC
$$u_1(0) = 0, \quad u_1(1) + u_2(1) = 0?$$

(e) What are inflow, outflow, and absorbing boundary conditions?

(f) Define the Godunov–Ryabenkii condition and explain its importance for stability analysis.

(g) What is particularly alarming about Figure 8.3(c)?

(h) The orders of discretization in Example 8.7 are equal to 4 both in time and in space. Yet the observed error rate in Table 8.2 is less than 4; in fact $rate_g = 2$. How can such an observed error rate reduction *not* violate theory?

(i) What are radiating boundary conditions?

1. Implement and test a numerical method of your choice for the problem of adsorption in a finite bath (Example 8.1). Justify your choice of method and try it for the following parameter values:

$$N_e = 2, \quad \lambda_1 = 1, \quad \lambda_2 = 1.15, \quad e_p = 0.7, \quad e_b = 0.5,$$

$$\alpha = \frac{1 - e_b}{e_b} \frac{e_p}{R_p} = 7, \quad k_i(\mathbf{y}) = \frac{a_{i0}}{e_p + a_{i1}y_1 + a_{i2}y_2},$$

$$a_{10} = 171.07, \quad a_{20} = 372.85, \quad a_{11} = 1400, \quad a_{21} = 1573,$$

$$a_{12} = 2835, \quad a_{22} = 2565, \quad u_{10} = 0.001, \quad u_{20} = 0.001,$$

$$R_p = 0.1, \quad \beta_1 = 100, \quad \beta_2 = 200.$$

Plot the solution at $t = 0, 0.01$ and 1. Discuss your observations.

2. Consider the classical wave equation of Section 7.2, namely, (7.6a) with $V(u) = 0$, and (7.6b).

    (a) Formulate the PDE as a first order hyperbolic system (8.8a) and explain how the initial conditions are specified.

    (b) Show that the initial-boundary value problem is well posed with the BCs
    $$\phi_x(t, 0) = g_0(t), \quad \phi_x(t, 1) = g_1(t).$$

    (c) Derive boundary conditions in order to solve the problem in first order form numerically.

    (d) Implement this using the leapfrog and the Lax–Wendroff schemes, and compare with the results in Example 7.3. What are your conclusions?

3. Consider the box scheme (Section 5.2.4) applied to the advection equation.

   (a) Show that no outflow BCs are needed here.

   (b) Explain how time-stepping can be achieved without inverting any linear system of algebraic equations.

   (c) Is there any difference expected in execution time between this suddenly turned explicit method and a "normal" explicit method, such as leapfrog or Lax–Wendroff?

   (d) Does the method remain explicit for a general hyperbolic PDE system?

4. In Section 3.1.3, natural and essential boundary conditions are derived for Poisson-like equations; see (3.18) and (3.19).

   The **biharmonic equation** is a fourth order elliptic PDE that arises in fluid flow modeling and reads

   $$\Delta \Delta u = q, \quad \mathbf{x} \in \Omega. \tag{8.17a}$$

   It, too, can be treated via a minimization functional as in (3.18).

   Of the following two popular BC pairs, determine which of the three different BCs is essential and which is natural:

   $$\text{(i)} \quad u|_{\partial\Omega} = 0, \quad \frac{\partial u}{\partial n}\Big|_{\partial\Omega} = 0, \tag{8.17b}$$

   $$\text{(ii)} \quad u|_{\partial\Omega} = 0, \quad \Delta u|_{\partial\Omega} = 0. \tag{8.17c}$$

5. Consider the linear boundary value ODE

   $$\frac{d^2 u}{dx^2} \equiv u'' = u, \quad 0 \leq x < \infty,$$
   $$u(0) = 1, \quad u(\infty) \text{ is bounded.}$$

   This is a (major) simplification of common situations where **radiating BCs** are considered.

   (a) Show that although the value of $u(\infty)$ is not specified, the problem has the unique solution

   $$u(x) = e^{-x}.$$

   (b) To solve this problem numerically one typically chooses a finite interval $[0, L]$ (where L stands for "long") and discretizes the problem on it. We next must determine which *homogeneous* BC to specify at L.

   Show (theoretically or computationally) that the radiating BC

   $$u(L) + u'(L) = 0$$

   is better than the more simpleminded

   $$u(L) = 0$$

   in the sense that a smaller L can be chosen to achieve the same accuracy with the same mesh width.

Chapter 9
Several Space Variables and Splitting Methods

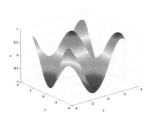

Most of the material covered in previous chapters concerns PDEs in one space dimension. Exceptions are Section 3.1 and Example 7.4, which involve the differential operator $\nabla \times$ and thus makes less sense in one dimension.

Of course, most mathematical models arising in practice involve more than one space variable. This chapter is devoted to extending our previous methods and analysis to the more general models. The following questions or issues naturally arise:

- Extending known schemes and examining their stability properties.
- Finding the solution at the next time level when implicit schemes are used.
- Handling BC on rectangular (tensor) meshes.
- Handling more generally shaped spatial domains.

As it turns out, many aspects of the methods and analysis that we have seen easily generalize to more dimensions, while some require more effort but eventually come under control. The most thorny issues seem to be the handling of solution discontinuities which are not aligned with the discretization mesh and of arbitrarily shaped domains.

An important basic observation related to theory is that for a hyperbolic PDE system,

$$\mathbf{u}_t + A\mathbf{u}_x + B\mathbf{u}_y = \mathbf{0}, \tag{9.1}$$

even with A and B constant matrices, complications may arise unless A and B commute or are **simultaneously diagonalizable**. The latter holds true when there is a nonsingular transformation T such that $T^{-1}AT$ and $T^{-1}BT$ are both diagonal. This assumption, which does not necessarily hold in practice, is needed to elevate results, such as Fourier stability and absorbing BC, from scalar equations to PDE systems.

In Section 9.1 we examine the extension of the basic methods developed earlier to several space dimensions. The material which follows is largely based on the books by Gustafsson, Kreiss, and Oliger [76] and Strikwerda [159]. Then, in Section 9.2 we consider the important issue of solving the large, sparse, linear algebraic systems arising from using implicit time discretization methods.

There are general methods for such linear systems of equations, and we briefly survey these in Section 9.4. This is really a review section, somewhere between Section 2.12 and the rest of the text in spirit. See Saad [149], Trottenberg, Oosterlee, and Schuller [171], and Elman, Silvester, and Wathen [59] for more. In Section 9.2.1 we demonstrate the application of techniques described in Section 9.4 in our time-dependent PDE context. The older ADI method and other variants of **dimensional splitting** are described in Section 9.2.2; see also Mitchell and Griffith [133].

Issues concerning the handling of boundary conditions are discussed in this chapter as they arise. See also Sections 3.1.3–3.1.5 and Exercise 3.7.

Dimensional splitting is a special case of **operator splitting**, giving rise to splitting methods, also known as *fractional step methods*. We have seen symplectic splitting methods for ODEs in Section 6.2.2 and for the KdV equation in Section 7.3. But the concept is more general, and splitting methods are important not only in the context of symplectic methods or dimensional splitting. For lack of a better place, we devote Section 9.3 to exploring further aspects of splitting methods, not just for several space variables.

Examples of applications are given to underscore the popularity and power of the splitting approach. We also briefly discuss related techniques. The latter include **additive**, or **implicit-explicit** methods, as well as certain **exponential integration** methods about which there has been considerable recent interest. All these involve a splitting of the differential operator.

9.1 Extending the methods we already know

Consider a Cauchy problem in two space variables. For instance, a system of conservation laws reads

$$\mathbf{u}_t + \mathbf{f}(\mathbf{u})_x + \mathbf{g}(\mathbf{u})_y = \mathbf{0}. \tag{9.2}$$

The extension from two dimensions to three dimensions turns out to be usually straightforward, at least in principle.

Extending the methods we have seen hitherto, such as the centered semi-discretization and its various subsequent time discretizations, or the Lax–Wendroff scheme, is straightforward in a simpleminded way. However, there are more variants now, with a larger variety of time step restrictions. The Fourier stability analysis extends directly, where the basic transform is given by (1.21). For two space variables we therefore have two wave numbers, ξ_1 and ξ_2. However, assessing situations under which the extension of the von Neumann condition for PDE systems is sufficient for stability is significantly harder than in one dimension.

Rather than writing down a general difference scheme, which looks tedious and involves multi-indices, let us consider some familiar instances.

The **Lax–Friedrichs** scheme for (9.2) naturally extends into

$$\mathbf{v}_{j,l}^{n+1} = \bar{\mathbf{v}}_{j,l}^n - \frac{\mu_1}{2}\left(\mathbf{f}(\mathbf{v}_{j+1,l}^n) - \mathbf{f}(\mathbf{v}_{j-1,l}^n)\right) - \frac{\mu_2}{2}\left(\mathbf{g}(\mathbf{v}_{j,l+1}^n) - \mathbf{g}(\mathbf{v}_{j,l-1}^n)\right),$$

$$\bar{\mathbf{v}}_{j,l}^n = \frac{1}{4}\left(\mathbf{v}_{j+1,l}^n + \mathbf{v}_{j-1,l}^n + \mathbf{v}_{j,l+1}^n + \mathbf{v}_{j,l-1}^n\right). \tag{9.3}$$

9.1. Extending the Methods We Already Know

Here $\mu_i = k/h_i$, where h_1 is the uniform step size in direction x and h_2 is the uniform step size in direction y.

Next, we write down the scheme for the case of a scalar PDE with constant coefficients (replacing $\mathbf{f}(\mathbf{u})$ by au and $\mathbf{g}(\mathbf{u})$ by bu), and apply a double Fourier transform,

$$v(t, x, y) = \frac{1}{2\pi} \int_{-\infty}^{\infty} \int_{-\infty}^{\infty} e^{\iota(\xi_1 x + \xi_2 y)} \hat{v}(t, \xi_1, \xi_2) d\xi_1 d\xi_2.$$

Substituting into the difference scheme we have

$$\int_{-\infty}^{\infty} \int_{-\infty}^{\infty} e^{\iota(\xi_1 x + \xi_2 y)} \hat{v}(t+k, \xi_1, \xi_2) d\xi_1 d\xi_2$$

$$= \int_{-\infty}^{\infty} \int_{-\infty}^{\infty} e^{\iota(\xi_1 x + \xi_2 y)} \left[\frac{1}{4} \left(e^{\iota \xi_1 h_1} + e^{-\iota \xi_1 h_1} + e^{\iota \xi_2 h_2} + e^{-\iota \xi_2 h_2} \right) \right.$$

$$\left. - \frac{\mu_1 a}{2} \left(e^{\iota \xi_1 h_1} - e^{-\iota \xi_1 h_1} \right) - \frac{\mu_2 b}{2} \left(e^{\iota \xi_2 h_2} - e^{-\iota \xi_2 h_2} \right) \right] \hat{v}(t, \xi_1, \xi_2) d\xi_1 d\xi_2.$$

The procedure is precisely the same as in Sections 1.2 and 4.1 for one spatial variable. We obtain stability iff

$$\left| \frac{1}{2} (\cos \zeta_1 + \cos \zeta_2) - \iota (\mu_1 a \sin \zeta_1 + \mu_2 b \sin \zeta_2) \right| \leq 1 \quad \forall |\zeta_1| \leq \pi, |\zeta_2| \leq \pi,$$

where $\zeta_i = \xi_i h_i$, $i = 1, 2$. Stability therefore holds if

$$\max \{\mu_1|a|, \mu_2|b|\} \leq 1/2.$$

More generally, for the system of conservation laws (9.2) we let $A = \frac{\partial \mathbf{f}}{\partial \mathbf{u}}$, $B = \frac{\partial \mathbf{g}}{\partial \mathbf{u}}$, freeze the coefficients, and obtain a *necessary* condition for stability involving the spectral radii of the two matrices A and B,

$$\max \{\mu_1 \rho(A), \mu_2 \rho(B)\} \leq 1/2. \tag{9.4}$$

In one dimension, to recall, we have a bound similar to (9.4), except that the constant at the right-hand side equals 1, not 1/2. Similarly, for the heat equation in two dimensions,

$$u_t = u_{xx} + u_{yy}, \tag{9.5}$$

applying the usual three-point discretization in each of the spatial variables and forward Euler in time we obtain the stability restriction

$$\max \left\{ \frac{k}{h_1^2}, \frac{k}{h_2^2} \right\} \leq \frac{1}{4},$$

which is again half the allowed step size compared to the one-dimensional case.

The stability restriction for all explicit schemes developed earlier for hyperbolic problems in one dimension, unless they are unconditionally unstable or use repetitive evaluations at each time step, coincides with the CFL condition. This gives the maximum time step size allowed based on physical considerations. In two dimensions, however, the CFL condition is more permissive and reads

$$\max\{\mu_1\rho(A), \mu_2\rho(B)\} \leq 1.$$

It is therefore conceivable (and, as it turns out, also true!) that other explicit schemes exist for the two-dimensional case which have a more generous stability restriction than (9.3) has, without resorting to multistage evaluations.

Another, less happy difference between one and two dimensions is that, in general, it is difficult to obtain a *sufficient* stability condition from (9.4). This is not because of a deficiency in the Fourier analysis but because of the passage from matrix norms to their eigenvalues. To see this better, write the Lax–Friedrichs scheme for the constant coefficient system, assuming also $\mu_1 = \mu_2 = \mu$ (or, equivalently, absorbing h_1/h_2 into B),

$$\mathbf{v}_{j,l}^{n+1} = \bar{\mathbf{v}}_{j,l}^n - \frac{\mu}{2}\left[A(\mathbf{v}_{j+1,l}^n - \mathbf{v}_{j-1,l}^n) + B(\mathbf{v}_{j,l+1}^n - \mathbf{v}_{j,l-1}^n)\right],$$

$$\bar{\mathbf{v}}_{j,l}^n = \frac{1}{4}\left(\mathbf{v}_{j+1,l}^n + \mathbf{v}_{j-1,l}^n + \mathbf{v}_{j,l+1}^n + \mathbf{v}_{j,l-1}^n\right),$$

and note that for a scalar PDE the condition (9.4) guarantees that all coefficients at time level n are nonnegative. Thus, $\|v^{n+1}\|_\infty \leq \|v^n\|_\infty$, which is a very strong stability statement, and no spurious oscillations occur either (recall Example 5.6 and Figures 5.4 and 5.5). We return to this property in Chapter 10. To elevate this to systems with constant coefficients, however, we must require that A and B be simultaneously diagonalizable. If they are, then we can envision a transformation to characteristic variables, as in (8.8), where the equations decouple and nonnegativity holds provided that (9.4) holds. Alas, in practice A and B do not usually commute and hence are not simultaneously diagonalizable.

The situation is theoretically better for *strictly dissipative* schemes where Theorem 5.2 may be extended to many space variables—the details can be found in [76].

The **leapfrog** scheme also extends to more dimensions in an obvious fashion. For the system of conservation laws (9.2), it reads

$$\mathbf{v}_{j,l}^{n+1} = \mathbf{v}_{j,l}^{n-1} - \mu_1\left(\mathbf{f}(\mathbf{v}_{j+1,l}^n) - \mathbf{f}(\mathbf{v}_{j-1,l}^n)\right) - \mu_2\left(\mathbf{g}(\mathbf{v}_{j,l+1}^n) - \mathbf{g}(\mathbf{v}_{j,l-1}^n)\right). \tag{9.6}$$

A Fourier analysis for the constant coefficient problem readily yields the necessary stability condition

$$\mu_1\rho(A) + \mu_2\rho(B) \leq 1. \tag{9.7}$$

If the step sizes in the spatial directions are chosen wisely, so that $\mu_1\rho(A) \approx \mu_2\rho(B)$, then the stability condition (9.7) is similar to that for Lax–Friedrichs (9.4), and it can be improved in principle. Such an improvement is considered in Exercise 2. However, unless A and B are simultaneously diagonalizable, it is hard to guarantee stability for the leapfrog scheme.

The **Lax–Wendroff** scheme generalizes directly, in principle, although in practice it is considerably messier than in one dimension, with mixed derivatives cropping up. Thus, we write the Taylor expansion for $\mathbf{u}(t+k, x, y)$ about $\mathbf{u}(t, x, y)$ and replace the first two time derivatives by their spatial equivalents according to the PDE, before commencing with

9.2. Solving for Implicit Methods

a symmetric, compact nine-point discretization in space. The variants which take advantage of the special form of conservation laws also generalize, the one by MacCormack (5.5c) being particularly direct,

$$\bar{\mathbf{v}}_{j,l} = \mathbf{v}_{j,l}^n - \mu_1 \left(\mathbf{f}_{j,l}^n - \mathbf{f}_{j-1,l}^n\right) - \mu_2 \left(\mathbf{g}_{j,l}^n - \mathbf{g}_{j,l-1}^n\right), \qquad (9.8)$$

$$\mathbf{v}_{j,l}^{n+1} = \frac{1}{2}\left(\mathbf{v}_{j,l}^n + \bar{\mathbf{v}}_{j,l}\right) - \frac{1}{2}\mu_1 \left(\mathbf{f}(\bar{\mathbf{v}}_{j+1,l}) - \mathbf{f}(\bar{\mathbf{v}}_{j,l})\right) - \frac{1}{2}\mu_2 \left(\mathbf{g}(\bar{\mathbf{v}}_{j,l+1}) - \mathbf{g}(\bar{\mathbf{v}}_{j,l})\right).$$

The necessary stability condition for the Lax–Wendroff scheme, which is sufficient for suitable variable-coefficient systems because of strict dissipativity, turns out to be

$$\max\{\mu_1 \rho(A), \mu_2 \rho(B)\} \leq \frac{1}{2\sqrt{3}}. \qquad (9.9)$$

The bound in (9.9) is not ideal, and several variants have been proposed to improve it up to 1; see Example 9.4 as well as [76, 133] and references therein.

Finally, the approach of applying a **Runge–Kutta** discretization in time to a centered semi-discretization in space generalizes directly. The classical RK4 method is dissipative and thus stable for strictly hyperbolic systems of variable coefficient PDEs, but with the maximal time step again cut to essentially half the corresponding one in one dimension.

The **Crank–Nicolson** and **box** schemes, together with their variants, and also the **BDF** methods we have seen for parabolic problems, are unconditionally stable.

9.2 Solving for implicit methods

When using an implicit discretization in time the solution values at the unknown level $n+1$ are coupled through the algebraic equations forming the discretization. This necessitates solving a large system of algebraic equations, typically the size of the mesh times the number of PDEs. The algebraic system would be nonlinear if nonlinear terms of the PDE are discretized implicitly, but here we consider at first a linear PDE, where the major departure from the one-dimensional case already appears, returning to nonlinear problems toward the end of the section.

Let us consider, for instance, the Crank–Nicolson discretization for the simplest heat equation (9.5) with the simplest mesh. Setting $\mu = k/h^2$ and $h_1 = h_2 = h$ the scheme reads

$$\left[1 - \frac{\mu}{2}D_{x+}D_{x-} - \frac{\mu}{2}D_{y+}D_{y-}\right]v_{j,l}^{n+1}$$
$$= \left[1 + \frac{\mu}{2}D_{x+}D_{x-} + \frac{\mu}{2}D_{y+}D_{y-}\right]v_{j,l}^n. \qquad (9.10)$$

Assume that this heat equation is defined on the unit square in space, i.e., $\Omega = \{0 \leq x, y \leq 1\}$, and that the boundary conditions are Dirichlet, i.e., $u|_{\partial\Omega} = g(t, \cdot, \cdot)$ for $t \geq 0$. Then the discretization (9.10) is defined for the J^2 unknowns $v_{j,l}^{n+1}$ with $j, l = 1, \ldots, J$, $(J+1)h = 1$, and with $v_{0,l}^{n+1}, v_{J+1,l}^{n+1}, v_{j,0}^{n+1}, v_{j,J+1}^{n+1}$ given by the boundary conditions g for all relevant j, l.

In the corresponding one-dimensional case we write the algebraic equations in matrix form, ordering the unknowns by their natural correspondence to mesh locations. This yields a tridiagonal system, which can be solved quickly using the algorithm described in Section 2.12.1. For a system of such PDEs we cannot expect a tridiagonal matrix, but in one dimension there is a narrow band (independent of J) which contains all nonzero elements of the matrix to be "inverted." For the KdV equation we have to deal with a penta-diagonal matrix (see Section 7.3), and moreover, for periodic BC there are additional terms. A variant of Gaussian elimination for the latter introduces additional zeros, but only $O(J)$ of them, during the decomposition stage. The overall work and storage required for all the one-dimensional cases remains $O(J)$, which is comparable to the asymptotic cost of an explicit time step, at least if issues of parallelization and vectorization are ignored.

This situation changes in more than one space variable. There is really no natural ordering of mesh unknowns into a vector in the multidimensional case. We can order them lexicographically by rows, or by columns, or even more innovatively in a nested dissection or minimal degree fashion—methods that we do not discuss here. Whichever ordering we use, in the end some unknowns that appear in (9.10) and are neighbors on the two-dimensional mesh are no longer neighbors in the vector of unknowns. Thus, while the obtained matrix is still sparse (a trademark of finite difference schemes, shared by finite volume and finite element techniques, but not shared by spectral methods or methods based on integral equations), its nonzero diagonals no longer huddle together next to the main diagonal. In the simple case of (9.10) for $J = 10$ we get the sparsity pattern depicted in Figure 9.1. This corresponds to ordering the unknowns either by row or by column. We

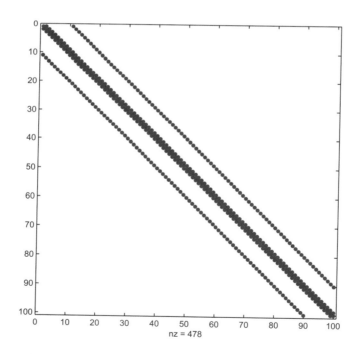

Figure 9.1. *Sparsity pattern of the matrix arising using an implicit method for the heat equation on a square.*

9.2. Solving for Implicit Methods

have five nonzero diagonals, three of which are the main-, the super-, and the subdiagonals. The remaining two diagonals are separated from the main diagonal by $O(J)$ zero diagonals.

If the system is treated as banded, and LU-decomposition is applied as described in Section 2.12.1, then the zero diagonals outside the outer band remain undisturbed, but those within the band get filled with nonzeros, in general. This is referred to as a **fill-in**. Such a Gaussian elimination algorithm therefore costs $O(J^3)$ storage locations and $O(J^4)$ operations, as compared to $O(J^2)$ operations and storage for an explicit scheme. Unfortunately, problems in several space variables on domains with complex geometries are precisely where explicit methods may require exceedingly small time steps due to small spatial widths that crop up near irregular boundaries; see Figure 9.2. Therefore, it is ill-advised precisely here to just ignore implicit methods.

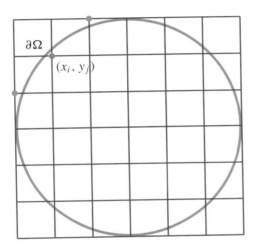

Figure 9.2. *Interior mesh points can get close to a boundary $\partial\Omega$ that does not align with the mesh.*

This sets the stage for possibly opting out of *direct methods* for solving linear algebraic systems and considering *iterative methods* as an alternative. Such methods usually have significantly lower storage requirements, they are often easier to set up, and they often require fewer flops as well. An added advantage here is that a good initial approximation for the unknowns v^{n+1} is given by the values at the current level v^n or an extrapolation of solutions at previous time levels. Matrices arising from the discretization of parabolic problems are particularly suited for the application of methods such as **multigrid, preconditioned conjugate gradient**, and other **Krylov space** methods surveyed in Section 9.4, because the matrices involved are typically (at least close to being) symmetric positive definite. When used correctly, such methods may require in some cases as little as one or two iterations per time step, bringing the operation count back to the optimal $O(J^2)$ per time step.

Before continuing let us remark that the above statements based on flop counts can be misleading in practice. Issues such as rapid memory access, which give the nod to algorithms that can be expressed in terms of vector operations and use high level BLAs wisely, can have a major impact on CPU time, and not only in MATLAB. Among explicit discretization methods for parabolic PDEs there is the **Runge–Kutta–Chebyshev** family, although the utility of these methods drops when considering more involved PDEs and domain geometry. Moreover, sparse matrix technology today is at such a level that many problems arising from implicit methods for PDEs in two dimensions can be solved rather rapidly using direct methods. See, for example, Davis [56]. There is a publicly available code called MUMPS [1]; see also the descriptions and user manuals of `SuperLU` and `ScaLAPACK` at Demmel's site,

http://www.cs.berkeley.edu/~demmel/

There are many such computations, especially but not only in finite element methods, where linear system solvers are not the bottleneck issue any more. The utility of iterative methods is really more crucial for PDE problems in three and more space variables or when very fine spatial meshes are involved, and when one does not possess a large machine architecture. What this amounts to is that in many situations the methods discussed next are the methods or choice, but often there is also some practical alternative breathing down their neck.

9.2.1 Implicit methods for parabolic equations

The methods described in Section 9.4 are directly applicable to discretized elliptic problems. This is what we face at each step of an implicit time discretization of a parabolic PDE with appropriate initial and boundary values prescribed.

Consider, for instance, the simple heat equation in the unit cube in d space dimensions

$$u_t = \sum_{l=1}^{d} \frac{\partial^2 u}{\partial x_l^2} + q(t, x_1, \ldots, x_d) \tag{9.11a}$$

with Dirichlet BC. The source function q is assumed known. Applying an implicit time discretization, and the standard three-point centered discretization on a uniform mesh with step size $h = 1/(J+1)$ in each space variable, yields an $m \times m$ system of linear equations,

$$A\mathbf{v}^n = \mathbf{q}^n, \quad n = 1, 2, \ldots, n_f. \tag{9.11b}$$

Thus we have a linear algebraic system for each time step n, where the mesh function v^n approximating the solution u at $t = t_n$ has been reshaped as a vector. The known right-hand side contains the effect of the time discretization as well. Note that $m = J^d$ and that the superscripts in (9.11b) are not powers.

The *condition number*[29] can be shown to satisfy

$$\text{cond}(A) = O(J^2).$$

---

[29]Recall that the *condition number* of a matrix A is defined by

$$\text{cond}(A) = \|A\| \|A^{-1}\|. \tag{9.12a}$$

Any induced matrix norm would do, but in the 2-norm we have

$$\text{cond}(A) = \frac{\max \sigma_i}{\min \sigma_i}, \tag{9.12b}$$

where σ_i are the singular values of A. Recall also that if A is symmetric positive definite, then $\sigma_i = \lambda_i$, i.e., eigenvalues replace singular values in (9.12b).

9.2. Solving for Implicit Methods

In particular, cond(A) grows unboundedly as $h \to 0$. For reasonable resolution values, say, $J = 100$ or even $J = 1000$, this condition number is not large enough to cause roundoff error trouble when using standard double precision floating point arithmetic. But it does slow down iterative methods such as conjugate gradients (CG) considerably, which is why we are bringing up the subject here and not earlier in the book. We urge you at this point to read Section 9.4: see in particular Example 9.12 and Figure 9.12.

Can the condition number be made to be bounded as resolution increases using another discretization method? This question and related ones are considered at length for ODE systems in [10]. A reformulation of a higher order ODE as a first order system prior to discretization or through it can help reduce the condition number, but unless the differential problem is reformulated as an integral equation, the discretization matrix will have a condition number that grows unboundedly with J.

To understand this better in the PDE context, consider (9.11a) in steady state. Recall from Review Section 1.3.3 that u is generally smoother than q, and thus u belongs to a Sobolev space which is a strict subspace of the space to which q belongs. But in (9.11b) both $\mathbf{v}^n$ and $\mathbf{q}^n$ belong to the same space, $\mathbb{R}^m$. The smoothing operation of the solution operator is grossly undervalued being approximated by the boundedness of $\|A^{-1}\|$, while the differentiation process leading from u to q does translate into unboundedness of $\|A\|$ in the limit.

The next natural question is, Which of the iterative methods of Section 9.4 are particularly suitable in the context of parabolic PDEs?

One advantage in the context of multigrid methods is that for parabolic problems relaxation is easy because the PDE is dissipative. Thus, a consistent discretization can naturally provide a smoother! See, e.g., [171, 175] and recall also Section 1.1. In some situations only one or two multigrid cycles are needed per time step [18, 17].

More generally, note that we have here a *family* of algebraic systems to solve, with time as the family parameter. These systems can be very similar to one another, in some cases differing only in their right-hand sides. This brings up the important question of a **warm start**: can the n_f linear systems (9.11b) sharing the same matrix be solved in much less effort than n_f times the effort required to solve one of them? The answer is affirmative. Using direct methods, an LU decomposition is applied to A only once, and then for each right-hand-side vector $\mathbf{q}^n$ only forward and backward substitutions are required; see Section 2.12.1.

In contrast, warm starting iterative methods in a similar way is generally an open research question. On the other hand, an obvious warm start effect not shared by direct methods can be obtained by using previous solutions for the initial guess in a PCG algorithm. The simplest is to set the initial iterate to the solution at the previous time level, $v^{n+1,0} = v^n$. Another possibility is to extrapolate from the previous time levels, $v^{n+1,0} = 2v^n - v^{n-1}$.

Some warm starting can be applied through the choice of a preconditioner, especially in case of a linear PDE with constant coefficients. If the preconditioner is a multigrid cycle, or an SSOR iteration, then not much sharing among systems with different n can be achieved. However, an ILU preconditioner can be shared, i.e., the same incomplete decomposition can be used for different linear systems (9.11b) to efficiently obtain the preconditioning effect. This can be done even for problems arising as linearizations for nonlinear PDEs, where A in (9.11b) does depend on n, so long as it does not vary too rapidly in time.

It is important to stress again that neither the Krylov space iteration nor the preconditioner (when done right) require forming A: only matrix-vector multiplications are performed. The number of iterations required at each time step should remain comfortably small given the excellent initial guess, or else the preconditioner is deemed ineffective.

Example 9.1 (Aquifer problem) This application is described in [69] and elsewhere. The motion of water in an aquifer is governed, under certain assumptions, by the PDE

$$b(x, y)u_t = (a(x, y)u_x)_x + (a(x, y)u_y)_y + q(t, x, y),$$

where u is the piezometric head height and q is a known function. The transmissivity a and the storage coefficient b are positive parameter functions which characterize the aquifer. Initial conditions, as well as a combination of Dirichlet and Neumann boundary conditions, are prescribed.

This problem must be solved for u in order to simulate an efficient water extraction from the aquifer. But first, the coefficient functions a and b must be found to calibrate the aquifer. To simplify, let us assume that $b \equiv 1$ and concentrate on finding a. This is done by observing values of u for certain initial and boundary data and solving the **inverse problem**, finding both a and u which would satisfy the PDE problem and fit the data well enough. What "well enough" means depends on the amount of noise in the measured data. The inverse problem, which we do not get into here, involves a regularization (see, e.g., [178, 60, 166]), and an iterative procedure in which u is solved for based on current iterate values of $a(x, y)$, and then the latter are updated. This requires solving the *forward problem* (i.e., finding u given a) efficiently and robustly.

Assuming a square spatial domain discretized by a uniform mesh with spacing h, we utilize a semi-discretization as in Example 3.4,

$$\frac{dv_{j,l}}{dt} = h^{-1}\left[a_{j+1/2,l}\frac{v_{j+1,l} - v_{j,l}}{h} - a_{j-1/2,l}\frac{v_{j,l} - v_{j-1,l}}{h}\right.$$
$$\left. + a_{j,l+1/2}\frac{v_{j,l+1} - v_{j,l}}{h} - a_{j,l-1/2}\frac{v_{j,l} - v_{j,l-1}}{h}\right] + q(t, x_j, y_l). \quad (9.13)$$

Assume for simplicity that the BC are Dirichlet all around a square boundary.[30] Then in (9.13), the subscripts are over the range $j, l = 1, \ldots, J$, and $v_{0,l}, v_{J+1,l}, v_{j,0}$ and $v_{j,J+1}$ are given by the BC for all t.

Next, we discretize (9.13) in time using an implicit method such as backward Euler, or a higher order BDF method, or the trapezoidal rule. In any of these cases we obtain at each time step a system of equations that is linear for a given $a(x, y)$. The system has the form (9.36) with $\mathbf{x}$ corresponding to the J^2 unknowns v^{n+1} ordered by row or by column and A is the corresponding matrix that has the sparsity structure of Figure 9.1.

For the solution of this system we can use a preconditioned CG algorithm, starting, e.g., from the initial guess v^n for v^{n+1}. For the preconditioner we can use either a multigrid cycle [171] (this gets trickier, although still possible, if a is bumpy), or an *incomplete Cholesky decomposition*, as described above. Indeed, here is a case where the same matrix appears at each time step as in (9.11), provided the time step is constant.

---
[30] In Section 3.1 we discuss in detail how Neumann BC can be discretized in a way compatible with the finite volume approach.

9.2. Solving for Implicit Methods

For a simple numerical example, let $a(x, y) = 1 + x + y$ on the unit square and construct the inhomogeneity $q(t, x, y)$ as well as initial and Dirichlet boundary data from the exact solution

$$u(t, x, y) = \sin(\pi x)(1 - e^{-y})e^{-t}.$$

Given a spatial step size h for the semi-discretization (9.13), a forward Euler time discretization requires $k \leq h^2/12$ to be stable. Using a Crank–Nicolson scheme of comparable accuracy instead, we can choose $k = h$, so the time step size is larger by a factor of $12/h$. Thus, a simple CG method without preconditioning would do well in comparison if the number of iterations it requires per time step is well below $12/h$.

The actual maximum absolute error at $t = 1$ for $h = .005$, using k as specified above for either time discretization method, is around 2.e-6. So we set the absolute convergence tolerance for the CG method at $tol = (h/.005)^2 10^{-6}$ and count the average number of iterations per step. The results are recorded in Table 9.1. Plots of the solution and its error are given in Figure 9.3.

Table 9.1. *Crank–Nicolson with conjugate gradients (CN-CG) for a diffusion problem. The maximum error at $t = 1$, recorded under Error, is $O(h^2)$ and dominates the CG convergence error. The average number of required CG iterations is listed under Iterations. The relative amount of work for a comparable purpose using forward Euler is listed under $12/h$. Even when no preconditioning is used the CN-CG method is far more efficient than the corresponding explicit time discretization method.*

$h = k$	Error	Iterations	$12/h$
.04	1.2e-4	25	300
.02	3.0e-5	44	600
.01	7.5e-6	74	1200
.005	1.9e-6	115	2400

We hasten to point out again that using a good preconditioner will further improve the efficiency of the CG method significantly. Our purpose here is to emphasize the *simplicity* of an implementation for the implicit method which already yields major gains. No matrix has been formed, let alone inverted, here, and the basic matrix-vector multiplication performed at each CG iteration is as simple as for the forward Euler step. ■

Example 9.2 (Inverting electromagnetic data) This case study briefly describes the work in [78], which in turn culminates work reported in several earlier papers by Haber and his collaborators. The problem is essentially to reconstruct, or at least image, the inhomogeneous **conductivity** in some survey area in earth from measurements of the electromagnetic field for a given source. It is posed and solved in time and three space dimensions. The purpose may be aiding a mining company to decide whether there is enough ore in a particular site that is worth extracting commercially. Or perhaps it is suspected that there is a gold box buried by pirates centuries ago under the sand in a particular beach. Or the

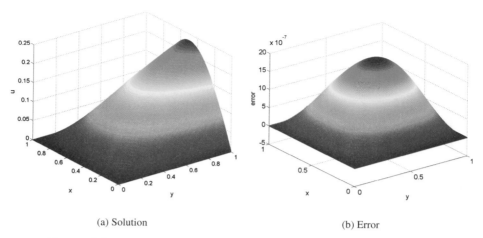

(a) Solution (b) Error

Figure 9.3. *Solution and error for the diffusion equation of Example 9.1 at $t = 1$ using the Crank–Nicolson scheme with $h = .005$. See Table 9.1.*

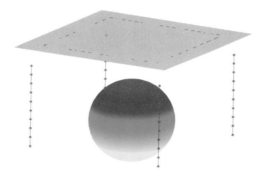

Figure 9.4. *Borehole geometry for Example 9.2. The dashed line at the surface is a loop source. All three components of the magnetic field are acquired in the boreholes.*

methodology can be possibly used to discover mines left during past wars in an area under examination. Such objects all have much higher conductivity than that of the sand or soft rock presumably surrounding them.

The inverse problem for Example 9.1, namely, estimating $a(x, y)$ there from given measurements on the field u, is of a somewhat similar nature mathematically, although it is rather simpler than the present problem. We have therefore allowed ourselves to only solve the forward problem numerically by simple means there, since we discuss here solving both the forward and the inverse.

An example for the experimental setting is depicted in Figure 9.4, which, like Figure 9.5, is taken from [78]. The source is the current along a square loop with dimensions 50×50 meters. The data are measured at 20 depths uniformly spaced down each of

9.2. Solving for Implicit Methods

four boreholes at 18 logarithmically spaced time instances between 10^{-4} and 10^{-1} seconds. The "true model" is that of a sphere (radius $15m$, conductivity $\sigma = 0.1 S/m$) buried in a uniform half-space that has conductivity $\sigma = 0.01 S/m$. Using this we generate data by solving the *forward problem* described below for the electromagnetic field, sampling all three magnetic components at the locations described above, adding 2% normally distributed random noise, and then "forgetting" the true model and trying to reconstruct it from these data. The results are depicted in Figure 9.5. For such sparse data and highly ill-posed inverse problem, this is actually a successful reconstruction. Our next step is then to see what is involved numerically.

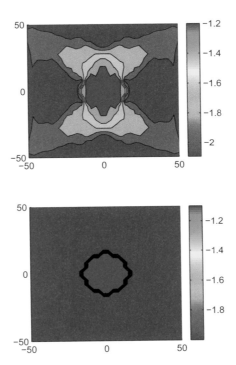

Figure 9.5. *Result of inversion of time domain electromagnetic data. The top panel is a horizontal slice of the recovered conductivity at a depth corresponding to the center of the sphere. The true conductivity is shown in the bottom panel.*

The forward model involves solving Maxwell's equations given in Example 3.6, except that there is an additional, known source $\mathbf{s}_r(t)$. The boundary conditions along the spatial domain boundary $\partial \Omega$ are

$$\mathbf{n} \times \mathbf{H} = \mathbf{0},$$

where $\mathbf{n}$ is the outward normal in three dimensions.

Given typical earth parameters these equations represent wave phenomena over very short time scales (up to 10^{-7} seconds, say). However, over longer times the equations tend to exhibit heavy dissipation, behaving more like stiff parabolic than like hyperbolic PDEs. There is an initial layer where the solution varies rapidly, but later on it is smoother in time. Moreover, the data measurements are not taken within this initial layer. Hence, stepping with a relatively large time step size that in particular skips the layer is sensible and desirable, and this gives rise to *L-stable* time discretization methods. There are additional reasons that rule out higher order methods in the inverse problem context, so the method of choice here is **backward Euler**. With $\alpha_n = (t_{n+1} - t_n)^{-1}$ being the inverse of the nth time step we have the semi-discretization in time

$$\nabla \times \mathbf{E}^{n+1} + \alpha_n \mu \mathbf{H}^{n+1} = \alpha_n \mathbf{H}^n \equiv \mathbf{s}_H \quad \text{in } \Omega, \quad (9.14a)$$

$$\nabla \times \mathbf{H}^{n+1} - (\sigma + \alpha_n \epsilon) \mathbf{E}^{n+1} = \mathbf{s}_r^{n+1} - \alpha_n \epsilon \mathbf{E}^n \equiv \mathbf{s}_E \quad \text{in } \Omega, \quad (9.14b)$$

$$\mathbf{n} \times \mathbf{H}^{n+1} = 0 \quad \text{on } \partial\Omega. \quad (9.14c)$$

Here solution quantities at $n + 1$ are unknown while those at n are known. Note that the time steps are such that data measurements are at mesh points. In the particular experiment shown we have 32 time levels spaced nonuniformly.

Note that we have deviated here from our usual semi-discretization treatment, which consists of applying spatial discretization first. Here we discretize in time first, because that choice is independent of the spatial discretization and yields an elliptic-like PDE system at each time level. What corresponds to discretizing in time first also arises in DAEs (see Section 2.8), where a set of algebraic equations resulting from the implicit discretization and the equations that do not contain a time derivative are simultaneously solved at each new time level. This also arises when differential equations are subjected to *inequality constraints*, for example, when simulating multibody systems with frictional contact, a problem that arises in mechanical engineering and computer graphics [2, 106].

To solve the initial-boundary value problem of which (9.14) describes one step, we apply a spatial discretization. Here, however, the plot thickens. Because the `curl` operator annihilates any function of the form $\nabla \phi$, it has a nontrivial null-space. A good, conservative finite volume method such as the one described in Example 3.6 may thus give rise to a large, sparse linear system of equations at each time level that is a challenge to solve using iterative methods. The challenge is, essentially, to restrict the iteration only to the active space, avoiding the null-space because there the iteration is totally ineffective. This removal of the null-space can indeed be done at the discrete level, e.g., inside a multigrid solver [91]; however, the resulting code can be complex and hard to modify. Another approach used in [78] is to remove the null-space first at the PDE level of (9.14), using a variable transformation called the *Helmholtz decomposition with Coulomb gauge*. The details can be found in [78] and further references therein and are too specific to this application to resurrect in full here. Suffice it to say for the present purpose that the resulting system at each time step is a discretization of four coupled Poisson equations of the form encountered in Example 3.4, and a finite volume method such as (3.15) is relevant.

To solve the resulting linear system, note that for α_n sufficiently small (time step sufficiently large!) the obtained system is dominated by its diagonal blocks, and therefore a good preconditioner can be obtained by solving for each Poisson equation separately. Based on this we have used one multigrid cycle and/or ILU to obtain effective preconditioners for a BICGSTAB iteration.

9.2. Solving for Implicit Methods

Combining all the time step discretizations together yields a system of the form

$$\begin{pmatrix} A_1(m) & & & & \\ B_2 & A_2(m) & & & \\ & B_3 & A_3(m) & & \\ & & \ddots & \ddots & \\ & & & B_s & A_N(m) \end{pmatrix} \begin{pmatrix} u_1 \\ u_2 \\ \vdots \\ \vdots \\ u_N \end{pmatrix} = \begin{pmatrix} q_1 \\ q_2 \\ \vdots \\ \vdots \\ q_N \end{pmatrix}$$

for N time steps. We write out explicitly the dependence on our model m, which can be, for instance, $m = \log \sigma$. This is our reconstruction goal in the inverse problem.

Obviously a block forward substitution can be performed, combined with the above iterative techniques for each A_n. We can write the system (before elimination) as

$$A(m)u = q.$$

This is then a very large, sparse system of algebraic equations that we know how to solve using iterative methods.

For the inverse problem, to recall, we have *observed* data b that correspond to certain components of the field u at particular points, denoted Qu. Thus we seek a model m such that the *predicted data* Qu, where $u = A(m)^{-1}q$, matches the observed data b to within the often unknown noise level. This problem turns out to be highly ill-posed, and some form of regularization is required in order to solve a saner, nearby problem and to incorporate a priori information such as that the solution model m is a smooth mesh function and not just an assortment of values at pixels. A Tikhonov-type regularization then seeks to minimize a data fitting term plus a penalty on the roughness of m

$$\min_{m,u} \frac{1}{2} \|Qu - b\|^2 + \beta R(m), \qquad (9.15)$$
$$\text{s.t.} \quad A(m)u = q.$$

The regularization functional can be a discretization of the integral

$$\frac{1}{2} \int_\Omega |\nabla m|^2 d\mathbf{x}.$$

But there are other choices, too, as briefly mentioned in Example 3.5. The parameter β balances the two terms of regularization and data fitting, being larger the higher the noise level.

To solve (9.15) we can eliminate u from the forward problem, yielding a minimization problem in m alone. The latter can be solved using a Gauss–Newton technique, employing a preconditioned CG procedure for the inner iteration. Or else we can solve the constrained formulation (9.15) without elimination. Thus we find ourselves iterating *all at once* for m, u and the Lagrange multiplier functions resulting upon stating the necessary conditions for the constrained minimization. However, the present description and the narrator are getting carried further and further away from the purpose of this example in its context, so let us bring this discussion to a halt, referring again to [78] and references therein for much more. ∎

Finally, we mention that there are packages available on NETLIB such as VODPK and DASPK, written by Hindmarsh, Petzold, and coworkers, which use preconditioned Krylov space methods in combination with the method of lines for systems of time-dependent PDEs.

Next, we proceed to briefly discuss a class of methods that make sense only in the present context and leads naturally to Section 9.3 as well.

9.2.2 Alternating direction implicit (ADI) methods

When ordering the unknowns corresponding to (9.10) by rows, the unknowns corresponding to consecutive mesh locations in x (i.e., $v_{j-1,l}$, $v_{j,l}$, $v_{j+1,l}$) remain close neighbors in the vector of unknowns, whereas the neighbors in the y direction are not. But when ordering the unknowns by column it is the neighbors in the y direction (i.e., $v_{j,l-1}$, $v_{j,l}$, $v_{j,l-1}$) which remain close neighbors. This suggests an iterative method where the ordering alternates according to rows or columns, and in which only the close neighbors are solved for. Such methods were proposed in the 1950s by Peaceman and Rachford, Douglas, and others, and they remain popular today among some practitioners for sufficiently simple, nonautomated situations.

Let us consider the Crank–Nicolson equations (9.10) on a simple domain. We can write them in operator form

$$\left[I - \frac{\mu}{2}(Q_L + Q_M)\right] v^{n+1} = \left[I + \frac{\mu}{2}(Q_L + Q_M)\right] v^n, \tag{9.16}$$

where v^n is the mesh function at level n: this is the collection of $(J+2)^2$ solution values corresponding to the spatial mesh at time t_n *not yet ordered* into a vector in one way or another. Note that v^{n+1} incorporates J^2 unknown interior mesh values, plus known boundary values.

Next, we **split** the operator, approximating (9.16) by

$$\left(I - \frac{\mu}{2}Q_L\right)\left(I - \frac{\mu}{2}Q_M\right) v^{n+1} = \left(I + \frac{\mu}{2}Q_L\right)\left(I + \frac{\mu}{2}Q_M\right) v^n. \tag{9.17a}$$

Note that the difference between (9.16) and (9.17a) is

$$k\frac{\mu^2}{4} Q_L Q_M \frac{v^{n+1} - v^n}{k},$$

where $\mu = k/h^2$. When we substitute the exact solution (with bounded second derivatives) this term is $O(k^3)$. Hence, (9.17a) retains the accuracy order $(2, 2)$. Writing the above equations as

$$\left(I - \frac{\mu}{2}Q_L\right) w = \left(I + \frac{\mu}{2}Q_M\right) v^n,$$
$$\left(I - \frac{\mu}{2}Q_M\right) v^{n+1} = \left(I + \frac{\mu}{2}Q_L\right) w \tag{9.17b}$$

(the formulations (9.17a) and (9.17b) are equivalent because the operators $I + \frac{\mu}{2}Q_L$ and $(I - \frac{\mu}{2}Q_L)^{-1}$ commute), we obtain two half-steps each corresponding to inversion in only

9.2. Solving for Implicit Methods

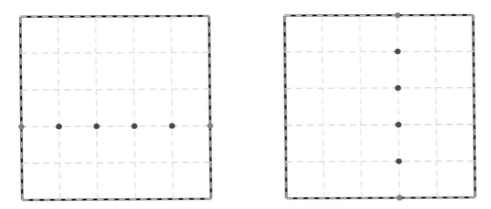

Figure 9.6. *ADI method. Solve tridiagonal systems in one direction, then in the other.*

one direction. Thus, we solve for w by ordering the unknowns horizontally, obtaining a set of tridiagonal systems (one for each mesh row, using boundary values at the vertical boundary segments), and then we solve another set of tridiagonal systems for v^{n+1}, ordering the unknowns vertically and using boundary conditions at the horizontal boundary segments; see Figure 9.6. Since each solution of a tridiagonal system has a cost which is linear in the number of unknowns, the total cost per step is the optimal $O(J^2)$, both in flops and in storage.

One question arising is, What are the appropriate boundary conditions for w? If the boundary values g depend on t, then one must avoid the temptation of simply setting the values of w corresponding to boundary locations to $g(t_{n+1/2}, \cdot, \cdot)$, which yields only a first order approximation. Instead, add the equations

$$\left(I - \frac{\mu}{2} Q_L\right) w = \left(I + \frac{\mu}{2} Q_M\right) v^n,$$
$$\left(I + \frac{\mu}{2} Q_L\right) w = \left(I - \frac{\mu}{2} Q_M\right) v^{n+1},$$

obtaining

$$w = \frac{1}{2}\left[\left(I + \frac{\mu}{2} Q_M\right) v^n + \left(I - \frac{\mu}{2} Q_M\right) v^{n+1}\right]. \tag{9.18}$$

Now substitute $g(t_n, \cdot, \cdot)$ and $g(t_{n+1}, \cdot, \cdot)$ for the boundary values of v^n and v^{n+1}, respectively, obtaining the corresponding boundary values of w.

There are other variants of dimensional splitting methods such as ADI in both two and three dimensions; see [133] and also the more recent but less detailed [137]. One persistent question with these methods is how to specify intermediate (or split-step) BC.

In Exercise 5 we ask our patient reader to extend and implement the ADI method for Example 9.1. Once the appropriate discretization, domain geometry, and BC combine well, a very simple solution procedure is obtained. More generally, however, the high degree of specialization required to use these methods gives the nod, in our opinion, to the more modern, general approaches described in Section 9.4 and utilized in Section 9.2.1.

9.2.3 Nonlinear problems

If the PDE system that is being discretized by an implicit method is nonlinear, then the large system of algebraic equations is also nonlinear. Bearing in mind the lessons of Section 2.7, in light of the mild stiffness present in parabolic problems we require an appropriate variant of Newton's method. The situation is not really different in this regard from that described in Section 2.7, and there is no reason to duplicate that exposition here. Instead we make two additional, important observations.

The first is that incorporating a Newton iteration with an iterative method for the resulting linear system yields an **inner–outer iteration**, with one of the methods described in Sections 9.2.1 and 9.4 providing the inner loop. A simple idea for gaining efficiency is to cut the inner iteration short when the outer, Newton iteration, is not yet close to the solution and thus not so important to resolve exactly. This leads to *inexact Newton* methods; see, e.g., [139, 78]. These methods can have central importance in some applications but are less crucial in the present context of solving a time-dependent PDE problem where we really don't expect to apply more than one or two Newton iterations per time step anyway. It is the tolerance for the convergence of the exact Newton method that can be set relatively loosely here, depending on the accuracy of the PDE discretization.

The second observation is that the Newton linearization (recall the review in Section 2.12.5) can be applied directly to the PDE rather than to the discretized algebraic equations. This often less messy approach to deriving Newton's method for PDE discretizations is also called **quasi-linearization**. See the description for the midpoint method in Section 7.3.1 for an instance.

9.3 Splitting methods

The splitting idea, utilized in a special way in Section 9.2.2 to obtain the ADI method, can be applied in much more general circumstances. It can be applied directly to the PDE, not just to its discretization, and it can be applied to split different PDE components, as discussed later in Section 9.3.1.

Example 9.3 The Cauchy problem for the simple, scalar advection equation

$$u_t + au_x + bu_y = 0,$$
$$u(0, x, y) = u_0(x, y)$$

has the solution

$$u(t, x, y) = u_0(x - at, y - bt)$$

(this can be verified directly). Further, treating each y as a parameter, let $\phi(t, x) = \phi(t, x; y)$ satisfy the one-dimensional advection equation

$$\phi_t + a\phi_x = 0,$$
$$\phi(0, x) = u_0(x, y).$$

Then from Section 1.1 we know that

$$\phi(t, x; y) = u_0(x - at, y).$$

9.3. Splitting Methods

Similarly, treating each $\hat{x} = x - at$ as a parameter, let $\psi(t, y) = \psi(t, y; \hat{x})$ solve the one-dimensional advection equation

$$\psi_t + b\psi_y = 0,$$
$$\psi(0, y) = u_0(\hat{x}, y) = \phi(t, x; y).$$

Then

$$\psi(t, y; \hat{x}) = u_0(x - at, y - bt) = u(t, x, y).$$

Thus, we can split the solution process of the two-dimensional advection equation into solutions of one-dimensional advection equations. Here there is no splitting error at all, which is unusual in the treatment of general splitting techniques. ∎

In terms of the solution operator we generalize (1.12) into

$$u(t, \cdot, \cdot) = \mathcal{S}(t)u_0, \qquad (9.19a)$$

where $\mathcal{S}(t)$ can be written as

$$\mathcal{S}(t) = \mathcal{S}^y(t)\mathcal{S}^x(t) \quad \forall t. \qquad (9.19b)$$

The solution operators for the simple advection equation in Example 9.3 are

$$\mathcal{S}^x(t) = e^{-ta\partial_x}, \quad \mathcal{S}^y(t) = e^{-tb\partial_y},$$
$$\mathcal{S}(t) = e^{-t(a\partial_x + b\partial_y)} = e^{-tb\partial_y}e^{-ta\partial_x},$$

where the exponent of a differential operator is to be understood, as for a matrix, in terms of its Taylor expansion; see Section 1.3.5.

While the splitting in the case of the scalar advection operator is exact, for matrix operators it is only approximate. For example, corresponding to the constant coefficient hyperbolic system (9.1) we obtain

$$e^{kA\partial_x}e^{kB\partial_y} - e^{k(A\partial_x + B\partial_y)} = \frac{1}{2}k^2(BA - AB)\partial_x\partial_y + O(k^3).$$

The same general lack of commutativity between A and B, which haunts us when trying to establish stability properties for conservative schemes, here reduces the approximation order to 2, leading in general to a first order accuracy in k.

For a semi-discretization of a constant coefficient PDE in two spatial variables,

$$\frac{dv}{dt} = (L + M)v,$$

the solution is

$$v(t + k) = e^{k(L+M)}v(t).$$

This suggests a natural splitting

$$e^{k(L+M)} \approx e^{kL}e^{kM}. \qquad (9.20)$$

The Crank–Nicolson scheme (9.10) for the heat equation corresponds to approximating e^z by the diagonal Padé approximation $\frac{2+z}{2-z}$. This splitting leads to **locally one-dimensional** (LOD) formulas, which are not as popular as the ADI splitting; see, e.g., [133, 137].

Generally we have as for the matrix PDE operators

$$e^{kL} e^{kM} - e^{k(L+M)} = \frac{1}{2} k^2 (ML - LM) + O(k^3),$$

and an overall first order accuracy in k may result if L and M do not commute.

Fortunately, a simple trick (in hindsight) due to Strang allows restoring an overall second order accuracy; see Exercise 10 and recall Section 6.2.2 as well as the semi-explicit method (7.27). The idea is to replace (9.20) by

$$e^{k(L+M)} \approx e^{\frac{k}{2}L} e^{kM} e^{\frac{k}{2}L}. \tag{9.21}$$

Now, approximate each exponent by a one-dimensional Lax–Wendroff scheme, say. For a system of conservation laws one of the corresponding variants of (5.5) may be used.

Denoting the corresponding full discretization operator by $Q(L, \frac{k}{2})$, etc., we obtain

$$\mathbf{v}^{n+1} = Q\left(L, \frac{k}{2}\right) Q(M, k) Q(L, \frac{k}{2}) \mathbf{v}^n = \cdots$$

$$= Q\left(L, \frac{k}{2}\right) Q(M, k) Q(L, k) Q(M, k) \cdots Q(L, k) Q(M, k) Q\left(L, \frac{k}{2}\right) \mathbf{v}^0.$$

Thus, the first fractional step of the next time step can be combined with the last fractional step of the current one, so there are essentially two one-dimensional half-steps per time step on a mesh that is staggered in time. Note that the application of one of these one-dimensional discretization operators carries with it the corresponding one-dimensional stability restrictions.

Example 9.4 Utilizing the splitting (9.21), the MacCormack scheme (9.8) becomes a four-stage prescription which reads

$$\bar{\mathbf{v}}_{j,l} = \mathbf{v}_{j,l}^{n-1/2} - \mu_1 \left(\mathbf{f}_{j,l}^{n-1/2} - \mathbf{f}_{j-1,l}^{n-1/2} \right), \tag{9.22}$$

$$\hat{\mathbf{v}}_{j,l} = \frac{1}{2} \left(\mathbf{v}_{j,l}^{n-1/2} + \bar{\mathbf{v}}_{j,l} \right) - \frac{1}{2} \mu_1 (\mathbf{f}(\bar{\mathbf{v}}_{j+1,l}) - \mathbf{f}(\bar{\mathbf{v}}_{j,l})),$$

$$\tilde{\mathbf{v}}_{j,l} = \hat{\mathbf{v}}_{j,l} - \mu_2 (\mathbf{g}(\hat{\mathbf{v}}_{j,l}) - \mathbf{g}(\hat{\mathbf{v}}_{j,l-1})),$$

$$\mathbf{v}_{j,l}^{n+1/2} = \frac{1}{2} (\tilde{\mathbf{v}}_{j,l} + \hat{\mathbf{v}}_{j,l}) - \frac{1}{2} \mu_2 (\mathbf{g}(\tilde{\mathbf{v}}_{j,l+1}) - \mathbf{g}(\tilde{\mathbf{v}}_{j,l})).$$

The advantage here is not in solving linear algebraic equations—the schemes are explicit—but in the restoration of the CFL time step restriction

$$\max\{\mu_1 \rho(A), \mu_2 \rho(B)\} \leq 1, \tag{9.23}$$

which is to be compared to the bound (9.9) obtained for the direct extension of the one-dimensional scheme. Remarkably, a significant improvement in the time step size restriction is obtained without much extra effort.

9.3. Splitting Methods

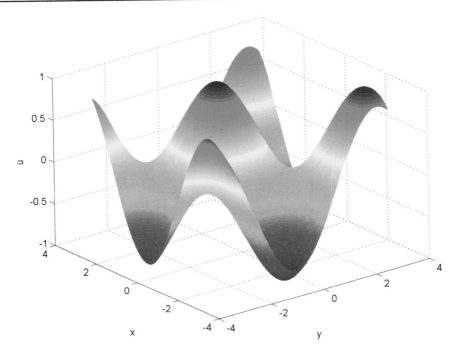

Figure 9.7. *Solution of an advection problem, Example 9.4.*

To see some numbers, consider the advection problem

$$u_t - u_x - 2u_y = 0, \quad (x, y) \in [-\pi, \pi] \times [-\pi, \pi], \ t \geq 0,$$
$$u(0, x, y) = \sin x \sin y,$$

under periodic BC. The exact solution is $u(t, x, y) = \sin(x + t) \sin(y + 2t)$. It is depicted for $t = 1$ in Figure 9.7. The meshes used below for the numerical experiments are fine enough that any of the *stable* configurations in Table 9.2 leads to a plot indistinguishable from Figure 9.7. Here we do not need to stagger the mesh in time to retrieve the second order accuracy of the Lax–Wendroff scheme.

We set $h_2 = 2h_1$, hence $\mu_2 = .5\mu_1$, so corresponding to the notation of (9.23), $\mu_1|a| = \mu_2|b| = \mu_1$. Table 9.2 records maximum errors at $t = 1$ for four methods:

1. Upwind, extending (1.15a) in an obvious fashion: see Exercise 3;

2. Lax–Wendroff as in (9.8);

3. Upwind split in an obvious fashion (see also Example 9.3); and

4. Lax–Wendroff split as in (9.22).

From the results in Table 9.2 it is clear that the upwind scheme is first order accurate and the Lax–Wendroff scheme is second order accurate in both split and unsplit forms.

Table 9.2. *Errors using explicit methods (upwind and Lax–Wendroff) in split and unsplit form for the advection equation in two dimensions. Unacceptably large error due to instability is denoted by* *.

h_1	μ_1	Upwind	L-W	Upwind split	L-W split
$.01\pi$	.45	4.1e-2	8.4e-4	4.2e-2	1.0e-3
$.005\pi$	.45	2.1e-2	2.1e-4	2.1e-2	2.6e-4
$.0025\pi$	.45	1.1e-2	5.3e-5	1.1e-2	6.5e-5
$.01\pi$	.9	*	*	7.7e-3	2.5e-4
$.005\pi$	.9	*	*	3.9e-3	6.2e-5
$.0025\pi$	.9	*	*	2.0e-3	1.6e-5

When $\mu_1 < .5$ the methods are all stable. But when $0.5 < \mu_1 < 1$ the original methods become unstable as predicted by (9.23), while the split methods remain stable.

We emphasize that the lack of splitting error in the computational part of this example is not general. ∎

Example 9.5 The restoration of second order accuracy by the Strang splitting is based on matching another term in the Taylor expansions of the corresponding matrix exponentials— see Exercise 10. Of course, if $\|kL\| \gg 1$ or $\|kM\| \gg 1$, then all bets are off, because the first few terms in the Taylor expansion of a rapidly decaying exponential function say little about the entire series. This situation occurs for stiff problems (recall Section 2.6). Consider the simple ODE system

$$\frac{d\mathbf{v}}{dt} = \begin{pmatrix} -\varepsilon^{-1} & 1 \\ 1 & 1 \end{pmatrix} \mathbf{v}$$

with $\mathbf{v}(t)$ a vector function of only two components and $\varepsilon \ll 1$ a positive parameter. We write this in split form in order to isolate the stiff component,

$$\frac{d\mathbf{v}}{dt} = (L+M)\mathbf{v}, \quad \text{where } L = \begin{pmatrix} -\varepsilon^{-1} & 0 \\ 0 & 1 \end{pmatrix}, \quad M = \begin{pmatrix} 0 & 1 \\ 1 & 0 \end{pmatrix}.$$

Further, we use the exact solution operators and consider the splitting error for step sizes $k \ll 1$ such that possibly $\varepsilon \ll k$. In this stiff situation the Strang splitting loses its second order status and becomes only first order accurate. In Table 9.3 we record absolute errors at $t = 1$ starting from the initial data $\mathbf{v}(0) = (1, 1)^T$. Note that the exact solution is $\mathbf{v}(1) = e^{L+M}\mathbf{v}(0)$. (In MATLAB, use the function expm for small matrices.) In addition to ε and k we record errors for the following splitting methods:

9.3. Splitting Methods

Table 9.3. *Errors in fast and slow solution components using four splitting methods for a simple ODE. The errors are due to splitting only.*

ε	k	$e^{kM}e^{kL}$		$e^{kL}e^{kM}$		$e^{\frac{k}{2}L}e^{kM}e^{\frac{k}{2}L}$		M-inhomog.	
10^{-3}	10^{-2}	2.46e-2	8.20e-3	2.72e-3	3.55e-2	2.54e-3	8.38e-3	3.53e-7	2.43e-2
	10^{-3}	1.58e-3	9.14e-4	1.14e-3	1.81e-3	1.10e-4	1.14e-4	1.02e-8	1.58e-3
	10^{-4}	1.38e-4	1.32e-4	1.34e-4	1.41e-4	1.13e-6	1.14e-6	8.34e-10	1.38e-4
10^{-6}	10^{-1}	2.85e-1	1.39e-1	2.72e-6	4.24e-1	2.72e-6	1.39e-1	2.46e-8	2.59e-1
	10^{-2}	2.73e-2	1.36e-2	2.72e-6	4.09e-2	2.72e-6	1.36e-2	2.69e-10	2.70e-2
	10^{-3}	2.72e-3	1.35e-3	2.72e-6	4.07e-3	2.72e-6	1.35e-3	2.72e-12	2.71e-3

1. $\mathbf{v}^{n+1} = e^{kM}e^{kL}\mathbf{v}^n$,
2. $\mathbf{v}^{n+1} = e^{kL}e^{kM}\mathbf{v}^n$,
3. $\mathbf{v}^{n+1} = e^{\frac{k}{2}L}e^{kM}e^{\frac{k}{2}L}\mathbf{v}^n$, and
4. $\mathbf{v}^{n+1} = [e^{kL} + (e^{kL} - I)L^{-1}M]\mathbf{v}^n$.

The last formula is an approximation arising from considering $M\mathbf{v}$ as an inhomogeneity and approximating the exact formula

$$\mathbf{v}(t+k) = e^{kL}\mathbf{v}(t) + \int_0^k e^{(k-s)L}M\mathbf{v}(t+s)ds. \qquad (9.24)$$

We will have more to say about this method in Section 9.3.3.

Let us concentrate first on the errors in v_2. It is clear that when $\varepsilon \ll k$ the errors in all these splitting methods are first order in k and hardly distinguishable from one another. On the other hand, when $k \leq \varepsilon$ the Strang splitting is more accurate than the other methods and displays a second order behavior.

In the first component v_1 there is more variety. The errors are usually smaller than those in v_2, for all but the first splitting method they decrease with ε, and for the last method they decrease with both ε and k. Note how significantly better it is to apply L after M, not before. Unfortunately, however, these effects are only observed when the "fast solution component" v_1 is essentially decoupled from the "slow solution component" v_2. If we consider the same problem, except that L is replaced by $S^T L S$ for some rotation or reflection matrix S, then the fast and slow solution components are mixed, and only errors such as those for v_2 in Table 9.3 are observed. ∎

9.3.1 More general splitting

From what has already been said it is clear that the idea of splitting can be generalized further. Splitting the operator does not necessarily have to be into only two parts (indeed, the direct generalization of the preceding methods to three spatial variables would involve splitting the spatial operator into three parts), and not necessarily because the problem is multidimensional. We have seen a more sophisticated instance of this sort in the semi-explicit, symplectic method (7.27) developed in Section 7.3.1 for the KdV equation.

In general, if the spatial differential operator contains (additively) parts of a different character then different discretization techniques may be applied to each part of the differential operator using the splitting idea. Candidates for this approach are multidimensional problems, problems with different time scales, nonlinear problems which contain linear leading differential terms, problems which combine parabolic and hyperbolic terms, etc. It must be understood, however, that by merely identifying a particular method as a splitting method not much is yet concluded about its accuracy or convergence properties.

Example 9.6 (Schrödinger equation) Let us recall again the nonlinear Schrödinger equation (NLS) from Exercise 1.7 and Example 5.10. Consider in particular the *cubic NLS*,

$$\psi_t = \iota(\psi_{xx} + |\psi|^2 \psi). \tag{9.25}$$

Note that $\psi(t, x)$ is complex-valued. From Example 5.10 we know that for the pure initial value problem the solution's $\mathcal{L}_2$-norm remains constant for all time, $\int \bar{\psi}(t, x) \psi(t, x) dx \equiv \|\psi(t)\|^2 = \|\psi(0)\|^2$. Moreover, this is a *Hamiltonian PDE* (see, e.g., [41]), which means that it can be written as

$$\mathbf{u}_t = \mathcal{D}\left(\frac{\delta \mathcal{H}}{\delta \mathbf{u}}\right), \quad \text{where}$$

$$\mathcal{H}[\mathbf{u}] = \int H(x, \mathbf{u}, \mathbf{u}_x, \ldots) dx,$$

$$\int \frac{\delta \mathcal{H}}{\delta \mathbf{u}} \mathbf{v} dx = \left(\frac{d}{d\varepsilon} \mathcal{H}[\mathbf{u} + \varepsilon \mathbf{v}]\right)_{\varepsilon=0} \tag{9.26}$$

with

$$\mathbf{u} = (\psi, \bar{\psi})^T, \quad \mathcal{D} = \iota, \quad H(\psi, \bar{\psi}) = \psi_x \bar{\psi}_x - \frac{1}{2} \psi^2 \bar{\psi}^2.$$

Note that the operator ι is skew-symmetric. There is also a multisymplectic structure here.

We can construct a symplectic method for this PDE by splitting the right-hand side of (9.25) in an obvious way. Note that the problem

$$u_t = \iota u_{xx}$$

is linear. We discretize it in space using the usual D_+D_- operator and in time using the (symplectic) midpoint method. The matrix that arises because the method is implicit does not depend on t and hence can be assembled and decomposed once using sparse *LU*-decomposition.

The nonlinear part

$$u_t = \iota |u|^2 u$$

is really an ODE with x as a parameter. It is easy to verify that its exact solution is

$$u(t) = u(t_0) \, e^{\iota(t-t_0)|u|^2}$$

9.3. Splitting Methods

with $|u|$ independent of t; see Exercise 11. Hence, stepping from t with any time step k we have

$$u(t+k) = u(t)\, e^{\imath k|u|^2}.$$

So, we use the Strang splitting (9.21) to compose these two solution operators, the midpoint for the linear PDE and the exact one for the ODEs at the spatial mesh points, to obtain a compact, symplectic method with accuracy order $(2, 2)$.

To see this method in action we use an example from [98]. Periodic BC are specified on the interval $[-20, 80]$, and the initial value function is

$$\psi(0, x) = e^{\imath x/2}\mathrm{sech}(x/\sqrt{2}) + e^{\imath(x-25)/20}\mathrm{sech}((x-25)/\sqrt{2}).$$

This yields two pulses which propagate to the right at different speeds, with their shapes unchanged except when they coalesce.

At first we run twice, for $k = h = .1$ and for $k = h = .01$, and at $t_f = 200$ compute the difference from the values at $t = 0$ in the discrete solution norm, $\sqrt{\sum_j |\psi_j|^2}$, and in the discrete Hamiltonian, $\sum_j h^{-2}|D_+\psi_j|^2 - \frac{1}{2}|\psi_j|^2$. These results are tabulated in the first two data rows of Table 9.4. The error indicators are pleasantly small. The solution profile is plotted in Figure 9.8(a). There is no significant phase error here, even for the coarser mesh. Nonetheless let us caution, with Example 6.5 in mind, that we may not conclude from this evidence that the solution is pointwise accurate.

Table 9.4. *Error indicators in Hamiltonian and norm for Example 9.6 measured at $t = 200$.*

t_f	k	h	Error-Ham	Error-norm
200	.1	.1	3.7e-5	4.3e-13
	.01	.01	3.9e-9	1.5e-11
1000	.1	.1	5.2e+2	2.9e-12
	.01	.1	3.3e-7	4.2e-12
	.01	.01	3.4e+2	7.7e-11

Next we increase the final time to $t_f = 1000$ and run again from $t = 0$. For $k = h = .1$ the solution after $t = 200$ develops an instability in ψ_x, i.e., fast oscillations with a bounded amplitude! The result recorded in the third row of Table 9.4 shows that the considerable pointwise error is reflected in the Hamiltonian deviation but not in the solution norm deviation. A similar phenomenon occurs for $k = h = .01$; see Figure 9.8(b). The intolerable spurious oscillation disappears when we choose $k = .01$, $h = .1$.

This surprising instability arises because we have for the semi-discretization $u_t = \imath h^{-2}D_+D_-u$ *imaginary* eigenvalues of size $O(h^{-2})$. When $k = h$ we are therefore in the highly oscillatory regime of the midpoint method, and the difficulties pointed out in Section 6.4 become relevant. Choosing $k = O(h^2)$ brings these eigenvalues back to the manageable regime. Splitting methods should not be trusted blindly! See also Part (b) of Exercise 11.

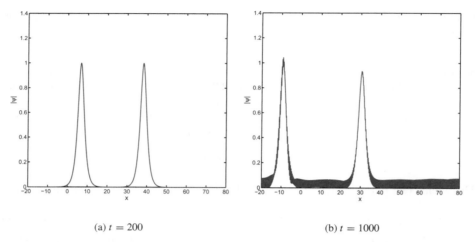

Figure 9.8. *Solution magnitude for the Schrödinger equation (Example 9.6) using $k = h = .01$. The two pulses look accurate at $t = 200$. But as integration proceeds an instability in the solution derivative arises, yielding sharp oscillations that in the figure look like a thick line. See Table 9.4.*

A popular scheme replaces the difference scheme for the linear PDE with a *spectral method* in the above splitting. See Section 7.4 and Review Section 2.12.6. The Fourier spectral method is applied to the constant coefficient problem $u_t = \iota u_{xx}$ subject to periodic boundary conditions. Let $\mathcal{F}$ denote applying the transform in space. Then we can write

$$u(t+k) = \mathcal{F}^{-1}\left(e^{-\iota \xi^2 k} \mathcal{F}(u(t))\right).$$

Spectral accuracy results upon replacing the transform by its discrete counterpart. The splitting error remains $O(k^2)$, though, and in fact k must often be taken rather small for this highly sensitive problem.

The tame-looking nonlinear Schrödinger equation (9.25) can thus cause serious surprises. In fact, it gets worse in more space dimensions where the PDE reads

$$\iota \psi_t + \Delta \psi = V'(|\psi|)\psi. \tag{9.27}$$

Incidentally, this equation arises in optics and water wave applications. In three dimensions there may be solution blowups and in two dimensions there may be "almost blowups" [41]. The splitting idea used above can be applied without change, although the solution of

$$u_t = \iota \Delta u$$

becomes more interesting. For an implicit time discretization the usual iterative methods may not work well (can you see why?), and direct methods using sparse matrix technology look attractive; see, e.g., [132]. Alternatively, a Fourier spectral method can be applied.

9.3. Splitting Methods

Another possibility is to replace the implicit midpoint scheme by the explicit RK4. No splitting is needed here. The method is not symplectic (recall Chapter 6), and it demands $k = O(h^2)$ even for integration intervals that are much shorter than $t_f = 200$. But if we are resigned to using $k = O(h^2)$, then this method is much faster than the implicit midpoint, and its relative advantage grows in more spatial dimensions. In one dimension for $k = .005$ and $h = .1$ it yields at $t_f = 1000$, Error-Ham = 3.2e-10, and Error-norm = 3.5e-12, which is quite respectable. A somewhat cheaper alternative to RK4, still with $k = O(h^2)$, would be the four-step Adams pair in PECE mode; see Chapters 2 and 6. ∎

Example 9.7 (Shallow water equations) This model is described in [110, 133, 76] and arises in meteorology and oceanography. The equations form a nonlinear hyperbolic system

$$\begin{pmatrix} u \\ v \\ \phi \end{pmatrix}_t + \begin{pmatrix} u & 0 & 1 \\ 0 & u & 0 \\ \phi & 0 & u \end{pmatrix} \begin{pmatrix} u \\ v \\ \phi \end{pmatrix}_x + \begin{pmatrix} v & 0 & 0 \\ 0 & v & 1 \\ 0 & \phi & v \end{pmatrix} \begin{pmatrix} u \\ v \\ \phi \end{pmatrix}_y = \mathbf{0}$$

and are subject to initial conditions and periodic boundary conditions. The unknown functions u and v denote velocities in the x- and y-directions, and $\phi > 0$ is a potential function. In many applications, it is realistic to assume that there is a mean value $\phi_0 > 0$ of ϕ, and $0 < \epsilon \ll 1$, such that

$$|u| + |v| \leq \epsilon \phi_0, \quad |\phi - \phi_0| \leq \epsilon \phi_0.$$

Thus, we have **two time scales** (as in Example 9.5 and the KdV equation of Section 7.3) and a small variation of ϕ about its known mean.

To investigate the problem first, it makes sense to symmetrize it, because more is known about symmetric hyperbolic systems. The change of variables $\phi \leftarrow 2\sqrt{\phi}$ achieves the desired result and yields the system which we'll assume from now on to be the given one,

$$\begin{pmatrix} u \\ v \\ \phi \end{pmatrix}_t + \begin{pmatrix} u & 0 & \tfrac{1}{2}\phi \\ 0 & u & 0 \\ \tfrac{1}{2}\phi & 0 & u \end{pmatrix} \begin{pmatrix} u \\ v \\ \phi \end{pmatrix}_x + \begin{pmatrix} v & 0 & 0 \\ 0 & v & \tfrac{1}{2}\phi \\ 0 & \tfrac{1}{2}\phi & v \end{pmatrix} \begin{pmatrix} u \\ v \\ \phi \end{pmatrix}_y = \mathbf{0}.$$

To investigate the problem further, freeze the coefficients at mean values U_0, V_0, Φ_0 and denote $\mathbf{w} = (u, v, \phi)^T$. Applying a Fourier transform (1.21) with $d = 2$ and $\boldsymbol{\xi} = (\xi_1, \xi_2)^T$ gives

$$\hat{\mathbf{w}}_t = \imath(\xi_1 A + \xi_2 B)\hat{\mathbf{w}},$$

and hence

$$\hat{\mathbf{w}}(t, \boldsymbol{\xi}) = T e^{\imath t \Lambda} T^{-1} \hat{\mathbf{w}}(0, \boldsymbol{\xi}),$$

where Λ and T are the eigenvalue and eigenvector matrices

$$\Lambda = \begin{pmatrix} \xi_1 U_0 + \xi_2 V_0 & 0 & 0 \\ 0 & \xi_1 U_0 + \xi_2 V_0 + \tilde{q} & 0 \\ 0 & 0 & \xi_1 U_0 + \xi_2 V_0 - \tilde{q} \end{pmatrix},$$

$$T = \begin{pmatrix} \xi_2/q & \xi_1/\tilde{q} & -\xi_1/\tilde{q} \\ -\xi_1/q & \xi_2/\tilde{q} & -\xi_2/\tilde{q} \\ 0 & 1 & 1 \end{pmatrix},$$

with $q = \|\xi\|_2$, $\tilde{q} = q\sqrt{\Phi_0}$.

Thus, we have two types of waves, a slow one moving at speed (U_0, V_0) and two fast ones with speeds depending on Φ_0. These correspond to the two time scales. There are practical situations in which we are interested in the slow wave only, but numerical stability may dictate discretization step sizes depending on the fast waves, unless we do something more clever.

For example, a direct application of the leapfrog scheme to this PDE yields the stability restriction

$$\mu \left(|U_0| + |V_0| + \sqrt{2\Phi_0} \right) < 1.$$

Nonlinear instability phenomena occur here for the leapfrog scheme, as in Example 5.1, and the introduction of some dissipativity into the numerical scheme is deemed necessary. Thus, we have again the same double incentive as for KdV to use an implicit scheme for this hyperbolic system, namely, to stabilize over a long time integration and to avoid a steep time step restriction.

Using the splitting approach of the current section, we can write this PDE system as

$$\mathbf{w}_t + (L_1 + L_2 + L_3 + L_4)\mathbf{w} = \mathbf{0}, \quad \text{where}$$

$$L_1 = \begin{pmatrix} 0 & 0 & \tfrac{1}{2}\phi_0 \\ 0 & 0 & 0 \\ \tfrac{1}{2}\phi_0 & 0 & 0 \end{pmatrix} \partial_x, \quad L_2 = \begin{pmatrix} u & 0 & \tfrac{1}{2}(\phi - \phi_0) \\ 0 & u & 0 \\ \tfrac{1}{2}(\phi - \phi_0) & 0 & u \end{pmatrix} \partial_x,$$

$$L_3 = \begin{pmatrix} 0 & 0 & 0 \\ 0 & 0 & \tfrac{1}{2}\phi_0 \\ 0 & \tfrac{1}{2}\phi_0 & 0 \end{pmatrix} \partial_y, \quad L_4 = \begin{pmatrix} v & 0 & 0 \\ 0 & v & \tfrac{1}{2}(\phi - \phi_0) \\ 0 & \tfrac{1}{2}(\phi - \phi_0) & v \end{pmatrix} \partial_y.$$

We can further approximate each e^{kL_j} in some way to second order accuracy and put these approximations together using an extension of the formula (9.21).

In this example, we have split the operators such that L_2 and L_4 are slow: they can be approximated using leapfrog with some dissipation added, or Lax–Wendroff, or the classical Runge–Kutta scheme RK4, with reasonably large step sizes in time. For the fast operators

9.3. Splitting Methods

L_1 and L_3 another method can be used: if ϕ_0 is constant, then an exact solution operator is possible. If not, then an implicit scheme can be used, now applied only for a linear operator. We can even use the same scheme as for L_2 and L_4, but with a smaller step size k for the simplified, fast operators only. ∎

Example 9.8 (Splitting applications) The appeal of the splitting approach, which provides a simple divide and conquer mechanism for complex models at the cost of an often small or at least amorphous additional error, has led to its use in many applications. Here we briefly describe three such applications, in addition to Examples 9.6 and 9.7, just to give an idea of the variety and appeal of splitting methods. There are many other papers which utilize similar techniques.

- **Diffusion-generated motion**

    This example concerns a numerical technique described in Merriman, Bence, and Osher [131] and Ruuth [147]. The task of simulating the motion of evolving surfaces with junctions according to some curvature-dependent speed is a major motivation for level set methods, briefly described in Section 11.3; for much more see [152, 140]. Motion by mean-curvature is a particularly popular instance of this sort. It is quite possible that a blob would separate into two blobs and more during such motion, or that several blobs will merge and become one, so a method that tracks the interface defining the surface of the blob too rigidly may fail at such singularities.

    An alternative method for this motion, which shares the topological flexibility of level set methods, is based on a sequence of steps alternately diffusing and sharpening the interface defining the surface. It may be considered as a splitting method for the *Allen–Cahn* equation

    $$u_t = \varepsilon \Delta u - \varepsilon^{-1} f'(u), \tag{9.28a}$$

    where f is a double well potential, e.g.,

    $$f(u) = u^2(u-1)^2, \quad \text{hence } f'(u) = u(u-1)(u-1/2), \tag{9.28b}$$

    in the limit $\varepsilon \to 0$. The function u is the *characteristic function* of the corresponding domain D whose shape we are tracking,

    $$u = \begin{cases} 1, & \mathbf{x} \in D, \\ 0 & \text{otherwise,} \end{cases} \tag{9.28c}$$

    so the initial value function $u_0(x)$ at $t = 0$ is determined by the initial shape.

    Each time step n of the method consists of two stages. In the first we turn off f and solve

    $$u_t = \Delta u,$$

    subject to homogeneous Neumann conditions and starting from the characteristic function that is the approximation for u from the previous step at t_n. This obviously has the effect of diffusing the interface. Call the result $\tilde{u}$. In the second stage we then sharpen the picture by turning off Δu in (9.28a) and solving at each $\mathbf{x}$

$$\varepsilon u_t = f'(u)$$

in the limit as $\varepsilon \to 0$. This is actually very simple, because we must set u to one of the zeros of $f'(u)$ in a stable way. But of the three zeros only $u = 0$ and $u = 1$ are stable. Thus, we set at each location $u = 0$ if $\tilde{u} \leq 1/2$ and $u = 1$ otherwise.

This method was proposed in [131], and its proved convergence does not follow from its interpretation as a splitting in the limit. Upon discretization, a finite difference method for the differential equation part turns out to be very slow, because small time steps must be taken and correspondingly the discretization step h must be fine enough so that some motion gets generated at this scale. A special *spectral method* was proposed in [147] to improve the efficiency of the resulting overall procedure. We will not dwell on this further here, but see Section 7.4 for a quick survey of spectral methods.

- **A semi-Lagrangian transport model**

This example is described in Manson, Wallis, and Hope [125]; see also [134]. The transport of reactive substances by groundwater flow is modeled by the differential system

$$\phi_t + (u\phi)_x = (dc_x)_x + \alpha a(s - c), \tag{9.29a}$$

$$\hat{a} s_t = -\alpha a(s - c), \tag{9.29b}$$

$$\phi = ac. \tag{9.29c}$$

Here the unknown functions are the concentration of chemical in the water, $c(t, x)$, and the concentration of the interacting rock, $s(t, x)$. The functions $u(x)$, $d(x)$, $a(x)$, $\hat{a}(x)$, and $\alpha(x)$ are all known. The domain is $0 < x < b, t \geq 0$. Initial conditions are given on both c and s. The boundary conditions are

$$c(t, 0) = g(t), \quad c_x(t, b) = 0,$$

where the function $g(t)$ is given.

The ODEs for s at each x, (9.29b), yield in principle a function $s = s(t, x; c)$ that can be substituted in (9.29a). This yields a reaction-diffusion-convection equation for the concentration c. In practice of course we do not eliminate s ahead of time but rather solve the system (9.29) simultaneously.

Denoting the time step size as usual by k, the method proposed in [125] reads at each time step n for each space location j

$$\tilde{\phi}_j = \phi_j^n + k\left[-(u\phi)_x\right]_{fb}^n,$$

$$\phi_j^{n+1} = \tilde{\phi}_j + k\left[(dc_x)_x\right]_j^{n+1} + k\left[\alpha a(s - c)\right]_j^{n+1},$$

$$\left[\hat{a}s\right]_j^{n+1} = \left[\hat{a}s\right]_j^n - k\left[\alpha a(s - c)\right]_j^{n+1},$$

9.3. Splitting Methods

where $[\cdot]_{fb}^n$ is evaluated at time t_n using spatial cubic interpolation for the foot of the characteristic of the convection part

$$\phi_t + (u\phi)_x = 0$$

arriving at (t_{n+1}, x_j). See also [25]. The ensuing spatial discretization of $(dc_x)_x$ is by now standard; see Section 3.1.2. Of course, $c = \phi/a$ may be obtained from ϕ at any given point.

This is a *semi-Lagrangian method*, where the convection part of the differential operator is split from the diffusion and reaction parts and receives a different numerical treatment. At first a Lagrangian method is applied to the convection part, and then a backward Euler discretization is applied to the rest. See [125] for further discussion and numerical results.

- **Fluid flow animation**

If there is one research area for which splitting methods seem just perfect, it is physics-based animations in computer graphics. The mathematical models describing in detail the complex scenes, for instance, of fluid flow, that one would like to have animated, say, for a large-budget movie or a computer game, can be very complicated. But the results need not correspond to true physics—they merely should look good. Hence replacing a combined action in such a model by a series of actions each involving part of the model is a major, natural simplification. The actual splitting error that results may be unimportant, provided that the order in which the subproblems are solved is correct and that the error is not allowed to accumulate unfavorably.

Typically, the *incompressible Navier-Stokes* equations in the inviscid limit, i.e., the **incompressible Euler** equations, model the flow. The equations read

$$\mathbf{u}_t + \mathbf{u} \cdot \nabla \mathbf{u} + \nabla p = \mathbf{g}, \quad (9.30a)$$

$$\nabla \cdot \mathbf{u} = 0, \quad (9.30b)$$

where $\mathbf{u}$ is the vector of velocities, e.g., $\mathbf{u} = (u, v)^T$ in two space variables, p is the pressure, and $\mathbf{g}$ are the forces, including in particular gravity. The component forms of the operators $\mathbf{u}_t + \mathbf{u} \cdot \nabla \mathbf{u}$ and $\nabla \cdot \mathbf{u}$ appear in Example 5.5.

An operator splitting method that goes back to [49] solves at each step the equation (9.30a) for the velocity $\mathbf{u}^{n+1}$ while holding $p = p^n$ at the known level and then adjusts the pressure to satisfy the incompressibility condition (9.30b) calling the result p^{n+1}. For the inviscid Euler equations often a marker-and-cell (MAC) technique is employed, which uses a staggered mesh as in Figure 9.9 and also has early roots [87, 33]. A semi-Lagrangian method as briefly described above in the context of the groundwater application (but in more space dimensions) is typically used.

An example employing such techniques is found in Feldman et al. [62]. Another paper where a multiple splitting method is massively applied is Gundelman et al. [74]. Here water, smoke, flexible bodies, and rigid bodies all interact. The figures and video clips that come with these papers look so good that it would be a shame to continue to describe them here in black and white with just words and formulas. ∎

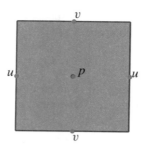

Figure 9.9. *The marker-and-cell staggered mesh for fluid flow, here in two dimensions. The velocity components u and v in the x and y directions, respectively, are located at the edge centers while the pressure p is at the cell center.*

9.3.2 Additive methods

These methods are not quite splitting methods in the form presented above, but they are often confused with the latter. More importantly, they do involve a splitting of the spatial differential operator, so here is a good place to discuss them.

Consider a semi-discretization of a PDE system which, as before, has a spatial differential operator consisting of a few additive parts, say,

$$\frac{d\mathbf{v}}{dt} = L(\mathbf{v}) + M(\mathbf{v}), \qquad (9.31a)$$

where $\mathbf{v}(t)$ is the mesh function turned vector and the operators L and M may be nonlinear (i.e., they may depend on $\mathbf{v}$ nonlinearly). We can write

$$\mathbf{v}(t) = \boldsymbol{\phi}(t) + \boldsymbol{\psi}(t), \qquad (9.31b)$$

where $\boldsymbol{\phi}$ and $\boldsymbol{\psi}$ are also time-dependent mesh functions satisfying

$$\frac{d\boldsymbol{\phi}}{dt} = L(\mathbf{v}) = L(\boldsymbol{\phi} + \boldsymbol{\psi}), \qquad (9.31c)$$

$$\frac{d\boldsymbol{\psi}}{dt} = M(\mathbf{v}) = M(\boldsymbol{\phi} + \boldsymbol{\psi}). \qquad (9.31d)$$

In (9.31) not only the operator but also the solution is being split. But the idea is to never compute values for $\boldsymbol{\phi}$ or $\boldsymbol{\psi}$, only for their sum $\mathbf{v}$. The partitioned form (9.31c)–(9.31d) does allow applying different time discretization methods to the two equations, and the results of these discretizations are combined into one formula. In particular, methods consisting of an implicit discretization for (9.31c) and an explicit one for (9.31d) (called **IMEX methods**) are possible. For instance, in Exercise 4.5 a backward Euler method for the diffusion term and an explicit one-sided method for the advection term of a simple advection-diffusion equation were combined. Examples of reaction-diffusion were studied in [146].

In general, IMEX linear multistep methods can be constructed, which allow for accuracy p in time when using a p-step method (Exercise 13). The construction of IMEX

9.3. Splitting Methods

Runge–Kutta schemes is also possible and leads to more generous time steps, although achieving higher order accuracy is more intricate; see [17]. Still, we are ahead in terms of high accuracy order when compared to simple operator splitting methods.

Example 9.9 For the advection-diffusion equation

$$u_t = \nu(u_{xx} + u_{yy}) + au_x$$

with the constants $\nu, a > 0$, a typical semi-discretization reads

$$\frac{dv_{j,l}}{dt} = \frac{\nu}{h^2}\left[D_{x+}D_{x-} + D_{y+}D_{y-}\right]v_{j,l} + \frac{a}{h}D_{x+}v_{j,l}.$$

The simplest IMEX method is to discretize diffusion by backward Euler and advection by forward Euler,

$$v_{j,l}^{n+1} = v_{j,l}^n + \frac{\nu k}{h^2}\left[D_{x+}D_{x-} + D_{y+}D_{y-}\right]v_{j,l}^{n+1} + \frac{ak}{h}D_{x+}v_{j,l}^n.$$

Like forward and backward Euler this method is first order, but now the stability condition is at worst $k \leq h/a$. Moreover, the solution of the implicit equations is much easier and faster without the advection term: using CG with a multigrid preconditioning cycle, we often need only one or two iterations per time step. See [18, 17] for numerical results.

Next, we can get a second order method by combining (carefully) an IMEX pair of two second order methods, e.g., the implicit Crank–Nicolson (or the two-step BDF) with the explicit two-step Adams–Bashforth. Additive two-stage Runge–Kutta methods of order two are not difficult to construct, either. ∎

Example 9.10 The following system of convection-diffusion equations arises in applications of fluid flow:

$$u_t + uu_x + vu_y + wu_z = \nu(u_{xx} + u_{yy} + u_{zz}),$$
$$v_t + uv_x + vv_y + wv_z = \nu(v_{xx} + v_{yy} + v_{zz}),$$
$$w_t + uw_x + vw_y + ww_z = \nu(w_{xx} + w_{yy} + w_{zz}).$$

(Recall Example 5.4.) These equations are defined on a three-dimensional domain Ω in space, for $t \geq 0$, and are equipped with appropriate initial and boundary conditions.

If the parameter ν is not small, then applying an explicit discretization would yield a severe restriction of the form $k \leq c\nu^{-1}h^2$ on the time step, where c is a constant of order unity. Thus, we look for an implicit discretization for this term. The terms from the material derivatives, however, are nonlinear and their explicit discretization would yield only a mild time step restriction of the form $k \leq ch$. The argument for an IMEX approach is therefore similar to the one made before for Example 9.9 and for the KdV equation in Section 7.3.

Unlike for the KdV equation, however, there is an additional argument here. Recall from Section 9.2 that when using an implicit scheme for a PDE in more than one space variable, the system of equations to be solved is large and sparse, which calls for iterative methods. In this respect it is also better that only the parabolic terms be discretized implicitly, as in Example 9.9, because they yield an algebraic system which is well suited to the application of methods like PCG and multigrid.

Another approach taken in the literature is to allow a semi-implicitness also in the first order terms, utilizing approximations such as $u^n u_x^{n+1}$. This still yields a linear algebraic system at each time step and allows taking a larger time step, which is advantageous especially when ν is not very large (i.e., as compared to the option of discretizing uu_x explicitly). However, in this case the iterative solution of the resulting system may be more costly as well. See, e.g., [172]. ∎

Example 9.11 Let us recall the cloth simulation case study, Example 2.9. Cloth is modeled there by a particle system in a mass-spring arrangement that should obey the equations of motion,

$$\mathbf{x}_t = \mathbf{v},$$
$$M\mathbf{v}_t = \mathbf{f}(\mathbf{x}, \mathbf{v}),$$

where $\mathbf{x}$ denotes the positions of the particles in three dimensions, i.e., three coordinates per particle, $\mathbf{v}$ are the corresponding velocities, M are the masses, and $\mathbf{f}$ are the forces. These forces include terms due to stretching, shearing, bending, gravity, damping, air drag, and particle collisions both with other particles and with solids. It is a good idea to refresh your knowledge of Example 2.9 at this point before going on.

Further, we have discussed the application of a semi-implicit backward Euler method in view of the stiff stretching forces. This numerical discretization essentially exchanges stiff spring energy with in-surface damping. What we have not discussed is how to efficiently solve the resulting system of algebraic equations at each step, which is a major issue if a realistic animation in real time is contemplated. We consider this next.

Just one full Newton step for backward Euler at each time step can already be too much because the Jacobian matrix is large and nonsymmetric. In [20] an approximate Jacobian matrix is used where terms that disrupt symmetry and positive definiteness are dropped. There is also the obvious benefit of sparsifying this matrix as much as possible before applying either a direct or an iterative solution method. Now, from Figure 2.13 we see that different forces connect between different unknowns and hence would yield nonzeros at different locations in the Jacobian matrix. This sets the stage for an IMEX method, where only stretching and occasionally shearing forces are discretized implicitly and thus contribute terms to the Jacobian matrix. See [31] and references therein.

The simplest IMEX schemes mix forward and backward Euler, so that (2.31) is discretized by

$$\begin{pmatrix} \mathbf{x}_{n+1} \\ \mathbf{v}_{n+1} \end{pmatrix} = \begin{pmatrix} \mathbf{x}_n \\ \mathbf{v}_n \end{pmatrix} + k \begin{pmatrix} \mathbf{v}_{n+1} \\ M^{-1} \left(\mathbf{f}^I(\mathbf{x}_n, \mathbf{v}_n) + \mathbf{f}^{II}(\mathbf{x}_{n+1}, \mathbf{v}_{n+1}) \right) \end{pmatrix},$$

where $\mathbf{f} = \mathbf{f}^I + \mathbf{f}^{II}$ is split into nonstiff and stiff force contributions. Note that $\mathbf{x}_{n+1} = \mathbf{x}_n + k\mathbf{v}_{n+1}$ can be eliminated, yielding a set of equations only for $\mathbf{v}_{n+1}$ at each time step. Which terms go into $\mathbf{f}^I$ and which into $\mathbf{f}^{II}$ can be determined adaptively by an absolute stability criterion borrowed from the Fourier stability analysis for a related PDE in three space variables (one which we do not wish to solve). Figure 9.10 demonstrates the potential gain from this strategy. There are further methods proposed in [31] for improving the sparsity structure, involving decomposition of the spatial mesh.

9.3. Splitting Methods

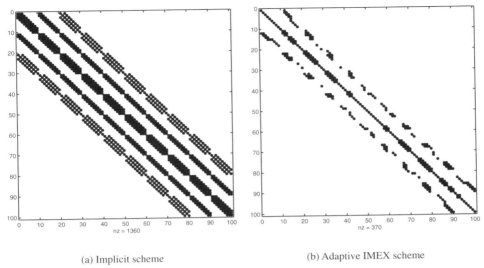

(a) Implicit scheme

(b) Adaptive IMEX scheme

Figure 9.10. *Sparsity gain when switching from a fully implicit time step to an IMEX step in Example 9.11. The matrices are both 300×300. Each point represents a 3×3 block, and* nz *denotes the number of nonzero blocks.*

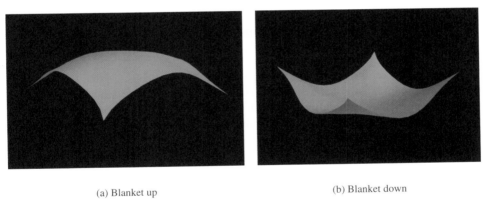

(a) Blanket up

(b) Blanket down

Figure 9.11. *Dusting a blanket by vigorously raising and lowering it at its four corners. The positions of the four corner particles are directly determined by external constraints.*

Before turning to solution methods for the linear algebra system there is another concern that needs to be addressed, namely, the incorporation of constraints. The above equations of motion apply to particles that are free to move about, not those that are directly constrained by an applied force. For instance, imagine two people holding the four ends of a blanket, raising and lowering it; see Figure 9.11.

It is possible to incorporate these constraints directly into the time-dependent system. Thus, the positions or velocities of the constrained particles are known and they appear at the right-hand side of the equations for a vector of unknowns $\hat{\mathbf{v}}_{n+1}$ which now contains only the unconstrained particle velocities. A direct solution method for the resulting sparse system is a competitive way to proceed, unless the particle mesh is very fine and the cloth cannot be easily decomposed as described in [31].

Another approach that is possible here is to describe the constraints in terms of a projection of the unconstrained system of equations. Let us write the unconstrained system as

$$A\mathbf{v}_{n+1} = \mathbf{b}.$$

We describe the constraints in terms of an **orthogonal projection** S. The generally rank-deficient matrix S satisfies

$$S^T = S^2 = S,$$

and it can be used to decompose vectors as a direct sum. The constrained system can then be written as

$$SA\mathbf{v}_{n+1} = S\mathbf{b},$$
$$(I - S)\mathbf{v}_{n+1} = (I - S)\mathbf{z},$$

where S is the block diagonal projection matrix and $\mathbf{z}$ are prescribed constraint values. Thus, for each particle the equations of motion hold only in the range subspace of S, range(S), whereas in the orthogonal subspace range($I - S$) the given values of $\mathbf{z}$ determine those of $\mathbf{v}_{n+1}$.

A *modified conjugate gradient* method for this constrained system was proposed in [20]. See [31] and references therein for subsequent improvements, including selecting a preconditioner for the constrained system. ∎

To summarize, the potential advantages of additive methods are that they naturally allow splitting of the spatial differential operator, or its discretization, without being confined to a low-order method and with less difficulty regarding handling of boundary conditions. However, their construction and certain resulting properties can be less obvious than those of some splitting methods.

> **Note:** Like the "purer" splitting methods of Section 9.3.1 and the exponential methods described in Section 9.3.3, the IMEX methods are not automatically best or even nontrivially applicable for any given problem. These methods all have their distinct advantages, though, and for the "right" application they can yield substantial improvements over other good alternatives.

9.3.3 Exponential time differencing

Another class of methods based on splitting the spatial differential operator extends the use of (9.24), which has been demonstrated in Table 9.3 under "M-inhomog." See Cox and Matthews [53]. In [105] many references are given and several methods are surveyed and compared in a particular context.

9.3. Splitting Methods

The essential observation leading to these methods is that the spatial differential operator of several important PDEs can be split into a *constant coefficient leading term* plus a lower order nonlinear term. Instances include the KdV equation, Boussinesq, NLS, and KS equations.

After spatial discretization we have the form

$$\frac{d\mathbf{v}}{dt} = L\mathbf{v} + M(\mathbf{v}), \qquad (9.32)$$

where L is a constant matrix containing the stiff, or leading, part of the PDE. Then we can write as in (9.24)

$$\mathbf{v}(t_n + k) = e^{kL}\mathbf{v}_n + \int_0^k e^{(k-s)L} M(\mathbf{v}(t_n + s))ds \qquad (9.33)$$

and obtain various method variants by discretizing the integral in (9.33) wisely. The time step restriction, if an explicit method is used, typically relates only to M, not to L, and the high accuracy achieved in sufficiently specialized circumstances can be rather impressive.

The simplest method of this sort replaces $M(\mathbf{v}(t_n + s))$ in (9.33) by $M(\mathbf{v}(t_n))$ à la forward Euler. Integrating the result exactly yields the time stepping method

$$\mathbf{v}_{n+1} = e^{kL}\mathbf{v}_n + (e^{kL} - I)L^{-1}M(\mathbf{v}_n). \qquad (9.34)$$

Higher order methods may result using higher order polynomial approximations of the integrand, either in a linear multistep fashion extending Adams methods or in a one-step fashion extending Runge–Kutta [53]. For example, replace $M(\mathbf{v}(t_n + s))$ by $M(\mathbf{v}_{n+1/2})$, where $\mathbf{v}_{n+1/2}$ is the average of $\mathbf{v}_n$ and the result of (9.34), to obtain a second order method from (9.33).

Of course an appropriately fast method for evaluating the exponential function, or more precisely $e^{kL}\mathbf{z}$ for any given vector $\mathbf{z}$, is crucial to the performance of methods based on this approach [135, 94, 95].

Further complications may arise because the function

$$\phi(z) = \frac{e^z - 1}{z}, \qquad (9.35)$$

if evaluated straightforwardly, yields cancellation error near $z = 0$. In that regime a truncated Taylor expansion works well, but the same approximation does not apply for large z. Of course, the reason for considering these methods in the first place is stiffness, which typically means that the matrix kL has both small and large eigenvalues, so we have a difficulty to deal with. If $\phi(kL)$ can be efficiently diagonalized, say, using a fast Fourier transform, then an appropriate treatment can be applied to individual eigenvalues, avoiding cancellation error. Alternatively, a fancy method involving integration in the complex plane is proposed in [105].

The exponential time differencing methods are somewhat related to the Rosenbrock methods (2.28), briefly introduced in Section 2.7, and to **exponential Runge–Kutta** methods. We refer the reader to [105, 94, 95] for further details and references.

9.4 Review: Iterative methods for linear systems

This section offers a review of iterative methods for linear systems of algebraic equations, a tad more detailed than those in Section 2.12. The reasons it is separated from the other reviews are that we use it essentially only from this chapter on, and because much of it is more advanced and less "canned" than the reviews of Section 2.12. This should become clear in what follows.

A full exposition of iterative methods for linear systems is well beyond the scope of this text, but there are several good sources around; see, e.g., [149, 21, 171, 59].

Consider the problem of solving a linear system of algebraic equations

$$A\mathbf{x} = \mathbf{b} \qquad (9.36)$$

with A an $m \times m$ real matrix. The notation in (9.36) is not really consistent with our notation in the rest of this chapter: the unknown $\mathbf{x}$ is our mesh function $\{v^{n+1}\}$ reshaped into a vector (say, column by column for the two-dimensional case), and this A is of course different from the small matrix A in Section 9.1. Rather, (9.36) corresponds to (9.11b), and the matrix A is like that depicted in Figure 9.1; if it arises from a PDE discretization in two space variables as in (9.10), then its dimension is roughly $J^2 \times J^2$, i.e., $m = J^2$. The reason we have switched is that (9.36) is the standard notation used in numerical linear algebra, and sticking to it here will make reading other texts easier.

The basic premise throughout this section is that to form a matrix-vector product $A\mathbf{y}$ is significantly cheaper than to form $A^{-1}\mathbf{y}$. In fact, A need not even be formed or stored as such: all we need is a function that returns $A\mathbf{y}$ for any input vector $\mathbf{y}$. This is the case if A is very sparse as, e.g., in Figure 9.1. However, the matrix A does not have to be sparse for the algorithms presented below to be worth considering.

The methods reviewed here all generate a sequence of iterates $\mathbf{x}_s$, $s = 1, 2, 3, \ldots,$ with $\mathbf{x}_0$ a given initial iterate.

9.4.1 Simplest iterative methods

This subsection reviews so-called **relaxation methods**. Given a *splitting* of A

$$A = M - (M - A),$$

the iteration reads

$$\begin{aligned} \mathbf{x}_{s+1} &= \mathbf{x}_s + M^{-1}\mathbf{r}_s, \quad \text{where} \\ \mathbf{r}_s &= \mathbf{b} - A\mathbf{x}_s, \quad s = 0, 1, \ldots. \end{aligned} \qquad (9.37)$$

Note that this is a fixed point iteration; see Section 2.12.5. The Jacobian matrix of the iteration is $T = I - M^{-1}A$, and since this is independent of $\mathbf{x}$ or s the iteration is called **stationary**. The trick is to choose M such that (i) unlike A it is easy to calculate $M^{-1}\mathbf{y}$ for any given vector $\mathbf{y}$, and (ii) the spectral radius of the iteration matrix, $\rho(T)$, is really small and in any case smaller than 1 to ensure convergence. The smaller ρ is the faster the convergence, but also the closer M must be to A in some sense.

9.4. Review: Iterative Methods for Linear Systems

None of the following methods should really be used in a standalone fashion in our applications because they are all too slow and specialized, but they are useful building blocks for other, faster methods. Let us denote

$$D = \begin{pmatrix} a_{11} & & & \\ & a_{22} & & \\ & & \ddots & \\ & & & a_{mm} \end{pmatrix}, \quad E = \begin{pmatrix} a_{11} & & & \\ a_{21} & a_{22} & & \\ \vdots & \vdots & \ddots & \\ a_{m1} & a_{m2} & \cdots & a_{mm} \end{pmatrix}.$$

- The **Jacobi** iteration uses $M = D = \operatorname{diag}(A)$, i.e.,

$$x_i^{s+1} = \frac{1}{a_{ii}} \left(b_i - \sum_{j=1, j \neq i}^{m} a_{i,j} x_j^s \right), \quad i = 1, \ldots, m, \qquad (9.38a)$$

where the superscript denotes the iteration, or **sweep**, counter.

This iteration converges if A is symmetric positive definite or diagonally dominant, but for the standard discretization problem (9.10) it does so rather slowly, requiring $O(J^2)$ iterations to reduce the error by a constant factor such as 10.

- The **Gauss–Seidel** (GS) sweep is defined by setting $M = E$, yielding a simple forward substitution,

$$x_i^{s+1} = \frac{1}{a_{ii}} \left(b_i - \sum_{j=1}^{i-1} a_{i,j} x_j^{s+1} - \sum_{j=i+1}^{m} a_{i,j} x_j^s \right), \quad i = 1, \ldots, m. \qquad (9.38b)$$

For a symmetric positive definite matrix the number of iterations is halved here compared to Jacobi's, although parallelization and vectorization become less straightforward.

If there is a mesh underlining the definition of A, then the neighboring relationship between the mesh unknowns defines the sparsity pattern of the matrix. This naturally leads to orderings such as *red–black* of the sweep in (9.38b) which allows restoring a higher degree of parallelism; see, e.g., [171].

- The ω-relaxation method is defined by setting

$$M = \frac{1}{\omega}(\omega \hat{M} + (1 - \omega) D) \qquad (9.38c)$$

in (9.37) with $\hat{M} = D$ for ω-Jacobi and $\hat{M} = E$ for ω-GS. Here ω is a parameter. The iteration is an *underrelaxation* if $0 < \omega < 1$ and an *overrelaxation* if $1 < \omega < 2$. The only combination that leads to a faster standalone method than before is overrelaxation with GS. This yields the successive overrelaxation (SOR). If ω is chosen well, then the number of iterations in our particular context may drop from $O(J^2)$ to $O(J)$. However, the choice of ω depends on the application, and some experimentation is required in practice.

- One somewhat bothersome aspect of Gauss–Seidel and SOR is that the direction of the *sweep* over the elements of **x** (i.e., over the mesh points in our applications of interest) matters. To make the iteration more symmetric, the present variant uses a sweep in one direction followed by one in the opposite direction, obtaining an iteration called symmetric SOR or SSOR.

9.4.2 Conjugate gradient and related methods

The methods considered here can be written as (9.37), except that M depends on the iteration s, so the methods are nonstationary.

Consider the problem of solving (9.36), where the real matrix A is symmetric positive definite. A natural way for viewing the *conjugate gradient* (CG) algorithm is as a method for minimizing the quadratic function

$$\phi(\mathbf{x}) = \frac{1}{2}\mathbf{x}^T A\mathbf{x} - \mathbf{b}^T\mathbf{x}. \tag{9.39}$$

The necessary and sufficient condition for minimizing (9.39) is our problem (9.36). We then look for a method which generates a sequence of iterates $\mathbf{x}_s$, $s = 1, 2, \ldots$, with $\mathbf{x}_0$ a given initial iterate and $\phi(\mathbf{x}_{s+1}) < \phi(\mathbf{x}_s)$. The update step is written as

$$\mathbf{x}_{s+1} = \mathbf{x}_s + \alpha_s \mathbf{p}_s, \tag{9.40}$$

where $\mathbf{p}_s$ is a *search direction* (in $\mathbb{R}^m$) and the scalar α_s is the *step length*.

Denote the residual as before by

$$\mathbf{r}_s = -\nabla\phi(\mathbf{x}_s) = \mathbf{b} - A\mathbf{x}_s. \tag{9.41}$$

The **steepest descent** direction is

$$\mathbf{p}_s = \mathbf{r}_s.$$

But the resulting method (with the step length defined by exact **line search**) tends to be slow. The CG direction is often better and is defined by

$$\mathbf{p}_s = \mathbf{r}_s + \beta_{s-1}\mathbf{p}_{s-1}, \tag{9.42}$$

where $\mathbf{p}_0 = \mathbf{r}_0$ and $\beta_{s-1} = \mathbf{r}_s^T\mathbf{r}_s / \mathbf{r}_{s-1}^T\mathbf{r}_{s-1}$. See the framed algorithm.

Note that only one matrix-vector multiplication involving A is needed per iteration. The matrix A need not be evaluated or stored; only the result of multiplying an arbitrary vector by it must be specified.

It can be shown that when the condition number of A is large the number of iterations s required to reduce the initial error by a constant factor is bounded by $O(\sqrt{\text{cond}(A)})$. For the steepest descent method a similar estimate yields $s = O(\text{cond}(A))$, which underscores the popularity of CG. In fact, if the eigenvalues of A are located in only a few narrow clusters, then the CG method requires only a few iterations to converge. The convergence is slower, though, if the eigenvalues are widely spread, as is typically but not always expressed

9.4. Review: Iterative Methods for Linear Systems

> **Algorithm: Conjugate gradients (CG).** Given an initial guess $\mathbf{x}_0$ and a tolerance tol, set at first $\mathbf{r}_0 = \mathbf{b} - A\mathbf{x}_0$, $\delta_0 = \mathbf{r}_0^T \mathbf{r}_0$, $b_\delta = \mathbf{b}^T \mathbf{b}$, $s = 0$, and $\mathbf{p}_0 = \mathbf{r}_0$. Then
> While $\delta_s > tol^2 \, b_\delta$
> $$\mathbf{s}_s = A\mathbf{p}_s,$$
> $$\alpha_s = \frac{\delta_s}{\mathbf{p}_s^T \mathbf{s}_s},$$
> $$\mathbf{x}_{s+1} = \mathbf{x}_s + \alpha_s \mathbf{p}_s,$$
> $$\mathbf{r}_{s+1} = \mathbf{r}_s - \alpha_s \mathbf{s}_s,$$
> $$\delta_{s+1} = \mathbf{r}_{s+1}^T \mathbf{r}_{s+1},$$
> $$\mathbf{p}_{s+1} = \mathbf{r}_{s+1} + \frac{\delta_{s+1}}{\delta_s} \mathbf{p}_s,$$
> $$s = s + 1.$$

by cond(A). In the context of implicit methods for time-dependent PDEs in two space variables, cond(A) = $O(J^2) = O(m)$, so the cost per time step using this algorithm for the Crank–Nicolson scheme (9.10) is $O(J^3)$ flops.

The **Krylov space** of degree s for A and $\mathbf{r}_0$ is defined as

$$K_s = \text{span}\{\mathbf{r}_0, A\mathbf{r}_0, A^2\mathbf{r}_0, \ldots, A^s\mathbf{r}_0\}.$$

Let us also define the *energy norm* $\|\mathbf{r}\|_B = \sqrt{\mathbf{r}^T B \mathbf{r}}$ for any symmetric positive definite matrix B. Then it is possible (and not even difficult) to show that CG is a subspace minimization method in the following sense: set

$$T_s = \mathbf{r}_0 + \text{span}\{A\mathbf{r}_0, A^2\mathbf{r}_0, \ldots, A^s\mathbf{r}_0\} \subset K_s;$$

then the defect in the sth iteration satisfies $\mathbf{r}_s \in T_s$ and

$$\|\mathbf{r}_s\|_{A^{-1}} = \min_{\mathbf{r} \in T_s} \|\mathbf{r}\|_{A^{-1}}. \tag{9.43}$$

Krylov space methods

What if A in (9.36) is not symmetric positive definite?

We can always consider solving the *normal equations*

$$A^T A \mathbf{x} = A^T \mathbf{b}$$

instead, using the CG method, but this squares the condition number of A and hence the number of required iterations. Other methods based on Krylov subspaces have been devised instead. These include *MINRES, GMRES, BICGSTAB,* and *QMR* [149, 21, 59]. Each of these is particularly good in some circumstances. Let us mention in our PDE context that if A is not symmetric positive definite simply because of handling boundary conditions, then possibly the BICGSTAB variant is most suitable.

The theory for these methods is not as complete as for CG, but the essential conclusion, that the methods converge rapidly if cond(A) is not too far from 1, holds in practice here as well.

Preconditioned conjugate gradient method

The performance of the CG algorithm and other Krylov space methods for our type of problems pales when compared against the (almost) constant number of iterations required for an alternative such as the multigrid method. Since the rate of convergence of the conjugate gradient method depends on the spread of the eigenvalues of A, we would like to redefine it so that the basic iteration is applied to a transformed problem, in which the matrix involved is better conditioned. This is the basic idea behind the *preconditioned conjugate gradient method* (PCG). If P is a symmetric positive definite matrix, the **preconditioner**, then we are essentially trying to solve the problem

$$P^{-1}A\mathbf{x} = P^{-1}\mathbf{b}. \tag{9.44}$$

This yields the algorithm stated below.

Algorithm: Preconditioned conjugate gradients (PCG). Given an initial guess $\mathbf{x}_0$ and a tolerance *tol*, set at first $\mathbf{r}_0 = \mathbf{b} - A\mathbf{x}_0$, $\mathbf{h}_0 = P^{-1}\mathbf{r}_0$, $\delta_0 = \mathbf{r}_0^T \mathbf{h}_0$, $b_\delta = \mathbf{b}^T P^{-1}\mathbf{b}$, $s = 0$, and $\mathbf{p}_0 = \mathbf{h}_0$. Then
   While $\delta_s > tol^2 \, b_\delta$

$$\mathbf{s}_s = A\mathbf{p}_s,$$

$$\alpha_s = \frac{\delta_s}{\mathbf{p}_s^T \mathbf{s}_s},$$

$$\mathbf{x}_{s+1} = \mathbf{x}_s + \alpha_s \mathbf{p}_s,$$

$$\mathbf{r}_{s+1} = \mathbf{r}_s - \alpha_s \mathbf{s}_s,$$

$$\mathbf{h}_{s+1} = P^{-1}\mathbf{r}_{s+1},$$

$$\delta_{s+1} = \mathbf{r}_{s+1}^T \mathbf{h}_{s+1},$$

$$\mathbf{p}_{s+1} = \mathbf{h}_{s+1} + \frac{\delta_{s+1}}{\delta_s}\mathbf{p}_s,$$

$$s = s + 1.$$

Similar rationale, modification, and improvements hold for other Krylov space methods designed for the more general case where A is not symmetric positive definite.

The choice of P should be such that the following seemingly contradictory requirements hold:

- It should be easy and efficient to solve problems of the form $P\mathbf{y} = \mathbf{c}$.

- The matrix $P^{-1}A$ should have a small condition number, hopefully not much larger than 1.

This echoes the requirements from a good splitting of A stated earlier, identifying P with M. Indeed, P^{-1} is often referred to as an **approximate inverse** of A. Of course, the problem (9.36) is *not* easy to solve, or else this entire section would not have been needed, so the reader may wonder if we are really ahead—how can P possess the property that A lacks and be close to it at the same breath?

The answer is yes, we are ahead, and the key word is "approximate." The beauty of preconditioners is that when designing them we no longer need to solve our problem exactly, which yields flexibility. For example, handling problematic BC on funny geometries in the PDE context can be much relaxed when the preconditioner is designed.

A simple, popular preconditioner is an SSOR sweep with $\omega = 1$. A more complex but more effective preconditioner is one multigrid cycle: we return to this below. Another class of preconditioners is based on *incomplete LU-decomposition* (ILU); see, e.g., [59] or [149]. The basic idea of the latter is to perform steps corresponding to the LU-decomposition of A, stopping short of replacing too many zeros in A by nonzeros in the factors L or U. What "too many" means is determined by a drop tolerance. A variant of this that is suitable for a symmetric positive definite A is *incomplete Cholesky decomposition*, which results in a symmetric decomposition without permutations, $L = U^T$.

Example 9.12 Let us compare the performance of the CG method, both preconditioned and not, and simple relaxation methods as standalone solvers. We consider the Poisson equation on the unit square with homogeneous Dirichlet BC and discretize it using the standard five-point formula,

$$-\frac{1}{h^2}\left[D_{x+}D_{x-} + D_{y+}D_{y-}\right]v_{j,l} = q_{j,l}, \quad j,l = 1,\ldots,J,$$

with $(J+1)h = 1$ and q some given right-hand side. This is obviously a special case of (9.11), and the same in essence as the system we need to solve at each time step of (9.10), but we start the iterations for each method with a zero guess, ignoring issues of warm start.

For SOR we pick $\omega = 1 + \frac{\cos(\pi h)}{(1+\sin(\pi h))^2}$. This is an optimal value available for this example but not in general, so SOR performs better than it may be expected to more generally. As a preconditioner for CG we use the incomplete Cholesky decomposition, such that no nonzeros are introduced where the original matrix has zeros.

The results are recorded in Figure 9.12. The most impressive performer, taking cost of iteration into consideration, is the CG method. The incomplete Cholesky preconditioner is not very impressive when taking the extra iteration cost into account. Better preconditioners can be tailored for this application upon taking the nature of the underlying PDE into account. ■

9.4.3 Multigrid methods

Next consider problems (9.36) where the matrix arises from a discretization on a two- or three-dimensional spatial mesh, or grid. The essential observation that has led to multigrid methods is that there exist simple and cheap **relaxation** schemes such as *Gauss–Seidel*, *underrelaxed Jacobi*, or the ILU method that reduce the high-frequency components of the residual (or the iteration error) in a given iteration much faster than they reduce the low-frequency ones. What this means is that the error, or the residual $\mathbf{r} = \mathbf{b} - A\mathbf{x}$, becomes

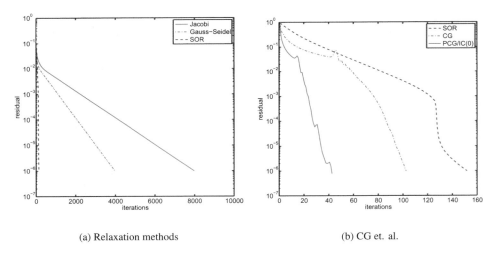

(a) Relaxation methods   (b) CG et. al.

Figure 9.12. *Relative performance of iterative methods for the Poisson equation with homogeneous Dirichlet BC on the unit square. A five-point discretization as in (9.10) is used with $J = 63$; hence $m = 63^2 = 3969$.*

smoother upon applying such a relaxation, when considered as a mesh function, well before it decreases significantly in magnitude. The relaxation operator is thus a **smoother**. Therefore, it is possible to represent the smoothed residual error well on a coarser grid,[31] thereby obtaining a smaller problem to solve, $A_C \mathbf{v}_C = \mathbf{r}_C$. The obtained correction $\mathbf{v}_C$ is then prolongated (interpolated) back to the finer grid and added to the current solution $\mathbf{x}$, and this is followed by additional relaxation. The process is repeated recursively: for the coarse grid problem the same idea is applied utilizing an even coarser grid. At the coarsest level the problem is so small that it can be rapidly solved "exactly," say, using a direct method.

The framed multigrid algorithm applies first ν_1 relaxations at the current finer level and then calculates the residual, coarsens it as well as the operator A, solves the coarse grid correction problem approximately γ times starting from the zero correction, prolongates (interpolates) the correction back to the finer grid, and applies ν_2 more relaxations. The whole thing constitutes one multigrid iteration, or a *cycle*, costing $O(m)$ operations. If $\gamma = 1$, then we have a V-cycle, while if $\gamma = 2$ we have a W-cycle. Typical relaxation values are $\nu_1 = \nu_2 = 2$. Although better theory requires a W-cycle, in practice the cheaper V-cycle is more popular.

Example 9.13 Let us compare the performance of a simple multigrid method (denoted MG) with that of the PCG method in three variants: not preconditioned (denoted CG), preconditioned with IC(0) (denoted PCG/IC(0)) and preconditioned by one multigrid iteration (denoted PCG/MG). We run these methods on the same very simple Poisson problem as in Example 9.12. Setting $J = 2^l - 1$ for $l = 5 : 9$, we record the number of iterations as well as the CPU time required on a garden variety desktop to satisfy a relative residual tolerance

---

[31] Throughout this book we use the term *mesh* rather than *grid*, viewing them as synonymous. However, in the multigrid context it feels like heresy to talk about coarser and finer meshes, so let us switch terminology, temporarily, to *grid*.

9.4. Review: Iterative Methods for Linear Systems

Algorithm: Multigrid method. Given three positive integer parameters ν_1, ν_2, and γ, start at $level = finest$:

    Function $\mathbf{x} = $ multigrid$(A, \mathbf{b}, \mathbf{x}, level)$
    If $level = coarsest$ then
      solve exactly $\mathbf{x} = A^{-1}\mathbf{b}$;
    else
      For j=1:ν_1
        $\mathbf{x} = $ relax$(A, \mathbf{b}, \mathbf{x})$;
      End
      $\mathbf{r} = \mathbf{b} - A\mathbf{x}$; $[A_c, \mathbf{r}_C] = $ restrict$(A, \mathbf{r})$;
      $\mathbf{v}_C = 0$;
      For l=1:γ
        $\mathbf{v}_C = $ multigrid$(A_C, \mathbf{r}_C, \mathbf{v}_C, level - 1)$;
      End
      $\mathbf{x} = \mathbf{x} + $ prolongate$(\mathbf{v}_C)$;
      For j=1:ν_2
        $\mathbf{x} = $ relax$(A, \mathbf{b}, \mathbf{x})$;
      End
    End

of $tol = 10^{-6}$. Our multigrid code uses an underrelaxed Jacobi with $\omega = .8$ (it turns out that $\omega = 1$ does not yield a smoother whereas $\omega = .8$ does!), and we set $\gamma = 1$, $\nu_1 = \nu_2 = 2$. The code generates A anew at each coarse level at each iteration, but MATLAB's command delsq is fast and simple so we didn't bother to optimize this. For the CG variants that do not involve MG, MATLAB commands pcg and cholinc(A,'0') are used. The results are recorded in Tables 9.5 and 9.6.

Table 9.5. *Relative performance in terms of iteration counts of iterative linear algebra methods for the Poisson equation with homogeneous Dirichlet BC on the unit square. A five-point discretization as in (9.10) is used with $J = 2^l - 1$, $l = 5:9$. For the largest J, the matrix order is $m \approx 260,000$.*

Iterations	$J = 31$	$J = 63$	$J = 127$	$J = 255$	$J = 511$
CG	52	102	202	396	771
PCG/IC(0)	23	42	69	119	229
MG	10	11	11	12	12
PCG/MG	27	22	16	13	13

Clearly the multigrid method offers great advantage here, both in pure form and as a preconditioner. This example is particularly good for the multigrid approach. The CG

Table 9.6. *Relative performance in terms of CPU time (secs) of iterative linear algebra methods for the Poisson equation with homogeneous Dirichlet BC on the unit square. A five-point discretization as in* (9.10) *is used with* $J = 2^l - 1$, $l = 5:9$.

CPU secs	$J = 31$	$J = 63$	$J = 127$	$J = 255$	$J = 511$
CG	4.7e-2	6.3e-2	6.3e-1	6.7	5.6e+1
PCG/IC(0)	4.7e-2	2.7e-1	3.1	4.4e+1	6.5e+2
MG	4.7e-2	6.3e-2	2.7e-1	1.4	6.1
PCG/MG	6.3e-2	1.6e-1	4.4e-1	1.6	7.6

method without preconditioning performs admirably well for a simple, basic method. The incomplete Cholesky preconditioner performs remarkably poorly. ∎

The multigrid mechanism, when it works, converges rapidly; indeed it can be spectacularly fast. But making it work precisely for a complex problem can lead to a hair-raising programming job. So, it is more customary to use one such multigrid iteration, a V-cycle to be precise, as a preconditioner for a Krylov space method, as we have done with the PCG/MG variant in Example 9.13. The number of iterations for the preconditioned Krylov space method is often pleasantly small and does not grow with the problem size. See [4, 59] for more vigorous applications.

Incidentally, the entire multigrid cycle is a stationary iteration of the form (9.37); however, the matrix that corresponds to M^{-1} is singular.

There are exciting methods of this type called **algebraic multigrid** (AMG) also for meshless problems, or problems that do not have a mesh that can be easily and logically coarsened. Modern, practical multigrid methods were started in the mid 1970s by Brandt [33]. A friendly, broad introduction to multigrid methods can be found in [184], and [171] is an excellent source for detailed information.

9.5 Exercises

0. **Review questions**

    (a) What are simultaneously diagonalizable matrices, and why is this important in the context of hyperbolic systems in more than one space variable?

    (b) Derive the CFL condition for the advection equation in three dimensions,
    $$u_t = au_x + bu_y + cu_z = 0,$$
    for a discretization on a given spatial-temporal mesh that is uniform in each direction with step sizes k and h_i, $i = 1, 2, 3$.

    (c) The relationship between the CFL condition and the stability restriction for the simplest, most basic discretizations for hyperbolic PDEs is different in one dimension than in more space dimensions. How?

9.5. Exercises

(d) What is a fill-in? How much fill-in may at worst occur during an LU-decomposition for the linear algebra problem defined in (9.10)?

(e) How do irregular boundaries relate to restrictions on the time step size for explicit methods?

(f) What is a warm start effect? Describe warm starting techniques for both direct and iterative methods.

(g) The ADI method is conceptually simple. Why is it rarely used nowadays?

(h) What are inner-outer iterations in the context of inexact Newton methods?

(i) What is a quasi-linearization method?

(j) What is an operator splitting method? Is ADI such a method?

(k) What is common to splitting methods, IMEX, and exponential time differencing methods? More subtly, distinguish among them.

1. Extend the method (9.3) to three dimensions and show stability for the scalar PDE $u_t + au_x + bu_y + cu_z = 0$, provided that

$$\max\{\mu_1|a|, \mu_2|b|, \mu_3|c|\} \leq 1/3.$$

2. Let $\bar{\mathbf{v}}_{j,l} = \frac{1}{2}(\mathbf{v}_{j,l+1} + \mathbf{v}_{j,l-1})$, $\hat{\mathbf{v}}_{j,l} = \frac{1}{2}(\mathbf{v}_{j+1,l} + \mathbf{v}_{j-1,l})$, and consider the following modification of the leapfrog scheme applied to (9.2):

$$\mathbf{v}_{j,l}^{n+1} = \mathbf{v}_{j,l}^{n-1} - \mu_1(\mathbf{f}(\bar{\mathbf{v}}_{j+1,l}^n) - \bar{\mathbf{f}}(\mathbf{v}_{j-1,l}^n)) - \mu_2(\mathbf{g}(\hat{\mathbf{v}}_{j,l+1}^n) - \mathbf{g}(\hat{\mathbf{v}}_{j,l-1}^n)).$$

Show that a Fourier analysis yields the CFL condition as the necessary condition for stability of this scheme [159]. This is twice as good as (9.4).

3. Extend the upwind scheme first introduced in Exercise 1.5 to the PDE

$$u_t + au_x + bu_y = 0,$$

where a and b are real constants. Find its stability restriction and compare it to the CFL requirement.

4. The hyperbolic PDE

$$\mathbf{u}_t + \begin{pmatrix} 2 & -1 \\ -1 & 2 \end{pmatrix} \mathbf{u}_x + \begin{pmatrix} 1 & 0 \\ 0 & \alpha \end{pmatrix} \mathbf{u}_y = \mathbf{0},$$

with periodic BC on $[-\pi, \pi] \times [-\pi, \pi]$ and suitable initial conditions, has the exact solution

$$\mathbf{u}(t, x, y) = \begin{pmatrix} \sin(x-t) + \sin(y-t) \\ \sin(x-t) + \cos(y-\alpha t) \end{pmatrix}.$$

The two matrices involved are simultaneously diagonalizable if $\alpha = 1$ and are not so if $\alpha = 3$. For both values of α, $\max\{\rho(A), \rho(B)\} = 3$.

Implement the leapfrog and the Lax–Friedrichs schemes, and run with different (moderate!) values of $h_1 = h_2$ and k, establishing experimentally (i) the methods' orders of accuracy, and (ii) the (approximate) maximum time step size k allowed for each of these two values of α. See [133].

5. The ADI method is described in Section 9.2.2 for the Crank–Nicolson discretization on the unit square of the PDE
$$u_t = u_{xx} + u_{yy}.$$

   (a) Staying with Crank–Nicolson on the unit square, extend the ADI method for the PDE semi-discretization (9.13).

   (b) Implement the method and carry out computations to parallel those described in Example 9.1. Compare your results to those in Table 9.1. Discuss.

6. The PDE
$$u_t = \Delta u - \frac{\lambda}{(1+u)^2}, \quad (x, y) \in \Omega, \tag{9.45a}$$

   subject to homogeneous Dirichlet BC and zero initial data ($u(0, x, y) = 0$ everywhere), is considered in [75]. It is related to the mathematical modeling of MEMS electric devices. Here λ is a parameter that is proportional to square the applied voltage of the idealized device.

   If λ is larger than a critical value λ_*, then u will reach the value of -1 at some point (x, y) in finite time, whereupon the solution derivative blows up. This is called *touchdown*. The voltage value corresponding to λ_* is called *pull-in voltage*.

   Your task is to integrate this problem numerically to determine touchdown time for a given $\lambda > \lambda_*$. Rather than tackling (9.45a) directly, however, it is a good idea to consider the transformation
$$w = \frac{1}{3\lambda}(1+u)^3 \tag{9.45b}$$

   for which the touchdown value is obviously $w = 0$; see [75].

   (a) Show that w satisfies the PDE
$$w_t = \Delta w - \frac{2}{3w}|\nabla w|^2 - 1 \tag{9.45c}$$

   with $w = \frac{1}{3\lambda}$ both initially and at the boundary for all t.
   The latter PDE (9.45c) reaches touchdown more gracefully than (9.45a); but you don't need to show this fact, just use it.

   (b) Devise a difference method of accuracy order (4, 2) to investigate the problem for the case where $\Omega = [0, \sqrt{\pi}] \times [0, \sqrt{\pi}]$ and $\lambda = 2$. The touchdown time T_* for this case should be around $t = .2$, but we want it precise to at least four digits. Plot $u(t, x, y)$ at a time t as close as you can to touchdown.
   Hint: Depending on your chosen method it may make sense to use smaller time steps as you get closer to touchdown.

9.5. Exercises

(c) Same as the previous item, but now the domain is the unit disk, $\Omega : x^2 + y^2 \leq 1$, and $\lambda = 1$. Your designed method is required to have accuracy order of only $(2, 2)$ here. You may want to use suitable coordinates, other than Cartesian.

(d) Optionally, investigate the determination of λ_* for a given domain Ω. You would want to consider the steady state case for this (why?), and a bifurcation analysis is involved, so this item is somewhat tangential to our course.

7. The notation in this exercise corresponds to that in Section 9.4, and we consider ODEs in time t for a solution $\mathbf{x}(t)$.

   (a) Consider applying forward Euler to the scalar ODE
   $$\frac{dx}{dt} = b - ax, \qquad x(0) \text{ given},$$
   with $a > 0$ and b constant. It reads
   $$x_{s+1} = x_s + k_s(b - ax_s), \qquad s = 0, 1, \ldots.$$
   Show that a step size $k = k_0$ can be chosen such that the steady state solution $x(\infty) = b/a$ is reached in one step, i.e., $x_0 = x(0)$ and $x_1 = x(\infty)$.

   (b) The same cannot be said, in general, about applying forward Euler to the corresponding system of size m,
   $$\frac{d\mathbf{x}}{dt} = \mathbf{b} - A\mathbf{x}, \qquad \mathbf{x}(0) \text{ given}, \qquad (9.46)$$
   with A a constant, symmetric positive definite matrix, and $\mathbf{b}$ a constant vector. Show that it is possible if working in infinite precision to select step sizes k_s such that after m steps the steady state $\mathbf{x}(\infty) = A^{-1}\mathbf{b}$ is reached. However, the resulting method for solving the linear system (9.36) is unstable, unless all eigenvalues of A are equal.

   (c) Consider applying forward Euler with a fixed step size to (9.46). A stable choice is $k = (\max_i \lambda_i)^{-1}$, where λ_i are the eigenvalues of A.
   Derive an expression for the number of steps required to reduce the "error" $\mathbf{x}_s - \mathbf{x}(\infty)$ by a factor of 10. Show that this number gets worse (i.e., larger) the higher the condition number $\text{cond}(A) = \frac{\max_i \lambda_i}{\min_i \lambda_i}$.

8. The *steepest descent* method for minimizing the quadratic function (9.39) is the "poorer cousin" of CG, where in (9.40) we take $\mathbf{p}_s = -\mathbf{d}_s$.

   (a) Show that the steepest descent method is equivalent to forward Euler for (9.46) with the "time step" $k_s = \alpha_s$.

   (b) An intelligent choice of time step for forward Euler, not requiring finding eigenvalues of A, is given by
   $$k_s = \frac{\mathbf{d}_s^T \mathbf{d}_s}{\mathbf{d}_s^T A \mathbf{d}_s}.$$

Justify and show that this is a stable choice.

More difficult: estimate the number of iterations required to reduce the error by a factor of 10.

9. Continuing the previous two exercises, consider applying forward Euler to the ODE system
$$M \frac{d\mathbf{x}}{dt} = \mathbf{b} - A\mathbf{x}, \qquad \mathbf{x}(0) \text{ given},$$
where the symmetric positive definite matrix M is easy to invert, unlike A.

   (a) Derive a corresponding preconditioned version of the steepest descent method. What step sizes would you choose?

   (b) How would you go about designing the matrix M? What properties should it have?

10. Show that
$$e^{k(L+M)} = e^{\frac{k}{2}L} e^{kM} e^{\frac{k}{2}L} + O(k^3)$$
for any $s \times s$ matrices L and M.

11. (a) Show that the nonlinear ODE
$$\iota \frac{du}{dt} = -|u|^2 u, \qquad t \geq t_0,$$
with the initial condition $u(t_0) = u_0$, has the exact solution
$$u(t) = u_0 \, e^{\iota(t-t_0)|u|^2}.$$
Furthermore, $|u(t)| = |u_0|$ for all $t \geq t_0$.

   (b) Carry out the calculations described in Example 9.6 using the full Crank–Nicolson method, i.e., no splitting. Do this for $k = h = .1$, or $k = h = .01$, and $t_f = 1000$. What are your observations?

12. Let us modify the cubic NLS (9.25) to read
$$\psi_t = \iota(\psi_{xx} + 2|\psi|^2 \psi),$$
and consider it, following [63], on the interval $x \in [-1, 1]$ with periodic BC and $\psi(0, x) = \psi_0(x) = \pi\sqrt{2}(1 + .1\cos(\pi x))$.

   (a) Derive the exact solution for the nonlinear part,
$$\iota \frac{du}{dt} = -2|u|^2 u, \qquad t \geq t_0.$$

9.5. Exercises

(b) Apply the splitting method of Example 9.6 for this problem, integrating from $t = 0$ to $t = 1$. Measure the error in norm and discrete Hamiltonian.

Set $h = .01$ and try $k = .01$ and $k = .0001$. For the latter you can also replace midpoint by RK4. What do you observe?

It's fun to watch $|\psi(t_n, x)|$ change with the time level n. If you use MATLAB, then something like the following (inside your time stepping loop) works:

```
if mod(n,10)==0,
    plot(x,abs(psi)); drawnow; pause(0);
end
```

13. Consider an ODE system, arising from a semi-discretization, of the form

$$\mathbf{v}' = \mathbf{f}(t, \mathbf{v}) + \mathbf{g}(t, \mathbf{v}), \qquad t \geq 0.$$

A linear, p-step IMEX method yields a full discretization and has the form

$$\sum_{l=0}^{p} \alpha_l \mathbf{v}^{n+1-l} = k \sum_{l=1}^{p} \beta_l \mathbf{f}^{n+1-l} + k \sum_{l=0}^{p} \gamma_l \mathbf{g}^{n+1-l}.$$

The combination of a two-step Adams–Bashforth method for $\mathbf{f}$ and a trapezoidal rule or a two-step BDF for $\mathbf{g}$ is common in the literature (especially in combination with spectral methods).

Show that

(a) The p-step method has order $\hat{p}$ if

$$\sum_{j=0}^{p} \alpha_j = 0,$$

$$\frac{1}{i!} \sum_{j=1}^{p} j^i \alpha_j = -\frac{1}{(i-1)!} \sum_{j=0}^{p} j^{i-1} \beta_j = -\frac{1}{(i-1)!} \sum_{j=0}^{p} j^{i-1} \gamma_j$$

for $i = 1, 2, \ldots, \hat{p}$ and such a condition does not hold for $i = \hat{p} + 1$.

(b) The $2\hat{p} + 1$ constraints above are linearly independent, provided that $\hat{p} \leq p$; thus, there exist p-step IMEX methods of order p.

(c) A p-step IMEX method cannot have order greater that p.

(d) The family of p-step IMEX methods of order p has p parameters.

See [18].

Chapter 10

Discontinuities and Almost Discontinuities

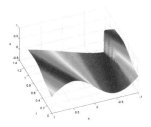

In Chapter 1 we have seen that the solution to the simple *advection equation*

$$u_t + au_x = 0, \qquad -\infty < x < \infty, \quad t \geq 0, \tag{10.1}$$

does not become smooth when starting from discontinuous initial data (recall Figure 1.3). In fact, starting from initial data $u_0(x)$ the solution is $u(t, x) = u_0(x - at)$ and thus consists of the initial values, discontinuities included, propagating along characteristics $dx/dt = a$. Furthermore, when solving nonlinear hyperbolic-type PDEs, discontinuities in the solution or its derivatives may arise even if the initial value function $u_0(x)$ is smooth.

However, usual discretization methods such as those considered in Chapters 3, 5, and 7 rely for their accuracy on the existence of bounded higher solution derivatives. When there are no such bounded derivatives the computed results can be unacceptably poor; see, e.g., Example 5.6. The present chapter is therefore devoted to designing methods that are appropriate for problems with discontinuities in their solutions.

While the advection equation causes propagation of initial data discontinuities, the solution of the *advection-diffusion equation*

$$u_t + au_x = \nu u_{xx}, \qquad -\infty < x < \infty, \quad t \geq 0, \tag{10.2}$$

with $\nu > 0$ and starting from the same initial conditions, is smooth for $t > 0$. See Figure 1.4, which displays the solution for the case $a = 0$, $\nu = 1$. And yet, this strong distinction between the properties of the two PDEs becomes weaker as ν gets smaller, $0 < \nu \ll \min\{|a|, 1\}$, with $a \neq 0$. In such a case we expect a sharp layer through which the solution varies rapidly, if smoothly. Recall Example 5.8, particularly Figure 5.7. This type of problem occurs in convection-dominated models of convection-diffusion type, and when simulating fluid flow governed by the Navier–Stokes equations with a large Reynolds number. Indeed, in the latter problems it is common to simplify the physical situation by sending the Reynolds number to ∞, obtaining in the inviscid limit the Euler equations, which are systems of conservation laws of the form (5.4) or (9.2). For the simple models above this would correspond to approximating (10.2) by (10.1) when ν is very small and positive. As Example 5.8 suggests, methods for problems with "almost discontinuities" corresponding to (10.2) with a small diffusion term naturally fit in the present chapter as well.

In Section 5.2.3 we discuss some basic ideas of upwinding and artificial viscosity, or diffusion, to solve the simple models above without encountering spurious wiggles. Recall that a general, explicit, consistent one-step scheme for linear, homogeneous PDEs,

$$v_j^{n+1} = \sum_{i=-l}^{r} \beta_i v_{j+i}^n,$$

is said to be **monotone** if $\beta_i \geq 0$, $i = -l, \ldots, r$. With such a scheme we have for each j

$$|v_j^{n+1}| \leq \max_{-l \leq i \leq r} |v_{j+i}^n| \sum_{i=-l}^{r} \beta_i = \max_{-l \leq i \leq r} |v_{j+i}^n|.$$

Hence $\|v^{n+1}\|_\infty \leq \|v^n\|_\infty$ and, moreover, no offshoots can arise in the numerical solution. In Section 5.2 we have seen that the Lax–Friedrichs and upwind schemes are monotone, and indeed they both produce smooth solution profiles for the advection equation, even for discontinuous initial data (recall Figures 5.4, 5.5, and 5.6).

However, extending these ideas to nonlinear problems can be nontrivial. In Section 10.1 we therefore discuss the theory for the simplest class of nonlinear problems which displays solution discontinuities, namely, scalar one-dimensional **conservation laws**.

For nonlinear problems we write a general explicit full discretization scheme as

$$v_j^{n+1} = \Phi(v^n; j), \tag{10.3}$$

where the dependence of Φ on the mesh spacing is suppressed. A scheme of the form (10.3) is called *monotone* if

$$\frac{\partial \Phi(v^n; j)}{\partial v_{j+i}^n} \geq 0 \quad \forall i. \tag{10.4}$$

Despite the imposing notation, this concept is very simple and extends the definition of monotone schemes for linear, constant coefficient problems.

In Section 10.2 we turn to basic, first order schemes for conservation laws. In addition to extending the upwind scheme we introduce **Godunov's method**, which is based on completely different principles but yields a similar scheme in the simplest cases.

The low-order schemes of Section 10.2 are all monotone, and this property not only provides stability but also ensures the prevention of spurious oscillations. However, higher order schemes cannot be monotone. In Section 10.3 we discuss methods which are at least second order accurate away from discontinuities and yet produce good resolutions of discontinuities as well. This long section introduces several important concepts and modern methods, including **high-resolution schemes**, **ENO**, and **WENO**.

In Section 10.4 we extend the methods introduced earlier to systems of conservation laws. Then in Section 10.5 we extend the methods to multidimensional problems. The theoretical backing available in one space dimension erodes significantly in two and three dimensions, but the practical utility of the methods often does extend.

10.1. Scalar Conservation Laws in One Dimension

Although discontinuities in hyperbolic-type problems may be regarded as extreme cases of parabolic-type problems with sharp transition layers, the passage to the extreme also yields some simplifications—a simple observation that is not always well understood in the community. In Section 10.6 we consider methods for more general problems with sharp transition layers. In some cases it is important to design a method that varies smoothly through layers.

A huge amount of work has been carried out on these topics in the past half century, especially on systems of conservation laws. Our main references here are LeVeque [116] and Shu [156]; see also Hundsdorfer and Verwer [98] and Hirsch [92]. The methods we describe fall under the general classification of **shock capturing**, where general methods are capable of handling also the appearance of shocks in the solution. Another class of exciting methods called **front tracking** is left out; see, e.g., Holden and Risebro [96].

10.1 Scalar conservation laws in one dimension

The examples of solution discontinuities in Chapters 1 and 5 arise from discontinuous initial data. One may wonder if the solution is smooth provided only that the initial plus boundary data are smooth. Unfortunately, we see next that solution discontinuities may develop for smooth data when solving nonlinear hyperbolic-type problems.

In this section we consider the Cauchy problem for a single conservation law,

$$u_t + f(u)_x = 0, \quad -\infty < x < \infty, \quad t \geq 0, \quad (10.5a)$$
$$u(0, x) = u_0(x).$$

The function f is called the **flux** and is assumed smooth.[32] Upon integration in x we have

$$\frac{d}{dt} \int_a^b u(t, x)dx = -(f(u(t, b)) - f(u(t, a))),$$

hence the interpretation of the conservation law is that the rate of change in u in a volume $[a, b]$ equals the difference between the flux in and the flux out.

This PDE can be written as

$$u_t + a(u)u_x = 0, \quad a(u) = \frac{df}{du}, \quad (10.5b)$$

and so u is constant along the characteristic curves $x = x(t)$ which are given by

$$\frac{dx}{dt} = a(u). \quad (10.5c)$$

We obtain the remarkable property that the characteristics for this nonlinear PDE are straight lines!

---

[32] By "smooth" we mean that f varies smoothly with u, *not* that $f(u(t, x))$ varies smoothly with t or x.

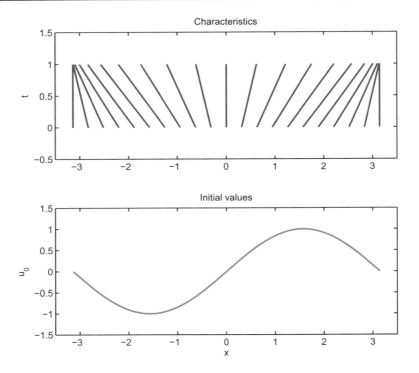

Figure 10.1. *Characteristic curves for the inviscid Burgers equation, and the corresponding initial data u_0.*

Example 10.1 The inviscid Burgers equation

$$u_t + \frac{1}{2}(u^2)_x = 0$$

is an instance of (10.5) with $f(u) = \frac{1}{2}u^2$ and $a(u) = u$. Figure 10.1 displays characteristic curves for the initial data function

$$u_0(x) = \sin x. \quad \blacksquare$$

While the characteristic curves for a conservation law are straight lines, their directions are not parallel, unlike for the simple advection equation. When two such characteristic curves collide (this would clearly eventually happen in Figure 10.1) they are bringing with them different solution values to the point of collision (t, x). Thus, a discontinuity forms and the solution no longer exists in the classical sense.

Example 10.2 The solution for the inviscid Burgers equation with initial data

$$u_0(x) = \frac{1}{4} + \frac{1}{2}\sin \pi x, \quad -1 \le x \le 1,$$

10.1. Scalar Conservation Laws in One Dimension

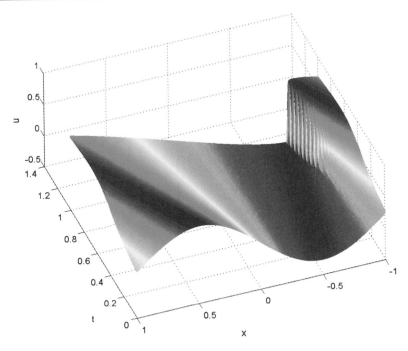

Figure 10.2. *Solution of the inviscid Burgers equation which, starting from a smooth sinusoidal function, develops a* **shock**.

and periodic boundary conditions is plotted in Figure 10.2. A discontinuity is seen to develop before t reaches the value 1. ∎

To be able to talk about solution existence for all time, even in the presence of discontinuities, we must admit *weak solutions*.

We multiply (10.5a) by an arbitrary differentiable function $w(t, x)$ which vanishes at ∞, and integrate over t and x. Integration by parts then transfers the derivatives from u to w and allows consideration of a piecewise continuous u. Thus, u is a **weak solution** of the PDE (10.5a) if it satisfies

$$\int_0^\infty \int_{-\infty}^\infty [w_t u + w_x f(u)] dx dt + \int_{-\infty}^\infty w(0, x) u_0(x) dx = 0 \qquad (10.6)$$

for all test functions $w(t, x) \in C^1$ with compact support.[33]

If u is a piecewise continuous weak solution of (10.5), then across the line of discontinuity

$$s(u_R - u_L) = f(u_R) - f(u_L), \qquad (10.7)$$

where s is the speed of propagation of the discontinuity and u_L and u_R are the states to the left and to the right of the discontinuity, respectively. This **jump condition** is also called

---
[33]A function with **compact support** vanishes outside a compact set.

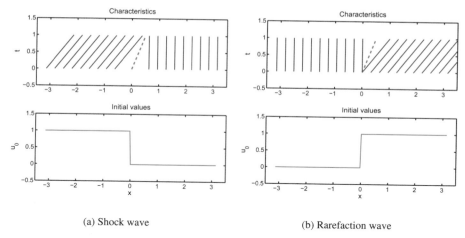

(a) Shock wave (b) Rarefaction wave

Figure 10.3. *Characteristic curves (solid blue) and wave speed (dashed green) for the inviscid Burgers equation with piecewise constant data.*

the *Rankine–Hugoniot* condition. It can be obtained by considering a test function $w(t, x)$ which vanishes everywhere except on a small volume which crosses the discontinuity line, and integrating (wisely) by parts. We leave the proof to Exercise 3.

Example 10.3 For the inviscid Burgers equation,

$$f(u_R) - f(u_L) = \frac{1}{2}(u_R^2 - u_L^2).$$

Thus, the jump condition (10.7) gives

$$s = \frac{u_R^2 - u_L^2}{2(u_R - u_L)} = \frac{1}{2}(u_R + u_L).$$

In Figure 10.3 we display the characteristics, as well as the curve with speed s, for two piecewise constant initial-value functions.

Note that in Case (a) of Figure 10.3, $s = 1/2$ and the unique weak solution is given by

$$u(t, x) = \begin{cases} 1, & x < t/2, \\ 0, & x > t/2. \end{cases}$$

In Case (b) of Figure 10.3, also $s = 1/2$ and a weak solution is given by

$$u(t, x) = \begin{cases} 0, & x < t/2, \\ 1, & x > t/2. \end{cases}$$

However, this latter weak solution is *not unique*. Physically, it is not very plausible either, because the characteristics do not really collide, so there is no reason to expect a shock. ∎

10.1. Scalar Conservation Laws in One Dimension

In Case (b) of Example 10.3 the conditions admitting a weak solution (10.6) plus the jump condition (10.7) do not yield a unique solution, and a question arises: What is the physically relevant weak solution? An answer is obtained, in the spirit of the previous section, by viewing the conservation law as a limit of a viscous problem,

$$u_t + f(u)_x = \varepsilon[b(u)u_x]_x, \qquad \varepsilon, b(u) > 0, \tag{10.8}$$

and considering the limit of the sequence of solutions of (10.8) as $\varepsilon \to 0$. This can be characterized by the **Oleinik entropy condition** stating that all discontinuities should have the property

$$\frac{f(u) - f(u_L)}{u - u_L} \geq s \geq \frac{f(u) - f(u_R)}{u - u_R} \tag{10.9a}$$

for all u between u_L and u_R. If f is convex (as it is for the inviscid Burgers equation) then (10.9a) is equivalent to the **Lax entropy condition**

$$f'(u_L) > s > f'(u_R). \tag{10.9b}$$

For Case (b) of Example 10.3 no value of s satisfies (10.9), hence there is no discontinuity in a solution satisfying the entropy condition, and the one prescribed at the end of Example 10.3 is neither the desirable **rarefaction wave** nor the one obtained using our upwind or Lax–Friedrichs methods. The corresponding continuous solution is obtained below using a *similarity transformation*.

10.1.1 Exact solution of the Riemann problem

A Cauchy problem consisting of a scalar conservation law, together with piecewise constant initial data with one discontinuity, is called a *Riemann problem*. In Example 10.3 we have seen the possible solutions of such a problem for the inviscid Burgers equation. Let us now generalize to a conservation law (10.5) with the initial data

$$u_0(x) = \begin{cases} u_L, & x < x_*, \\ u_R, & x > x_*. \end{cases}$$

Assume for notational simplicity that $[f'(u_R) - f'(u_L)](u_R - u_L) > 0$.

1. If there is a shock, then its speed is given by (10.7), i.e.,

$$s = \frac{f(u_R) - f(u_L)}{u_R - u_L}.$$

Thus, if $u_L > u_R$, then the solution of the Riemann problem is given by

$$u(t, x) = \begin{cases} u_L, & x - x_* < st, \\ u_R, & x - x_* > st. \end{cases} \tag{10.10a}$$

2. Otherwise, we have $u_L < u_R$. Then we look for a **similarity solution**

$$u(t, x) = g(\xi), \qquad \xi = \frac{x - x_*}{t}.$$

Substituting into (10.5a) gives

$$-\xi g' + f'g' = 0.$$

Thus, $g(\xi)$ is determined as the solution of

$$f'(g(\xi)) = \xi. \tag{10.10b}$$

We have the rarefaction wave

$$u(t, x) = \begin{cases} u_L, & x - x_* \le f'(u_L)t, \\ g(\frac{x-x_*}{t}), & f'(u_L)t < x - x_* \le f'(u_R)t, \\ u_R, & x - x_* > f'(u_R)t. \end{cases} \tag{10.10c}$$

In (10.10a) and (10.10c) the solution of the Riemann problem is completely specified. A similar construction may be applied for the case $[f'(u_R) - f'(u_L)](u_R - u_L) < 0$, and the two cases may then be combined. In particular, at $x = x_*$, $t > 0$, we obtain a simple expression for the solution as a function of time,

$$u(t, x_*) = \begin{cases} u_L, & u_L > u_R, \ s > 0, \\ u_R, & u_L > u_R, \ s \le 0, \\ u_L, & u_L \le u_R, \ 0 \le f'(u_L), \\ g(0), & u_L \le u_R, \ f'(u_L) < 0 \le f'(u_R), \\ u_R, & u_L \le u_R, \ 0 > f'(u_R). \end{cases} \tag{10.11}$$

Observe that at $x = x_*$ the solution is constant for all $t > 0$.

For the inviscid Burgers equation, $f'(\xi) = \xi$, so g is the identity, $g(\frac{x-x_*}{t}) = \frac{x-x_*}{t}$. Specifically, for Case (b) of Example 10.3 we have

$$u(t, x) = \begin{cases} 0 & \text{if } x \le 0, \\ x/t & \text{if } 0 < x \le t, \\ 1 & \text{if } x > t. \end{cases}$$

This solution is continuous throughout, but its derivative is not. See also Example 8.9.

10.2 First order schemes for scalar conservation laws

The simple schemes presented for the simple advection equation in Sections 5.2.2 and 5.2.3 generalize to handle scalar conservation laws; but care must be exercised, especially with the upwind scheme. The good news is that their interpretation as higher order approximations for

10.2. First Order Schemes for Scalar Conservation Laws

a corresponding modified PDE with an added artificial diffusion term, allows for entertaining the hope that they yield a solution which approximates a weak solution satisfying the entropy condition. On the other hand, not every generalization of (5.11) is guaranteed to approximate the discontinuity speed s correctly. To achieve that, *we must approximate the conservation form*.

To see this, consider the inviscid Burgers equation written in nonconservative form,

$$u_t + u u_x = 0.$$

Multiplying this PDE by u and denoting $\psi = u^2$ we have

$$\psi_t + \psi^{1/2} \psi_x = 0.$$

This can be written as a new conservation law,

$$\psi_t + \frac{2}{3}(\psi^{3/2})_x = 0,$$

and the jump condition yields

$$s_\psi = \frac{2}{3} \left[\frac{u_R^3 - u_L^3}{u_R^2 - u_L^2} \right].$$

This speed of propagation for the discontinuity differs from the speed obtained for the inviscid Burgers equation in conservation form. But a numerical scheme based on the nonconservative form may well approximate any of these speeds, or none of them!

> **Note:** Recall that in Sections 5.3 and 7.3 the discretization of a particular combination of conservative and nonconservative forms has proved advantageous. Here in the presence of discontinuities the story is different, and we should discretize the conservation form! Our advice to the reader is to keep alert.

Let us next integrate the conservation law $u_t + f(u)_x = 0$ over the control volume $V = [t_n, t_{n+1}] \times [x_{j-1/2}, x_{j+1/2}]$ and apply the Gauss divergence formula. See Figure 10.4. This yields

$$\bar{u}_j^{n+1} - \bar{u}_j^n + \mu \left(\bar{f}_{j+1/2}^{n+1/2} - \bar{f}_{j-1/2}^{n+1/2} \right) = 0, \qquad (10.12a)$$

where, e.g.,

$$\bar{u}_j^n = h^{-1} \int_{x_{j-1/2}}^{x_{j+1/2}} u(t_n, x) dx,$$

$$\bar{f}_{j+1/2}^{n+1/2} = k^{-1} \int_{t_n}^{t_{n+1}} f(u(t, x_{j+1/2})) dt. \qquad (10.12b)$$

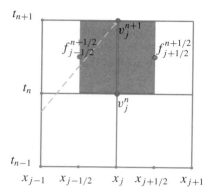

Figure 10.4. *A control volume (shaded). See* (10.12) *and* (10.13).

The formula (10.12) is exact and may be considered a better representation of the conservation law in the presence of low-smoothness solutions. Note that $\bar{u}_j^n$ is the average of u over the space interval $[x_{j-1/2}, x_{j+1/2}]$, and $\bar{f}_{j+1/2}^{n+1/2}$ is the *average flux* over the time step at $x = x_{j+1/2}$. Thus, (10.12a) expresses flux conservation across the finite volume cell.

A comparison of the above to the formulas used for deriving the box scheme in Sections 5.2 and 7.3 is inevitable. The major difference is that there we have tried our best to treat time and space in the same way. Here the propagation of solution information along characteristic curves, such as the dashed arrow in Figure 10.4, forward in time is of utmost importance. In fact, we seek to approximate the average flux $\bar{f}_{j+1/2}^{n+1/2}$ by explicit schemes.

Thus, consider three-point explicit schemes in conservation form

$$v_j^{n+1} = v_j^n - \mu \left(f_{j+1/2}^{n+1/2} - f_{j-1/2}^{n+1/2} \right), \tag{10.13a}$$

where

$$f_{j+1/2}^{n+1/2} = \hat{f}(v_j^n, v_{j+1}^n) \tag{10.13b}$$

is the **numerical flux** function required to approximate (and be at least consistent with) the average flux. The conservation form (10.13) can be shown to yield qualitatively correct approximations to discontinuity speeds.

The Lax–Friedrichs scheme can be written in this form:

$$\begin{aligned} v_j^{n+1} &= \frac{1}{2}(v_{j+1}^n + v_{j-1}^n) - \frac{\mu}{2}(f(v_{j+1}^n) - f(v_{j-1}^n)) \\ &= v_j^n - \mu \left(\left[\frac{f(v_j^n) + f(v_{j+1}^n)}{2} + \frac{v_j^n - v_{j+1}^n}{2\mu} \right] - \left[\frac{f(v_j^n) + f(v_{j-1}^n)}{2} + \frac{v_{j-1}^n - v_j^n}{2\mu} \right] \right). \end{aligned}$$

The Lax–Wendroff scheme and its variants mentioned in (5.5) are also in conservation form. For instance, in the Richtmyer formula (5.5b) we have $f_{j+1/2}^{n+1/2} = f(\bar{v}_j)$, where in fact $\bar{v}_j = v_{j+1/2}^{n+1/2}$ is an explicit Runge–Kutta-style approximation for $u(t_{n+1/2}, x_{j+1/2})$.

10.2. First Order Schemes for Scalar Conservation Laws

To extend the upwind scheme we follow the work of Engquist and Osher [61]. First we write (5.11) as

$$v_j^{n+1} = v_j^n - \mu[a^+ D_- + a^- D_+]v_j^n,$$

where

$$a^+ = \max(a, 0) = \frac{1}{2}(a + |a|), \quad a^- = \min(a, 0) = \frac{1}{2}(a - |a|). \quad (10.14)$$

Now extend to a conservation law (10.5)

$$v_j^{n+1} = v_j^n - \mu[D_- f^+(v_j^n) + D_+ f^-(v_j^n)], \quad (10.15a)$$

where

$$f^+(u) = \int_0^u a^+(\xi) d\xi, \quad f^-(u) = \int_0^u a^-(\xi) d\xi. \quad (10.15b)$$

Notice that for any u (10.15b) satisfies

$$f(u) = f^+(u) + f^-(u), \quad \frac{df^+(u)}{du} \geq 0, \quad \frac{df^-(u)}{du} \leq 0. \quad (10.16)$$

A flux decomposition that satisfies (10.16) is called **flux splitting**.

Example 10.4 For the inviscid Burgers equation, the definitions (10.15b) simplify into

$$f^+(u) = \frac{1}{2}[\max(u, 0)]^2, \quad f^-(u) = \frac{1}{2}[\min(u, 0)]^2.$$

Then the scheme (10.15a) reads as expected,

$$v_j^{n+1} = v_j^n - \frac{\mu}{2} \begin{cases} [(v_{j+1}^n)^2 - (v_j^n)^2] & \text{if } v_j^n < 0, \\ [(v_j^n)^2 - (v_{j-1}^n)^2] & \text{if } v_j^n \geq 0. \end{cases}$$

In Figure 10.5 we display the results using the Lax–Friedrichs and upwind schemes, both in conservative and in nonconservative form. The initial data are the same as in Case (a) of Figure 10.3, and the step sizes are $h = .01\pi$, $k = 0.5h$.

The best scheme for this example is clearly the upwind scheme in conservative form. The nonconservative form is a disaster. The Lax–Friedrichs variants both smear the discontinuity significantly. In Exercise 5 we see that the nonconservative form of the Lax–Friedrichs scheme can lead to a disaster as well. ∎

The flux splitting (10.15b) is not a smooth function of u, and determining it can become a challenge in general. A simpler flux splitting is given by

$$f^{\pm}(u) = \frac{1}{2}(f(u) \pm Mu), \quad M = \max_u |a(u)|. \quad (10.17)$$

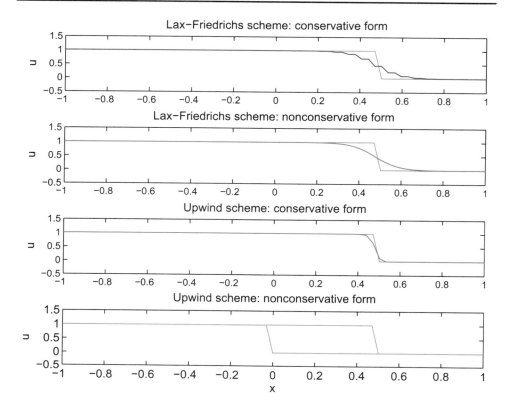

Figure 10.5. *Solving the inviscid Burgers equation, starting from a step function leading to a shock, and employing conservative and nonconservative forms of the Lax–Friedrichs and upwind schemes.*

Since we are considering explicit schemes in time the practical calculation of M by maximizing over all mesh points v_j^n at each time t_n is straightforward. It is easy to see that (10.17) satisfies (10.16), i.e., it is indeed a flux splitting. Moreover, this is a *smooth splitting*, unlike (10.15b). For the simple advection equation where a is constant, $M = |a|$ and the same upwind scheme (5.11), (10.14) is retrieved. But for nonlinear conservation laws the two flux splittings can yield significantly different results. Generally, (10.17) tends to smear the discontinuity more.

10.2.1 Godunov's scheme

This is another basic, low-order scheme which is based on a different approach than the other methods in this section, although for the simple advection equation it, too, coincides with the upwind scheme (5.11).

In a finite volume approach we can view the approximate solution at a mesh point, v_j^n, as approximating the *cell average* $\bar{u}_j(t_n)$ defined in (10.12b). This then suggests using

10.2. First Order Schemes for Scalar Conservation Laws

v_j^n to define a piecewise constant function

$$\tilde{v}^n(x) = v_j^n, \qquad x_{j-1/2} \le x < x_{j+1/2}. \tag{10.18a}$$

The basic idea is to use this piecewise constant function as initial data for the next step and to solve the resulting Riemann problem on each mesh subinterval $[x_j, x_{j+1}]$ exactly. This is possible, according to Section 10.1.1, so long as the time step k is small enough that waves from neighboring subintervals do not begin to interact.

Denote the solution so constructed from the sequence of Riemann problems by $\tilde{v}^n(t, x)$. Then at the end of the time step we set

$$v_j^{n+1} = h^{-1} \int_{x_{j-1/2}}^{x_{j+1/2}} \tilde{v}^n(t_{n+1}, x) dx. \tag{10.18b}$$

Despite the different approach, the method can be seen to be in the conservative form (10.13). Indeed, noting that $\tilde{v}^n(t, x)$ satisfies the conservation law (10.5a) exactly, we integrate this PDE over the volume $[t_n, t_{n+1}] \times [x_{j-1/2}, x_{j+1/2}]$ and insert (10.18b) and (10.18a) to obtain

$$v_j^{n+1} = v_j^n - \mu \left(f_{j+1/2}^{n+1/2} - f_{j-1/2}^{n+1/2} \right) \tag{10.18c}$$

with the flux

$$f_{j+1/2}^{n+1/2} = k^{-1} \int_{t_n}^{t_{n+1}} f(\tilde{v}^n(t, x_{j+1/2})) dt \tag{10.18d}$$

(see (10.12b)).

It remains to calculate the integral in (10.18d). But this is easy, because $\tilde{v}^n(t_n, x)$ is piecewise constant on $[x_j, x_{j+1}]$, with the discontinuity at $x_* = x_{j+1/2}$. The solution is given by (10.10a) or (10.10c) with $u_L = v_j^n$ and $u_R = v_{j+1}^n$. In particular, at $x = x_{j+1/2}$ the solution is constant in t and is given by (10.11). Thus

$$f_{j+1/2}^{n+1/2} = f(\tilde{v}_{j+1/2}^n), \tag{10.19}$$

where $\tilde{v}_{j+1/2}^n$ is the value of $u(t, x_*)$ according to (10.11). The working formula for Godunov's method thus becomes very simple.

Recall that the above procedure works if waves from neighboring cells do not interact. This happens if k is small enough to satisfy

$$\mu |a(v_j^n)| \le 1 \quad \forall j. \tag{10.20}$$

We get a natural extension of the CFL condition.

Example 10.5 Consider the inviscid Burgers equation with the initial conditions

$$u_0(x) = \frac{1}{4} + \frac{1}{2} \sin \pi x, \quad -1 \le x \le 1.$$

The exact solution is smooth up to $t = \frac{2}{\pi}$; then it develops a moving shock which interacts with the rarefaction waves [88]. At $t = 1.1$ there is a shock, with a monotone solution

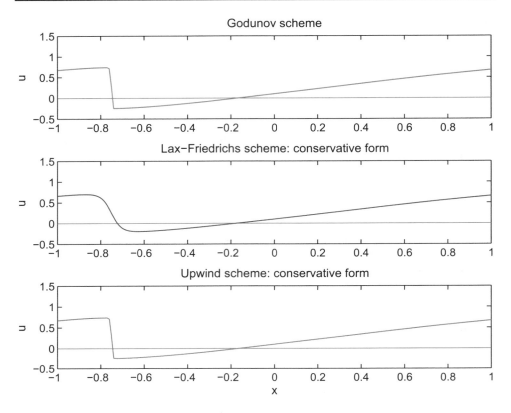

Figure 10.6. *Solving the inviscid Burgers equation, starting from a smooth sinusoidal function leading to a shock. The schemes employed are Godunov, Lax–Friedrichs in conservative form, and the upwind scheme in conservative form.*

elsewhere—see Figure 10.2. Employing the schemes we have seen so far in this chapter, the results using $h = 0.01$ and $\mu = 0.9$ are displayed in Figure 10.6. The Godunov results are qualitatively similar to those obtained by the upwind scheme and considerably better than those obtained by the Lax–Friedrichs scheme. ∎

The Lax–Friedrichs scheme does have one potential advantage over the upwind and Godunov schemes, in that its coefficients vary smoothly, even through **sonic points**, i.e., points where $f'(u) = 0$. Higher order schemes based on upwind or Godunov as building blocks without smoothing may lose accuracy at sonic points because of the sharp transition in these schemes.

10.3 Higher order schemes for scalar conservation laws

The schemes featured in Section 10.2 are all first order accurate and monotone (i.e., they satisfy (10.4)). For a monotone scheme, if $v_j^n \leq w_j^n \; \forall j$, then $v_j^{n+1} \leq w_j^{n+1} \; \forall j$. This

10.3. Higher Order Schemes for Scalar Conservation Laws

gives stability and prevents the possibility of spurious oscillations resulting from solution discontinuities. However, as it turns out, monotone schemes can only be first order accurate, essentially because they require $O(h)$ artificial diffusion (recall Section 5.2.3): a scheme that approximates with high accuracy another problem that is $O(h)$ away from our problem obviously cannot have accuracy better than $O(h)$ for our problem.

Therefore, weaker properties than monotonicity are sought. For a spatial mesh function v, define the **total variation** as

$$TV(v) = h \sum_j |v_{j+1} - v_j|. \tag{10.21a}$$

A numerical scheme (10.3) is called **total variation diminishing** (TVD) if for all mesh functions

$$TV(v^{n+1}) \leq TV(v^n) \quad \forall n \geq 0. \tag{10.21b}$$

The scheme is called **total variation bounded** (TVB) if for all mesh functions

$$TV(v^{n+1}) \leq B \quad \forall n \geq 0, \tag{10.21c}$$

where B is a constant depending only on $TV(v^0)$. The scheme is called **monotonicity preserving** if $v_j^0 \leq v_{j+1}^0 \ \forall j$ implies $v_j^n \leq v_{j+1}^n \ \forall j$ for all $n \geq 0$.

As it turns out, these properties are related as follows.

Theorem 10.6.

$$\text{monotone} \implies \text{TVD} \implies \text{monotonicity preserving}.$$

The search is therefore on for second order TVD methods.

10.3.1 High-resolution schemes

Unfortunately, a linear monotonicity-preserving scheme is at most first order accurate. This was shown already in 1959 by Godunov. Note that it is the scheme, not the PDE, which is linear in this negative theoretical result. More depressingly, in two space dimensions a TVD scheme is at most first order accurate except in certain trivial cases [70]. In one dimension it is possible to construct second order TVD (nonlinear) schemes, and such methods work well in practice also in two dimensions, despite the negative theory. We will return to the latter point in Section 10.4; now we proceed with methods for scalar, one-dimensional problems.

The basic observation here is that the excessive amount of artificial diffusion (or viscosity) which causes the scheme to be only first order accurate is needed only in the vicinity of a discontinuity. Elsewhere, a higher order method with less artificial viscosity can be used.

In **flux limiter** schemes we combine a higher order flux approximation $f_{H,j+1/2}(v)$ such as Lax–Wendroff, which works better away from discontinuities, with a first order flux $f_{L,j+1/2}(v)$ such as that of any of the monotone schemes of the previous section. For instance, with an upwind method the combined scheme can be seen as a higher order centered method with upwinding introduced only if and where needed. The scheme then has the form (10.13) with

$$f_{j+1/2}^{n+1/2} = f_{L,j+1/2}(v) + \phi(\theta_j)[f_{H,j+1/2}(v) - f_{L,j+1/2}(v)], \quad (10.22a)$$

where ϕ is a function which is near 0 in the vicinity of a discontinuity and near 1 when the solution varies slowly. The parameter θ_j measures the "smoothness" of the solution, e.g.,

$$\theta_j = \frac{v_j^n - v_{j-1}^n}{v_{j+1}^n - v_j^n}. \quad (10.22b)$$

Various possibilities have been proposed for ϕ, e.g., the "superbee,"

$$\phi(\theta) = \max(0, \min(1, 2\theta), \min(\theta, 2)). \quad (10.22c)$$

In [76] and [116] there are numerical comparisons of various variants on a simple model problem.

Such schemes have been shown to be TVD. We refer to [116] and references therein for much more detail.

10.3.2 Semi-discretization and ENO schemes

"Higher order" can be more than 2! However, the TVD schemes of the previous section are not. If we give up satisfying the TVD property orthodoxly, but still require a careful control of the oscillations, then we can aspire to an even higher order of accuracy. Moreover, we now introduce a pleasing semi-discretization setting which allows construction of higher order schemes in time and space separately, a fact that becomes even more important in more than one space variable. The pioneering work here was done by Harten and his collaborators [88].

This again is a finite volume approach. Consider the cell averages that have inspired the Godunov scheme. Independent of the time step, we define

$$\bar{u}_j(t) = h^{-1} \int_{x_{j-1/2}}^{x_{j+1/2}} u(t,x)dx. \quad (10.23)$$

Integrating the conservation law over the cell $[x_{j-1/2}, x_{j+1/2}]$ we obtain

$$\frac{d\bar{u}_j}{dt} + h^{-1}[f(u(t,x_{j+1/2})) - f(u(t,x_{j-1/2}))] = 0 \quad (10.24a)$$

with no approximation utilized as yet. Now, if $v(t)$ is a mesh function approximating $u(t,\cdot)$, then we can construct a numerical flux $\hat{f}_{j+1/2}(v(t))$ approximating $f(u(t,x_{j+1/2}))$ based on spatial considerations alone. Suppose that $\bar{v}_j(t)$ correspondingly approximates $\bar{u}_j(t)$. Then we have the semi-discretization

$$\frac{d\bar{v}_j}{dt} + h^{-1}[\hat{f}_{j+1/2}(v(t)) - \hat{f}_{j-1/2}(v(t))] = 0. \quad (10.24b)$$

10.3. Higher Order Schemes for Scalar Conservation Laws

If we discretize these equations in time using forward Euler, then a conservative scheme of the form (10.13a) is obtained. But we can discretize in time more accurately. Runge–Kutta methods come to mind, although some care must be exercised, as explained in Section 10.3.3.

To figure out the order of such a scheme, let us concentrate first on the semi-discretization (10.24b). The unusual feature in (10.24a)–(10.24b) is that these formulas involve both cell average and pointwise values of the solution. Let $\bar{v}(t)$ be the mesh function composed of the values $\bar{v}_j(t)$, and likewise denote by $\bar{u}(t)$ the corresponding mesh function for the exact cell averages. The relationship between $\bar{u}$ and the exact solution u is given by (10.23). For the approximate solution we construct in the sequel local interpolations for pointwise values v_{j+i} in terms of the values $\bar{v}_j$. Denoting the result of such a construction by $v(t) = \Psi(\bar{v}(t))$, we can write (10.24b) as

$$\frac{d\bar{v}_j}{dt} + h^{-1}[\hat{f}_{j+1/2}(\Psi(\bar{v}(t))) - \hat{f}_{j-1/2}(\Psi(\bar{v}(t)))] = 0. \tag{10.24c}$$

Thus, the local truncation error for (10.24), considered as a formula for the cell averages, is

$$\tau = h^{-1}[(\hat{f}_{j+1/2}(\Psi(\bar{u}(t))) - f(u(t, x_{j+1/2})))$$
$$- (\hat{f}_{j-1/2}(\Psi(\bar{u}(t))) - f(u(t, x_{j-1/2})))].$$

If $\tau = O(h^p)$ $\forall t$, and $\hat{f}$ is Lipschitz continuous, then upon subtraction of (10.24b) from (10.24a) and integration we obtain

$$\bar{u}_j(t) - \bar{v}_j(t) = O(h^p).$$

If we now use a Runge–Kutta method of order $\tilde{p}$ to integrate the ODE system (10.24c), then the overall order of the resulting full discretization is $(\tilde{p}, p)$.

Reconstruction of v, and ENO schemes

The integration in time is for cell averages $\bar{v}_j$. However, to obtain the formula and the final solution reconstruction we need pointwise values v_j and $v_{j+1/2}$. This can be developed elegantly using **primitive functions**. For a fixed t, let w be defined by

$$w'(x) = u(t, x), \qquad w(x_{1/2}) = 0. \tag{10.25a}$$

Then

$$w_{j+1/2} = w(x_{j+1/2}) = \int_{x_{1/2}}^{x_{j+1/2}} u(t, \xi) d\xi = h \sum_{i=1}^{j} \bar{u}_i(t). \tag{10.25b}$$

The values of $w(x_{j+1/2})$ are known exactly in terms of the cell averages, so we can apply a high-order polynomial interpolation and recover $w(x)$ to a high order! Thus, if p_j in a polynomial of degree at most q which interpolates $q+1$ values $w_{j+i+1/2}$ nearby x_j, then (see Section 2.12.2)

$$p_j(x) = w(x) + O(h^{q+1}), \qquad x_{j-1/2} \leq x \leq x_{j+1/2}, \tag{10.25c}$$

provided that w is smooth over the interpolated area, i.e., provided that the span of the points $x_{j+i+1/2}$ involved does not cross a discontinuity.

Now, we don't know the $\bar{u}_i$, but if we know the $\bar{v}_i$ accurate to $O(h^p)$, $p \geq q$, then we can use those in place of $\bar{u}_i$ in (10.25b) to define $w_{j+1/2}$. Then we set

$$v_j = p'_j(x_j), \quad v^R_{j-1/2} = p'_j(x_{j-1/2}), \quad v^L_{j+1/2} = p'_j(x_{j+1/2}), \qquad (10.25d)$$

obtaining $O(h^q)$ approximations to $u(t, x_j)$, $u(t, x^+_{j-1/2})$, and $u(t, x^-_{j+1/2})$, respectively. In the notation of (10.24c), this defines $\Psi(\bar{v}(t))$.

Example 10.7 Let us demonstrate the construction of v at mesh points from cell averages. Choose $q = 3$, i.e., we consider a local cubic interpolation of the primitive function w near x_j. Suppose, say, that the interpolation points are $x_{j-1/2}, x_{j+1/2}, x_{j+3/2}$, and $x_{j+5/2}$.

To construct the polynomial interpolant we form the divided differences

$$w[x_{j-1/2}, x_{j+1/2}] = \frac{w_{j+1/2} - w_{j-1/2}}{h} = \bar{v}_j,$$

$$w[x_{j+1/2}, x_{j+3/2}] = \bar{v}_{j+1}, \quad w[x_{j+3/2}, x_{j+5/2}] = \bar{v}_{j+2},$$

$$w[x_{j-1/2}, x_{j+1/2}, x_{j+3/2}] = \frac{\bar{v}_{j+1} - \bar{v}_j}{2h}, \quad w[x_{j+1/2}, x_{j+3/2}, x_{j+5/2}] = \frac{\bar{v}_{j+2} - \bar{v}_{j+1}}{2h},$$

$$w[x_{j-1/2}, x_{j+1/2}, x_{j+3/2}, x_{j+5/2}] = (3h)^{-1} \left[\frac{\bar{v}_{j+2} - \bar{v}_{j+1}}{2h} - \frac{\bar{v}_{j+1} - \bar{v}_j}{2h} \right]$$

$$= \frac{\bar{v}_{j+2} - 2\bar{v}_{j+1} + \bar{v}_j}{6h^2}.$$

So, the cubic

$$p_j(x) = w(x_{j-1/2}) + \bar{v}_j(x - x_{j-1/2}) + \frac{\bar{v}_{j+1} - \bar{v}_j}{2h}(x - x_{j-1/2})(x - x_{j+1/2})$$

$$+ \frac{\bar{v}_{j+2} - 2\bar{v}_{j+1} + \bar{v}_j}{6h^2}(x - x_{j-1/2})(x - x_{j+1/2})(x - x_{j+3/2})$$

approximates $w(x)$ to fourth order and, more importantly, the quadratic

$$p'_j(x) = \bar{v}_j + \frac{\bar{v}_{j+1} - \bar{v}_j}{2h}\left[(x - x_{j-1/2}) + (x - x_{j+1/2})\right] + \frac{\bar{v}_{j+2} - 2\bar{v}_{j+1} + \bar{v}_j}{6h^2} \cdot$$

$$\left[(x - x_{j-1/2})(x - x_{j+1/2}) + (x - x_{j-1/2})(x - x_{j+3/2}) + (x - x_{j+1/2})(x - x_{j+3/2})\right]$$

approximates v near x_j to third order.

In particular, substitute to get the third order approximations

$$v_j = p'_j(x_j) = \frac{1}{4}(-\bar{v}_{j+2} + 6\bar{v}_{j+1} - \bar{v}_j),$$

$$v^L_{j+1/2} = p'_j(x_{j+1/2}) = \frac{1}{6}(-\bar{v}_{j+2} + 5\bar{v}_{j+1} + 2\bar{v}_j), \qquad (10.26a)$$

$$v^R_{j-1/2} = p'_j(x_{j-1/2}) = \frac{1}{6}(2\bar{v}_{j+2} - 7\bar{v}_{j+1} + 11\bar{v}_j). \qquad (10.26b)$$

10.3. Higher Order Schemes for Scalar Conservation Laws

Exercise 6 shows that similarly, using the interpolation points $x_{j-3/2}, x_{j-1/2}, x_{j+1/2}$, and $x_{j+3/2}$ we have another local third order interpolant yielding

$$v^L_{j+1/2} = \frac{1}{6}(2\bar{v}_{j+1} + 5\bar{v}_j - \bar{v}_{j-1}), \tag{10.26c}$$

$$v^R_{j-1/2} = \frac{1}{6}(-\bar{v}_{j+1} + 5\bar{v}_j + 2\bar{v}_{j-1}), \tag{10.26d}$$

and using the interpolation points $x_{j-5/2}, x_{j-3/2}, x_{j-1/2}$, and $x_{j+1/2}$ yields

$$v_j = \frac{1}{4}(-\bar{v}_j + 6\bar{v}_{j-1} - \bar{v}_{j-2}),$$

$$v^L_{j+1/2} = \frac{1}{6}(11\bar{v}_j - 7\bar{v}_{j-1} + 2\bar{v}_{j-2}), \tag{10.26e}$$

$$v^R_{j-1/2} = \frac{1}{6}(2\bar{v}_j + 5\bar{v}_{j-1} - \bar{v}_{j-2}). \tag{10.26f}$$

During an ENO reconstruction, one of these formula pairs is used at each mesh point j.

Note that the tedious algebra needs to be done just once, and what comes out are simple formulas defining Ψ. ∎

Next, we use these obtained values to define $\hat{f}$. We need to approximate $f(u(t, x_{j+1/2}))$, and we have two values at this point, $v^L_{j+1/2} = p'_j(x_{j+1/2})$ and $v^R_{j+1/2} = p'_{j+1}(x_{j+1/2})$. A Riemann solver comes to mind. We then use the formula (10.11) to define $v_{j+1/2}$ and set $\hat{f}_{j+1/2}(\Psi(\bar{v}(t))) = f(v_{j+1/2})$.

Clearly, we must choose the points over which to interpolate carefully. It is well known that high-order polynomial interpolation may produce some oscillations even for smooth data (hence we give up the notion of strict monotonicity preservation when we go to orders higher than 2), but we can realistically hope to achieve good results, provided that no region with large gradients is crossed. The main idea behind **essentially nonoscillatory schemes** (ENO) is to choose the set of neighbors of x_j for the jth polynomial interpolation automatically, in such a way that the oscillation is minimized.

For this purpose, the Newton form of polynomial interpolation is ideal—for a review see Section 2.12.2. Since the construction adds the interpolation points one at a time we can start with the pair $x_{j-1/2}, x_{j+1/2}$, noting that $w[x_{i-1/2}, x_{i+1/2}] = \bar{v}_i$. Then, at some stage where we have already interpolated at $x_{j-l+1/2}, \ldots, x_{j+r-1/2}$, do the following:

- Form the divided differences $w[x_{j-l-1/2}, \ldots, x_{j+r-1/2}]$ and $w[x_{j-l+1/2}, \ldots, x_{j+r+1/2}]$ according to (2.37).

- Choose the point $x_{j-l-1/2}$ or $x_{j+r+1/2}$ which yields the divided difference that is smaller in magnitude.

- Form the next term of the interpolating polynomial, increasing its degree from $r+l-1$ to $r+l$.

As in Example 10.7, no values of w, only values of $\bar{v}_i$, are needed to form the divided difference tables. Once a step is completed in choosing the next neighbor, we add the corresponding contribution to the differentiated polynomial at the points of interest using coefficient tables that have been prepared in advance.

Example 10.8 Continuing with Example 10.5 we display results using first and second order schemes of the type described above. The "exact solution" is calcuated using a second order method with small values of k and h, and this is used to calculate errors in results obtained using much coarser step sizes. The time discretization used together with the first order spatial discretization is the forward Euler method. This yields essentially the Godunov scheme. For the second order spatial discretization, explicit trapezoidal (a two-stage Runge–Kutta method of order 2), given by

$$\begin{array}{c|cc} 0 & 0 & 0 \\ 1 & 1 & 0 \\ \hline & \frac{1}{2} & \frac{1}{2} \end{array}$$

is used. Thus, we are looking essentially at two schemes, one of order (1, 1) and the other of order (2, 2).

In Table 10.1 we record solution errors $\|error\|$ in three (normalized) norms: (i) at $t = 0.3$, where the solution is still smooth, and (ii) at $t = 1.1$, where the solution contains a shock layer, as in Figure 10.5. In all cases we took $\mu = k/h = 0.5$. The results in Table 10.1 clearly indicate the expected order of accuracy of these methods for the smooth solution at $t = 0.3$. At $t = 1.1$ we have not only a shock but also a relatively flat profile away from the shock, which gives the first order method an advantage. The results differ significantly depending on the norm in the latter case, an observation which you should be able to explain.

In Figure 10.7 we display the mesh function values (marked with ×) together with the exact solution, for the following cases: (i) first order method, $h = 0.05$, (ii) second order method, $h = 0.05$, (iii) first order method, $h = 0.1$, and (iv) second order method, $h = 0.1$. The second order method is more accurate, but the improvement is not amazing in this particular instance.

The solution displayed in Figure 10.2 was calculated using our current second order method, with $h = 0.02$ and $k = 0.01$, for $t \in [0, 1.5]$. ∎

Note that ENO can naturally be applied for outflow boundary conditions by taking the appropriate one-sided stencils that use only mesh points inside the domain.

10.3.3 Strong stability preserving methods

The idea of separating the considerations for the spatial and time discretizations is very appealing. In the previous section we achieve that, but at a price. Generally, many families of methods such as (10.13) (including the high-resolution methods of Section 10.3.1) look as if the forward Euler method is utilized in time, and theoretical results have been derived

10.3. Higher Order Schemes for Scalar Conservation Laws

Table 10.1. *Errors for ENO schemes applied to the inviscid Burgers equation. At $t = 0.3$ the solution is still smooth, but at $t = 1.1$ there is a shock. The measured errors are generally acceptable, and the higher order method is more accurate, but not spectacularly so.*

t	Method	h	$\|error\|_\infty$	$\|error\|_2$	$\|error\|_1$
0.3	(1, 1)	0.1	5.21e-2	2.14e-2	1.69e-2
		0.05	3.25e-2	1.19e-2	9.13e-3
		0.02	1.57e-2	5.11e-3	3.90e-3
		0.01	8.03e-3	2.63e-3	2.00e-3
		0.005	4.07e-3	1.34e-3	1.01e-3
	(2, 2)	0.1	2.97e-2	1.18e-2	8.03e-3
		0.05	1.24e-2	3.77e-3	2.30e-3
		0.02	3.71e-3	7.76e-4	3.78e-4
		0.01	1.47e-3	2.38e-4	9.94e-5
		0.005	5.52e-4	7.24e-5	2.55e-5
1.1	(1, 1)	0.1	3.24e-1	7.76e-2	3.66e-2
		0.05	1.93e-1	4.30e-2	2.01e-2
		0.02	2.96e-1	3.08e-2	7.32e-3
		0.01	1.44e-1	1.35e-2	3.20e-3
	(2, 2)	0.1	2.89e-1	6.52e-2	2.14e-2
		0.05	1.16e-1	2.42e-2	6.66e-3
		0.02	2.81e-1	2.82e-2	3.43e-3
		0.01	1.05e-1	1.03e-2	1.09e-3

for such formulations. It is natural, then, to want to "lift" such results up for more accurate time discretizations.

Let us consider an ODE system

$$\mathbf{v}' \equiv \frac{d\mathbf{v}}{dt} = \mathbf{g}(\mathbf{v}), \qquad (10.27)$$

and *assume* that the forward Euler method satisfies, in some norm, the *monotonicity property*

$$\|\mathbf{v}^n + k\mathbf{g}(\mathbf{v}^n)\| = \|\mathbf{v}^{n+1}\| \leq \|\mathbf{v}^n\|, \quad n = 0, 1, \ldots. \qquad (10.28)$$

Then an ODE method for (10.27) is said to be *strong stability preserving* (SSP) if it, too, satisfies $\|\mathbf{v}^{n+1}\| \leq \|\mathbf{v}^n\|$. See [72, 98].

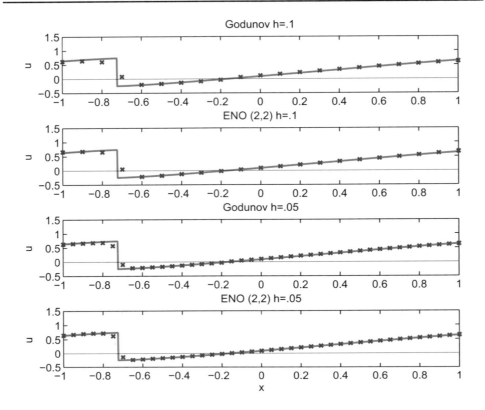

Figure 10.7. *Solving the inviscid Burgers equation, starting from a smooth sinusoidal function leading to a shock, and employing Godunov's scheme as well as a second order ENO scheme.*

Next, consider an explicit s-stage Runge–Kutta method (2.10) for (10.27), which we write as

$$\mathbf{Y}_1 = \mathbf{v}^n,$$
$$\mathbf{Y}_{i+1} = \mathbf{v}^n + k \sum_{j=1}^{i} a_{i+1,j} \mathbf{g}(\mathbf{Y}_j), \quad i = 1, \ldots, s,$$
$$\mathbf{v}^{n+1} = \mathbf{Y}_{s+1} \tag{10.29}$$

(where we set $a_{s+1,j} = b_j$). This can be written (nonuniquely, in general) as

$$\mathbf{Y}_1 = \mathbf{v}^n,$$
$$\mathbf{Y}_{i+1} = \sum_{j=1}^{i} \alpha_{i,j} \mathbf{Y}_j + k \sum_{j=1}^{i} \beta_{i,j} \mathbf{g}(\mathbf{Y}_j), \quad i = 1, \ldots, s,$$
$$\mathbf{v}^{n+1} = \mathbf{Y}_{s+1} \tag{10.30}$$

10.3. Higher Order Schemes for Scalar Conservation Laws

with $\alpha_{i,j} \geq 0$, $1 \leq j \leq i \leq s+1$. Now, if $\beta_{i,j} \geq 0$, all $1 \leq j \leq i \leq s+1$, then the higher order RK method is just a convex combination of forward Euler steps, and therefore the SSP property holds.

Example 10.9 The explicit trapezoidal method employed in Example 10.8 is SSP. This can be seen by writing it as

$$\mathbf{Y}_1 = \mathbf{v}^n,$$
$$\mathbf{Y}_2 = \mathbf{Y}_1 + k\mathbf{g}(\mathbf{Y}_1),$$
$$\mathbf{v}^{n+1} = \mathbf{Y}_3 = \frac{1}{2}(\mathbf{Y}_1 + \mathbf{Y}_2) + \frac{k}{2}\mathbf{g}(\mathbf{Y}_2).$$

The explicit midpoint method

$$\begin{array}{c|cc} 0 & 0 & 0 \\ \frac{1}{2} & \frac{1}{2} & 0 \\ \hline & 0 & 1 \end{array}$$

is not SSP, and we leave verifying this as Exercise 7. Note, however, that the results of running this variant as in Example 10.8 are not significantly different from those of the explicit trapezoidal method. ∎

Higher order RK methods that have the SSP property have been constructed [157, 72]. Unfortunately, this turns out not to be a simple matter [148]. Explanations of this concept in terms of **contractivity** are given in [90]. Linear multistep SSP methods have also been devised. The most popular method of this sort in practice is based on Runge–Kutta time discretization and has accuracy order (3,3). The reconstruction in space uses the formulas in Example 10.7. The corresponding RK method is given by

$$\begin{array}{c|ccc} 0 & 0 & 0 & 0 \\ 1 & 1 & 0 & 0 \\ \frac{1}{2} & \frac{1}{4} & \frac{1}{4} & 0 \\ \hline & \frac{1}{6} & \frac{1}{6} & \frac{2}{3} \end{array}$$

10.3.4 WENO schemes

In Chapter 7 we take extra pains to apply centered, *conservative* methods such as the leapfrog and box schemes in order to discretize hyperbolic PDEs. Such schemes are more accurate than one-sided ones and preserve solution structure better. However, in Section 5.2 we have already seen that in the presence of discontinuities such methods perform poorly, hence upwind methods are introduced there and studied further in Section 10.2, considerably lowering expectation of both high accuracy and structure preservation: survival takes precedence over fancy footwork.

In Section 10.3.1 we again bring back more accurate, centered discretizations to be used away from discontinuities through the mechanism of flux limiters. But in Sections 10.3.2 and 10.3.3 we are back with a purer upwind idea, slightly violated but improved in accuracy through the ENO methods.

The next stage is then to realize that ENO may occasionally also be "too one-sided," unnecessarily so in regions where the solution is smooth. Moreover, ENO methods may change too abruptly through a sonic point where $a(u)$ changes sign. It then makes more sense, after forming those local divided differences from which ENO chooses the smallest in magnitude, to use a weighted average of them. The weights are chosen so as to yield essentially an ENO method near a discontinuity and a higher order method in regions where the solution is smooth. This yields a modification of ENO called WENO (for **weighted** ENO). See Shu's detailed paper [156].

To see how this works in detail let us restrict attention to the case where local cubics are used to interpolate the primitive function $w(x)$ of (10.25), as in Example 10.7. In (10.26) there are three sets of interpolation formulas, and we denote the corresponding interpolants $v^{(1)}$, $v^{(0)}$, and $v^{(-1)}$ respectively. Next, consider a weighted sum of the three interpolants,

$$v_{j+1/2} = d_{-1} v_{j+1/2}^{(-1)} + d_0 v_{j+1/2}^{(0)} + d_1 v_{j+1/2}^{(1)}. \tag{10.31a}$$

Since all five values from $\bar{v}_{j-2}$ to $\bar{v}_{j+2}$ are involved, one may reasonably hope that there are weights d_l that yield a fifth order approximation where the approximated solution is smooth. Indeed, it turns out that such values exist and are given by

$$d_{-1} = 0.1, \quad d_0 = 0.6, \quad d_1 = 0.3. \tag{10.31b}$$

Similarly we have

$$v_{j-1/2} = d_1 v_{j-1/2}^{(-1)} + d_0 v_{j-1/2}^{(0)} + d_{-1} v_{j-1/2}^{(1)}. \tag{10.31c}$$

Therefore, in the WENO method we choose weights adaptively, so that in smooth areas they are close to those given by (10.31b). By adjusting the weights appropriately, the order of the obtained method away from discontinuities can be increased from 3 to 5 [121, 102].

Specifically, the interpolation formula is

$$v_{j+1/2}^L = w_{-1} v_{j+1/2}^{(-1)} + w_0 v_{j+1/2}^{(0)} + w_1 v_{j+1/2}^{(1)}, \tag{10.31d}$$

$$w_r = \frac{\alpha_r}{\alpha_{-1} + \alpha_0 + \alpha_1}, \quad \alpha_r = \frac{d_r}{(\epsilon + \beta_r)^2},$$

where $\epsilon = 10^{-6}$ is there to ensure no division by zero. The β_r are smoothness indicators calculated for each j, where $\beta_r = O(h^2)$ if the rth stencil covers a smooth area and $\beta_r = O(1)$ otherwise. Following [156] we set

$$\beta_1 = \frac{13}{12}(\bar{v}_j - 2\bar{v}_{j+1} + \bar{v}_{j+2}) + \frac{1}{4}(3\bar{v}_j - 4\bar{v}_{j+1} + \bar{v}_{j+2})^2,$$

$$\beta_0 = \frac{13}{12}(\bar{v}_{j-1} - 2\bar{v}_j + \bar{v}_{j+1}) + \frac{1}{4}(\bar{v}_{j+1} - \bar{v}_{j-1})^2, \tag{10.31e}$$

$$\beta_{-1} = \frac{13}{12}(\bar{v}_j - 2\bar{v}_{j-1} + \bar{v}_{j-2}) + \frac{1}{4}(3\bar{v}_j - 4\bar{v}_{j-1} + \bar{v}_{j-2})^2.$$

10.3. Higher Order Schemes for Scalar Conservation Laws

Similarly

$$v_{j-1/2}^R = \tilde{w}_{-1} v_{j-1/2}^{(-1)} + \tilde{w}_0 v_{j-1/2}^{(0)} + \tilde{w}_1 v_{j-1/2}^{(1)}, \tag{10.31f}$$

$$\tilde{w}_r = \frac{\tilde{\alpha}_r}{\tilde{\alpha}_{-1} + \tilde{\alpha}_0 + \tilde{\alpha}_1}, \quad \tilde{\alpha}_r = \frac{d_{-r}}{(\epsilon + \beta_r)^2}.$$

Note that for the advection equation, where $a(u)$ is a nonzero constant, we have in smooth solution regions $f(v_{j\pm1/2}) = av_{j\pm1/2}$. Examining the above formulas we can easily see that away from discontinuities the method inches toward a centered one. The obtained scheme thus typically boils down to something close to a centered discretization in space. This makes the SSP concept irrelevant for WENO schemes, because the condition (10.28) may not be expected to hold (recall the unstable scheme in our very first Example 1.2). The higher order explicit Runge–Kutta methods that are typically used to match the order of the spatial discretization then double up as a savior from forward Euler instability [179]. The classical RK4 method shines once again, yielding a method of order $(4, 5)$ in combination with the semi-discretization described above.

The finite volume version of WENO schemes described above may be too fancy, or too expensive, to apply in practice, especially when extensions to systems and to more than one space variable are considered. A more straightforward, **finite difference** variant applies a similar methodology directly to a *flux splitting* of f. Unfortunately, to be able to retrieve the high accuracy order we must have a smooth splitting, i.e., $f^+(u)$ and $f^-(u)$ should have at least as many derivatives as appear in the high order interpolation bounds. This gives advantage to the splitting (10.17).

The procedure is as follows [179, 156]:

1. For all i identify $\bar{v}_i$ with $f^+(v_i)$; carry out the reconstruction procedure described in (10.31) and (10.26) to obtain $v_{j+1/2}^L$ for each j.

2. For all i identify $\bar{v}_i$ with $f^-(v_i)$; carry out the reconstruction procedure described in (10.31) and (10.26) to obtain $v_{j-1/2}^R$ for each j.

3. For each j calculate the *numerical flux*

$$\hat{f}_{j+1/2} = \hat{f}_{j+1/2}^+ + \hat{f}_{j+1/2}^-$$
$$= v_{j+1/2}^L + v_{j+1/2}^R.$$

4. For each j form the right-hand side of the semi-discretization,

$$\frac{dv_j}{dt} = -\frac{1}{h}\left(\hat{f}_{j+1/2} - \hat{f}_{j-1/2}\right).$$

Currently, WENO schemes are probably the most popular methods in practical use for conservation laws when higher order accuracy away from discontinuities is desired. Therefore, rather than demonstrating them here, we ask for your help in carrying out the implementation and testing the scheme as described in Exercises 8 and 9.

10.4 Systems of conservation laws

In this section we extend the methods developed earlier to systems of conservation laws, still in one space dimension. Our essential approach is similar to the one used in earlier chapters, especially Section 8.2.1. Thus, we (i) transform to characteristic variables, (ii) apply ideas and methods to each resulting scalar PDE, and (iii) transform back to the original variables.

At first, consider a constant coefficient hyperbolic PDE system,

$$\mathbf{u}_t + A\mathbf{u}_x = \mathbf{0}. \tag{10.32a}$$

As in (8.8), A can be diagonalized

$$T^{-1}AT = \Lambda = \begin{pmatrix} \lambda_1 & & \\ & \ddots & \\ & & \lambda_s \end{pmatrix}, \tag{10.32b}$$

and the eigenvalues λ_i are real. We further define **characteristic variables**,

$$\mathbf{w} = T^{-1}\mathbf{u}, \tag{10.32c}$$

and obtain the decoupled set of s advection equations,

$$\mathbf{w}_t + \Lambda \mathbf{w}_x = \mathbf{0}. \tag{10.32d}$$

Each solution $w_i(t, x)$ is a wave propagating at speed λ_i, and our solution

$$\mathbf{u} = T\mathbf{w}, \tag{10.32e}$$

which is what gets observed eventually, is therefore a *superposition* of s waves.

Next consider a Riemann problem for (10.32a), i.e., with the initial data

$$\mathbf{u}_0(x) = \begin{cases} \mathbf{u}_L, & x < x_*, \\ \mathbf{u}_R, & x > x_*. \end{cases}$$

We can apply a similar transformation, defining

$$\mathbf{w}_L = T^{-1}\mathbf{u}_L, \quad \mathbf{w}_R = T^{-1}\mathbf{u}_R,$$

obtain s scalar Riemann problems which we know how to solve (see Section 10.1.1), and compose the solution $\mathbf{u}$ using the back-transformation (10.32e). It is important to realize that s is small and the transformation T is a constant $s \times s$ matrix that can be obtained ahead of the game. The process above is therefore not just for theoretical interest.

Similarly, we can extend the simple upwind and Lax–Friedrichs methods. The extension of the Lax–Friedrichs method is immediate and is a special case of (9.3).[34] For the

---

[34] Indeed, if you get lost in our treatment for systems of conservation laws and for problems in several space variables, it is useful to always remember that the Lax–Friedrichs method (9.3) is an option, if one can live with its significant inaccuracy.

10.4. Systems of Conservation Laws

simple upwinding method we again apply the decoupling transformation (10.32). Define

$$|\Lambda| = \begin{pmatrix} |\lambda_1| & & \\ & \ddots & \\ & & |\lambda_s| \end{pmatrix}, \quad |A| = T|\Lambda|T^{-1}. \tag{10.33}$$

Then (5.12b) extends into

$$\mathbf{v}_j^{n+1} = \mathbf{v}_j^n - \frac{\mu}{2}A(\mathbf{v}_{j+1}^n - \mathbf{v}_{j-1}^n) + \frac{\mu}{2}|A|(\mathbf{v}_{j+1}^n - 2\mathbf{v}_j^n + \mathbf{v}_{j-1}^n). \tag{10.34}$$

As before, the same method (10.34) is also the one obtained when extending Godunov's scheme for the case of constant coefficient systems. The stability condition for this upwind method involves the spectral radius of A, $\rho(A)$, in an obvious manner.

We can also write (10.34) as

$$\mathbf{v}_j^{n+1} = \mathbf{v}_j^n - \mu[A^+ D_- + A^- D_+]\mathbf{v}_j^n, \tag{10.35a}$$

where

$$A^+ = \frac{1}{2}(A + |A|), \quad A^- = \frac{1}{2}(A - |A|), \tag{10.35b}$$

which better reflects the scheme's upwind nature.

Example 10.10 The simplicity of the Lax–Friedrichs scheme, which on the face of it does not require transformation to characteristic variables or construction of $|A|$, is attractive. Let us therefore consider a simple sobering example.

We construct an advection system of two components by

$$\Lambda = \begin{pmatrix} -1 & 0 \\ 0 & -0.1 \end{pmatrix}, \quad T = \begin{pmatrix} \cos(\pi/3) & -\sin(\pi/3) \\ \sin(\pi/3) & \cos(\pi/3) \end{pmatrix}, \quad A = T\Lambda T^{-1}.$$

With $\mathbf{u} = T\mathbf{w}$ the Lax–Friedrichs scheme discretizes $\mathbf{u}_t + A\mathbf{u}_x = \mathbf{0}$ directly, whereas for the upwind scheme we must consult the eigenvalues first before realizing that for this particular example, since both eigenvalues are negative, the scheme simply uses forward differencing for $\mathbf{u}$ in space, as in (1.15a).

Using periodic BC on $[-\pi, \pi]$ and initial conditions $\mathbf{u}_0 = T\mathbf{w}_0$, with

$$w_0^1(x) = \begin{cases} 1, & .25 \leq x < .75, \\ 0 & \text{otherwise}, \end{cases} \quad w_0^2(x) = \begin{cases} 2, & .5 \leq x < .7, \\ 0 & \text{otherwise} \end{cases}$$

(the superscripts denote the components of $\mathbf{w}$), we run the two methods with $h = .01\pi$ and $\mu = .9$. The results are plotted in Figure 10.8, where the filter (5.10) has been applied at the end of the integration process as well.

The two square waves move intact in time for the characteristic variables $\mathbf{w}$ and get mixed through the transformation T to obtain the exact $\mathbf{u}$. The performance of the

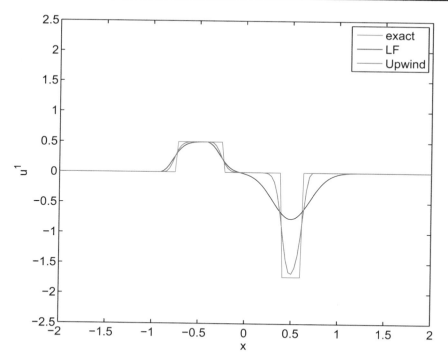

Figure 10.8. *Exact and approximate solutions for Example 10.10. The first solution component u_1 and its approximations are displayed. It is a superposition of the square waves w^1 and w^2. With $\mu = 0.9$ the upwind and Lax–Friedrichs methods have comparable accuracy for w^1, but for w^2 the upwind method is significantly better.*

two schemes for w^1 is comparable, although the upwind scheme is slightly better; for w^2, however, the upwind scheme is much better and the Lax–Friedrichs scheme is unacceptably poor.

This is not difficult to understand if we recall the comparison made in Section 5.2.3 between the diffusion terms added by the two methods: the closer $\mu|a|$ is to 1, the closer Lax–Friedrichs is to the performance of the upwind scheme. But here we have two wave speeds, $a = \lambda_1$ and $a = \lambda_2$! The time step is limited by the larger one, and so with $k = .9h$ the methods perform fairly similarly for the faster wave w^1. But for w^2 the wave speed is only $\lambda_2 = -.1$, and for it $\mu|\lambda_2|$ is much smaller than 1. So, the Lax–Friedrichs scheme introduces way too much dissipation for the slower wave.

A simple remedy is to apply the transformation to characteristic variables and then apply the Lax–Friedrichs method using *different time steps* for the different characteristic variables. But this is no longer simpler than upwinding. ∎

Next, consider a system of nonlinear conservation laws,

$$\mathbf{u}_t + \mathbf{f}(\mathbf{u})_x = \mathbf{0}, \quad A(\mathbf{u}) = \mathbf{f}'(\mathbf{u}) \equiv \frac{\partial \mathbf{f}}{\partial \mathbf{u}}. \tag{10.36}$$

10.4. Systems of Conservation Laws

As in Section 10.2, it is better to consider the integrated form corresponding to (10.12)–(10.13), which extends directly.

The Godunov method does extend if we only know how to solve the resulting Riemann problem. Indeed, the symbols used in (10.18)–(10.19) need only be boldfaced, and we obtain

$$\mathbf{v}_j^{n+1} = \mathbf{v}_j^n - \mu \left(\mathbf{f}_{j+1/2}^{n+1/2} - \mathbf{f}_{j-1/2}^{n+1/2} \right), \qquad (10.37a)$$

$$\mathbf{f}_{j+1/2}^{n+1/2} = \mathbf{f}(\tilde{\mathbf{v}}_{j+1/2}^n), \qquad (10.37b)$$

see (10.19) and (10.11). Note again that these Riemann solutions are used only at the cell edges, $x_{j+1/2}$ $\forall j$. The connection between (10.37a) and (10.35a) is made by writing (10.37a) as

$$\mathbf{v}_j^{n+1} = \mathbf{v}_j^n - \mu \left(\left(\mathbf{f}(\mathbf{v}_j^n) - \mathbf{f}_{j-1/2}^{n+1/2} \right) + \left(\mathbf{f}_{j+1/2}^{n+1/2} - \mathbf{f}(\mathbf{v}_j^n) \right) \right)$$

and identifying these **flux fluctuations** in the special constant coefficient case.

However, the process of solving the Riemann problem for nonlinear systems is prohibitively expensive. Thus, **approximate Riemann solvers** are sought. An obvious approach is to replace the nonlinear problem by a linearized one defined for each time step n at each cell interface $j + 1/2$,

$$\hat{\mathbf{u}}_t + \hat{A}_{j+1/2}^n \hat{\mathbf{u}}_x = \mathbf{0}. \qquad (10.38)$$

The matrix $\hat{A}_{j+1/2}^n$ depends on two states $\mathbf{v}_j^n$ and $\mathbf{v}_{j+1}^n$ and is required to satisfy the following:

1. It should be diagonalizable with real eigenvalues, so the local PDE (10.38) is hyperbolic.

2. It should satisfy consistency

$$\hat{A}_{j+1/2}^n \to A(\mathbf{u}) \quad \text{as } \mathbf{v}_j^n, \mathbf{v}_{j+1}^n \to \mathbf{u}.$$

3. Preferably

$$\hat{A}_{j+1/2}^n (\mathbf{v}_{j+1}^n - \mathbf{v}_j^n) = \mathbf{f}(\mathbf{v}_{j+1}^n) - \mathbf{f}(\mathbf{v}_j^n).$$

This turns out to be a tall order. But the first two requirements are satisfied by using the midpoint rule,

$$\hat{A}_{j+1/2}^n = A\left(\frac{1}{2}(\mathbf{v}_j^n + \mathbf{v}_{j+1}^n) \right). \qquad (10.39)$$

For more sophisticated choices satisfying all three requirements for important special cases, see Section 15.3 in [116].

Extension of the high-order schemes of Section 10.3 is possible. For flux limiter schemes a natural candidate for the higher order method is one of the variants of Lax–Wendroff aimed at directly approximating conservation laws, namely, the Richtmyer scheme (5.5b) or the MacCormack scheme (5.5c). The ENO and WENO schemes also extend in a similar fashion; see, e.g., [156].

10.5 Multidimensional problems

Without a doubt the simplest way to extend the methods of the previous sections to the case of systems of conservation laws in more than one space variable is using **dimensional splitting**. See Section 9.3.1. Thus, for the PDE system

$$\mathbf{u}_t + \mathbf{f}(\mathbf{u})_x + \mathbf{g}(\mathbf{u})_y = \mathbf{0}, \tag{10.40}$$

we first solve for each mesh value in y, $y = y_l$, the one-dimensional system

$$\mathbf{u}_t + \mathbf{f}(\mathbf{u})_x = \mathbf{0},$$

starting from $\{\mathbf{v}_{j,l}^n\}$ and obtaining a solution $\{\mathbf{v}_{j,l}^*\}$. Then we solve for each mesh value in x, $x = x_j$, the one-dimensional system

$$\mathbf{u}_t + \mathbf{g}(\mathbf{u})_y = \mathbf{0},$$

starting from $\{\mathbf{v}_{j,\cdot}^*\}$ and obtaining a solution $\{\mathbf{v}_{j,\cdot}^{n+1}\}$.

Care must be taken in specifying appropriate boundary conditions for $\mathbf{v}^*$. Moreover, the resulting method is in principle only first order accurate. But the BC question can be resolved by carrying out the fractional step integration in x also for ghost unknowns which then provide BC for the next split-step, and the accuracy is often much better than a first order designation would indicate.

Example 10.11 The equations governing compressible fluid flow, called the Euler equations for gas dynamics, form a system of conservation laws. In two dimensions they have the form (10.40) with

$$\mathbf{u} = \begin{pmatrix} \rho \\ \rho u \\ \rho v \\ E \end{pmatrix}, \quad \mathbf{f} = \begin{pmatrix} \rho u \\ \rho u^2 + p \\ \rho u v \\ (E+p)u \end{pmatrix}, \quad \mathbf{g} = \begin{pmatrix} \rho v \\ \rho u v \\ \rho v^2 + p \\ (E+p)v \end{pmatrix}, \tag{10.41a}$$

where u and v are velocity components in the x and y directions, respectively, ρ is the density, E the total energy, and p the pressure given by the equation of state for an ideal polytropic gas,

$$E = \frac{p}{\gamma - 1} + \frac{1}{2}\rho(u^2 + v^2). \tag{10.41b}$$

The known constant γ is the *ratio of specific heats*, or *adiabatic exponent*, e.g., in air $\gamma \approx 1.4$. You should be able to figure out how the system (10.41) looks like in one dimension and in three dimensions.

10.5. Multidimensional Problems

The Jacobian matrix $A(\mathbf{u}) = \mathbf{f}'(\mathbf{u})$ has the eigenvalue and eigenvector matrices (Exercise 11)

$$\Lambda^x = \begin{pmatrix} u-c & & & \\ & u & & \\ & & u & \\ & & & u+c \end{pmatrix}, \quad T^x = \begin{pmatrix} 1 & 1 & 0 & 1 \\ u-c & u & 0 & u+c \\ v & v & 1 & v \\ H-uc & \frac{1}{2}(u^2+v^2) & v & H+uc \end{pmatrix},$$

where $c = \sqrt{\gamma p/\rho}$ is the sound speed. There are similar expressions for the eigenvalues and eigenvectors of $B(\mathbf{u}) = \mathbf{g}'(\mathbf{u})$.

Godunov's method brought a breakthrough, back in the late 1950s, in succeeding to solve these equations in one dimension. For the present problem we can apply dimensional splitting, relying on the method described in Sections 10.4 and 10.2 for the one-dimensional case. ■

Let us return to the basic scalar advection equation

$$u_t + au_x + bu_y = 0$$

and consider methods other than dimensional splitting. The basic upwind scheme (5.12) or (10.34) generalizes directly into

$$v_{j,l}^{n+1} = \left(1 - \mu_1[a^+ D_{x,-} + a^- D_{x,+}] - \mu_2[b^+ D_{y,-} + b^- D_{y,+}]\right) v_{j,l}^n. \tag{10.42}$$

For simplicity of notation assume $a, b > 0$. Then we have the scheme

$$v_{j,l}^{n+1} = v_{j,l}^n - a\mu_1(v_{j,l}^n - v_{j-1,l}^n) - b\mu_2(v_{j,l}^n - v_{j,l-1}^n). \tag{10.43}$$

As discussed in Section 9.1, however, we expect this first order scheme to be stable only under the more restrictive condition

$$\max\{\mu_1|a|, \mu_2|b|\} \le 1/2.$$

Indeed, there is something not entirely satisfactory in (10.43) beyond just the halving of the allowable step size. Recall the Lagrangian interpretation of the upwind scheme in one dimension, namely, that the characteristic curve hitting (t_{n+1}, x_j) is traced back to time level t_n and its value is then linearly interpolated by neighboring values. Here, in two dimensions we would expect an interpolation based on the four nodes of a mesh cell at $t = t_n$ (see Figure 10.9) to participate in the action, and yet only three of these nodes appear in (10.43). Recall that whereas on a triangle a general linear function is given by $\alpha + \beta x + \gamma y$, on a rectangle we must consider the bilinear function $\alpha + \beta x + \gamma y + \delta xy$. In other words, it requires the participation of all four nodes to get an interpolation corresponding to the linear interpolation of the Lagrangian method in one dimension. The resulting formula is then

$$\begin{aligned} v_{j,l}^{n+1} = v_{j,l}^n &- a\mu_1(v_{j,l}^n - v_{j-1,l}^n) - b\mu_2(v_{j,l}^n - v_{j,l-1}^n) \\ &+ \frac{1}{2}\left[a\mu_1[b\mu_2(v_{j,l}^n - v_{j,l-1}^n) - b\mu_2(v_{j-1,l}^n - v_{j-1,l-1}^n)] \right. \\ &\quad \left. + b\mu_2[a\mu_1(v_{j,l}^n - v_{j-1,l}^n) - a\mu_1(v_{j,l-1}^n - v_{j-1,l-1}^n)]\right]. \end{aligned} \tag{10.44}$$

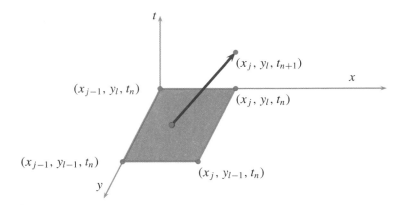

Figure 10.9. *Tracing back the characteristic curve in two dimensions.*

It is called **corner-transport upwind**, in contrast to (10.43), which is called **donor-cell upwind** [116]. Its theoretical stability restriction based on Fourier analysis is now back to the CFL condition.

The method is next generalized to variable coefficient problems (the coefficients in (10.44) beceome localized) and to conservation laws. In addition to the previous one-dimensional Riemann solvers, a transverse Riemann solver is required. Flux limiters can also be introduced to obtain high resolution methods. See Chapter 20 of [116] for details.

Finally, let us note that the semi-discrete approach which leads in Section 10.3 to ENO, WENO, and SSP methods can be generalized to problems in more than one dimension as well, not only through dimensional splitting. Note, however, that the high-order methods can become prohibitively complex for systems in more than one space variable, and many people restrict themselves to methods of order at most two.

The software package CLAWPACK available at

$$\text{http://www.amath.washington.edu/}{\sim}\text{claw/apps.html}$$

implements methods that are described in detail in LeVeque's book [116]. There are also several numerical examples on that site.

10.6 Problems with sharp layers

If the original PDE to be solved is of the form (5.13b), but with $\nu > 0$ very small, we may expect spurious oscillations in the solution upon applying leapfrog or Lax–Wendroff, unless $\nu \geq \frac{h|a|}{2}$. Thus, if ν is not sufficiently large in this relative sense, then artificial diffusion or viscosity may have to be added. The number $R = \frac{h|a|}{\nu}$ is sometimes referred to as the **cell Reynolds number**. Artificial viscosity may be added, then, if the cell Reynolds number exceeds 2, and the amount to be added in such a case should bring this number down to at most 2 to avoid spurious oscillations.

Example 10.12 Throughout this chapter we have been using the inviscid Burgers equation as an example of a conservation law. Let us now consider the Burgers equation with a small

10.6. Problems with Sharp Layers

amount of viscosity,

$$u_t + \frac{1}{2}(u^2)_x = \sigma u_{xx},$$

where $\sigma \geq 0$ is a given parameter.

The above considerations suggest extending the upwind scheme of Example 10.4 to read

$$v_j^{n+1} = v_j^n + \hat{\sigma}_j^n \frac{\mu}{h} D_+ D_- v_j^n - \frac{\mu}{2} \begin{cases} D_+(v_j^n)^2 & \text{if } v_j^n < 0, \\ D_-(v_j^n)^2 & \text{if } v_j^n \geq 0, \end{cases} \quad (10.45a)$$

where $\mu = k/h$ and

$$\hat{\sigma}_j^n = \max(\sigma - .5h|v_j^n|, 0). \quad (10.45b)$$

This is clearly an explicit, first order scheme in time and space. To obtain an indication of the stability restriction we freeze coefficients and obtain, as in Example 5.8, that the time step size $k = k_n$ should satisfy

$$k < h / \max_j \left\{ |v_j^n| + \frac{2\hat{\sigma}_j^n}{h} \right\}. \quad (10.45c)$$

Thus, where σ is significantly smaller than $.5h|v_j^n|$ we have essentially the upwind scheme for the inviscid case. The scheme switches smoothly to the parabolic regime with the forward Euler stability restriction, as σ grows with the numerical method mesh fixed.

An alternative scheme for this problem is obtained by a splitting, semi-Lagrangian approach; see Section 9.3.1 and especially Example 9.8. We apply a fractional step using (10.45a) with $\hat{\sigma} = 0$, calling the result $\{\hat{v}_j\}$, say. This is followed by a Crank–Nicolson step for $u_t = \sigma u_{xx}$, starting from $\{\hat{v}_j\}$ to obtain v^{n+1}. The time step size restriction is now (10.45c) with $\hat{\sigma}_j^n = 0$, for any σ.

We tried out these two schemes for the same initial and (periodic) boundary conditions as in Example 10.4. There is no difficulty with boundary conditions for the split step here, and no fancy splitting is required as the scheme is first order accurate anyway. Figure 10.10 records solution profiles at $t = 1$ for various values of σ. These are all calculated using the splitting scheme with $k = h = .001$.

We then varied k and h for the different σ values to investigate what happens when $h \gg \sigma$, $h \approx \sigma$, and $h \ll \sigma$.

When $h \ll \sigma$ the results are not qualitatively different from those displayed in Figure 10.10. However, the explicit upwind scheme requires exceedingly small time steps for reasons of stability. This handicap is not shared by the splitting scheme.

When $h \geq \sigma$ both schemes work fairly well. However, again the splitting scheme is better in localizing the region of fast solution variation. Near the sonic point, where v^n is slow and changes sign, the presence of σ that satisfies $0 < \sigma \ll h$ is suddenly felt in (10.45), and this produces a jarring effect that is not apparent when using the other scheme.

Of course the explicit scheme step is simpler, and it can be significantly cheaper to carry out in more than one space variable. ■

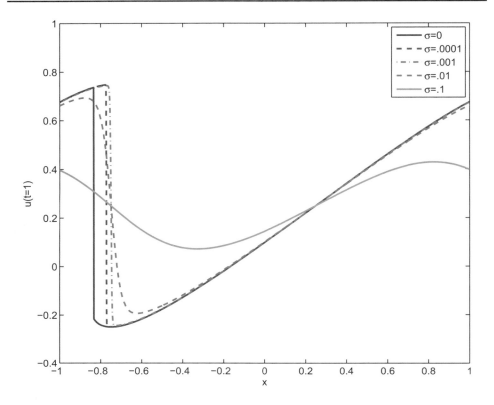

Figure 10.10. *Solving the slightly viscous Burgers equation, with initial and boundary conditions as in Figure* 10.2. *Solution profiles at* $t = 1$ *are sharp when* σ *is small and get more diffused as* σ *is increased. All solutions were computed using the semi-Lagrangian scheme with* $k = h = .001$.

The orderly progression to the limit $\sigma \to 0$ demonstrated in Example 10.12 does not hold automatically for any PDE that looks like a conservation law plus a higher order differential term. The most obvious example of this is the KdV equation, which may be viewed as adding a third derivative term νu_{xxx} to the Burgers equation. Here the exact solution under similar initial and boundary conditions is infinitely smooth and bounded, and it remains wiggly so long as $\nu > 0$. In particular, for Examples 5.1 and 7.6, where $\nu \approx .004 \ll 1 = \alpha$, the initial conditions were chosen such that the corresponding Burgers equation develops a shock (without any oscillations) much before any of the KdV schemes become unstable. See Exercise 7.17.

10.7 Exercises

0. **Review questions**

    (a) Define monotonicity for a one-step difference scheme and explain its importance.

10.7. Exercises

(b) Show that the characteristic curves for a conservation law are straight lines even when the PDE is nonlinear. Explain how solution discontinuities may arise even if the initial data function of a pure IVP is smooth.

(c) What is a weak solution? What is a jump condition?

(d) Explain the difference between a shock wave and a rarefaction wave.

(e) The inviscid Burgers equation can be written as $u_t + \frac{1}{2}(u^2)_x = 0$ and as $u_t + uu_x = 0$. Which of these forms is more appropriate to seek discretizations of in the presence of solution discontinuities?

(f) What is a flux splitting and why is it important?

(g) The Godunov scheme is first order accurate and delivers results similar to those of the upwind scheme in conservation form. Are these schemes one and the same?

(h) Define the properties of total variation diminishing and total variation bounded and explain their importance.

(i) What is a flux limiter?

(j) Explain the basic ideas behind the ENO scheme.

(k) Explain the basic ideas behind the WENO scheme.

(l) Define the SSP property. Explain why it is important for ENO schemes and why it is unimportant for WENO schemes.

(m) What is an approximate Riemann solver?

(n) How does dimensional splitting help make most of this chapter's material relevant to practical problems?

(o) What is the difference between corner-transport upwinding and donor-cell upwinding?

1. Devise an upwind scheme for the advection-diffusion equation (10.2) which avoids spurious oscillations by being monotone and adds a minimal amount of artificial diffusion. Your scheme should automatically adjust for any values of $\nu \geq 0$ and a.

2. Consider the advection equation with variable coefficient [76]

$$u_t - (\sin x)\, u_x = 0$$

with initial condition

$$u(0, x) = \sin x$$

and periodic BC on $[-\pi, \pi]$.

(a) Would you expect the solution to this problem to develop a discontinuity or remain smooth for all time?

(b) Propose an appropriate discretization method, implement it, and report your results at $t = 1$ and $t = 10$. These results don't have to be perfect, but your discussion of them should be.

3. Prove the Rankine–Hugoniot jump condition (10.7).

4. Obtain the form of the rarefaction wave for Case (b) of Example 10.3 by plugging in a solution of the form $u(t, x) = g(x/t)$.

5. Consider the inviscid Burgers equation with initial data given by

$$u_0(x) = \begin{cases} 0 & \text{if } -1 \leq x < -2/3, \\ 1 & \text{if } -2/3 \leq x < 0, \\ 0 & \text{if } 0 \leq x < 1 \end{cases}$$

and with periodic boundary conditions on $[-1, 1]$.

   (a) Find the exact solution for $t \leq 1$.

   (b) Implement the four schemes of Example 10.5. Find the solutions using $h = 0.01$ and $h = 0.001$ with $\mu = 0.9$. Compare against the exact solution at $t = 1$. What are your observations?

6. Verify the interpolation formulas for third order ENO and fifth order WENO reconstructions given in (10.26).

   Find the corresponding formulas for v_j.

7. Show that the explicit midpoint scheme in the context of Section 10.3.3 is not SSP.

8. Implement either the finite volume or the finite difference version of the fifth order WENO scheme in one space variable, using RK4 in time. Test your program as follows.

   (a) Construct a table of errors corresponding to Table 10.1. At least for $t = .3$ you should get much smaller errors. Then construct also two plots for $h = .1$ and $h = .05$ as in Figure 10.7. What are your observations?

   (b) Solve the inviscid Burgers equation with periodic BC on $[-1, 1]$ starting from the initial data,

$$u(0, x) = \sin(10\pi x) e^{-10x^2}.$$

   Plot the solution every few time steps and watch it change shape. At $t = 1$ you should obtain a profile that matches that of Figure 10.11.

9. The **Buckley–Leverett flux** [116] for water is given by

$$f(u) = \frac{u^2}{u^2 + a(1-u)^2}. \tag{10.46}$$

Use your WENO program from the previous exercise above, or another program you wrote implementing another suitable method, to integrate this conservation law with the same initial conditions and periodic boundary conditions as in Exercise 8. You should be able to retrieve (as far as the eye can tell) the results in Figure 10.12.

10.7. Exercises

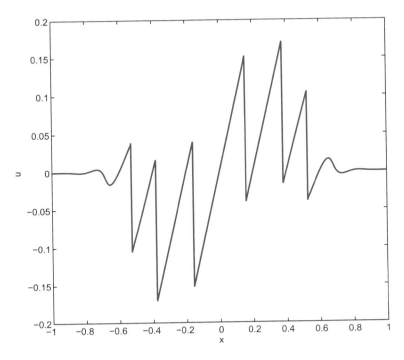

Figure 10.11. *Solving the inviscid Burgers equation, starting from $u(0, x) = \sin(10\pi x)e^{-10x^2}$. Several discontinuities are observed at $t = 1$.*

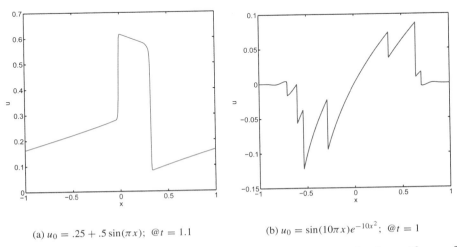

(a) $u_0 = .25 + .5\sin(\pi x)$; @$t = 1.1$ (b) $u_0 = \sin(10\pi x)e^{-10x^2}$; @$t = 1$

Figure 10.12. *Solution profiles for the Buckley–Leverett conservation law with $a = .5$.*

10. Repeat the calculations of Example 10.10, and produce corresponding observations and considerations, for the case where the eigenvalue $\lambda_1 = -1$ of A is replaced by $\lambda_1 = 1$. The other eigenvalue remains the same. Thus

$$\Lambda = \begin{pmatrix} 1 & 0 \\ 0 & -0.1 \end{pmatrix}, \quad T = \begin{pmatrix} \cos(\pi/3) & -\sin(\pi/3) \\ \sin(\pi/3) & \cos(\pi/3) \end{pmatrix}, \quad A = T\Lambda T^{-1}.$$

11. Consider the Euler equations for gas dynamics, (10.41).

    (a) Write explicitly $A(\mathbf{u}) = \mathbf{f}'(\mathbf{u})$ and $B(\mathbf{u}) = \mathbf{g}'(\mathbf{u})$.

    (b) For each matrix find the eigenvalues and eigenvectors. Verify that the expressions for T^x and Λ^x in Example 10.11 are correct, and write the expressions for B.

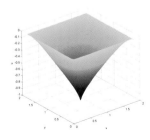

Chapter 11

Additional Topics

11.1 What first: Optimize or discretize?

Many differential systems arise as necessary conditions for a stationary or critical point of a functional involving temporal and spatial derivatives. For instance, consider the Poisson problem in two spatial variables with a mix of Dirichlet and Neumann conditions,

$$-\Delta u = q \quad \text{in } \Omega, \tag{11.1a}$$

$$u|_{\partial\Omega_1} = 0, \quad \frac{\partial u}{\partial n}|_{\partial\Omega\setminus\partial\Omega_1} = 0,$$

where $\partial\Omega_1$ is part of the boundary $\partial\Omega$. Recall Section 3.1.3, and in particular equations (3.18)–(3.20), as well as Example 3.8. This PDE problem is the Euler–Lagrange equation (which is, to recall, the necessary condition) for the minimization of the functional

$$\int_\Omega \left[|\nabla u|^2 - 2qu\right] d\Omega \tag{11.1b}$$

over all functions u satisfying $u|_{\partial\Omega_1} = 0$ that are bounded together with their gradient in the $\mathcal{L}_2(\Omega)$-norm. This steady state problem is clearly related to a parabolic diffusion one.

Another instance is the family of optimal control problems, where a functional depending on a control function and state variables is to be minimized subject to satisfying an ODE system. This is further described below; see Section 11.1.3. A third instance is the variational principles briefly discussed in Section 6.2.3.

Our usual approach in this text has been to ignore the source of a given differential system and consider discretization methods directly. Thus, the discretization of (11.1) occurs *after* the optimization, in the sense that necessary conditions for the optimal or critical solution are formed first. However, we have occasionally seen in Sections 3.1, 6.2.3, and elsewhere that good numerical methods may be obtained by discretizing the functional first, and only then proceeding to generate the necessary conditions. The discretization of the variational principle corresponding to a functional such as (11.1b) is routine in finite element methods: indeed, the piecewise linear basis functions described in Section 3.1.4 and Exercise 3.5 do not have the bounded second derivative required in (11.1a), but can be readily used to discretize (11.1b) where only first derivatives appear. Finite volume techniques, which in general attempt to mimic finite element methods without the overhead of operating with basis functions, likewise lead to a discretize-first approach.

The class of methods that are built on the principle of reasonably discretizing first is generally smaller (because the eventual result can always be interpreted as a discretization of the necessary conditions) and more structured than the optimize-first class. This often allows for avoiding some pitfalls and concentrating on "the good ones," as we demonstrate below. The *downside* is in situations where some phenomenon or property of the exact solution cannot be reproduced at the discrete variational level. An instance is bang-bang control in optimal control problems. Another instance is the theoretically appealing finite element method of streamline diffusion; see [101] and references therein. This method is in a sense more faithful to the principle of multidimensional upwinding, and yet it is typically not nearly as effective in practice as the methods described in Section 10.5 and [116]. Upwinding methods that use the differential equations locally to decide on the discretization itself do not lend themselves to being cast as the discretization of a general optimization or variational formulation. Finally, in many situations a practical decision about discretizing or optimizing first may depend on what adequate software is available for the optimization problem or the differential system that result.

Let us next consider specific instances of the discretize-first approach.

11.1.1 Symmetric matrices for nonuniform spatial meshes

As an opener for Section 11.2 consider discretizing the boundary value ODE,

$$-(au')' = q, \quad 0 < x < 1, \quad (11.2)$$
$$u(0) = 0, \quad u'(1) = 0,$$

where $a(x) > 0$ and $q(x)$ are given, using a general, nonuniform mesh.

If we try any reasonable-looking 3-point discretization for (11.2), coupled with a reasonable-looking discretization for the BC involving u', then we could end up with a tridiagonal *nonsymmetric* matrix problem to solve. On the other hand, if we discretize instead the corresponding functional as specified in Exercise 3.4, then the obtained matrix (for the obtained second order discretization) is symmetric positive definite!

This is even more important in several space variables, because iterative methods are typically used for the matrix problem. If the matrix is symmetric positive definite, then a preconditioned CG method can be applied, while the loss of symmetry obtained by using a general finite difference discretization mandates use of a Krylov space method such as preconditioned BICGSTAB, a method that is less safe and potentially less efficient; see Section 9.4.

11.1.2 Efficient multigrid and Neumann BCs

Consider the usual 5-point formula for Poisson's equation (11.1a), as in Example 3.4 with $a \equiv 1$. In the early 1980s it was already known that in the case where there are no Neumann BCs, i.e., $\partial \Omega_1 = \partial \Omega$, a simple multigrid method solves the resulting linear system of equations rather accurately in but a few sweeps over the grid (or mesh) points. However, this high efficiency, demonstrated in Table 9.5, deteriorated in the case where Neumann BCs were present and discretized in certain simple ways. Special mechanisms were proposed to improve the relaxation phase of the multigrid cycle for the latter case [33].

Yet, the problem of slowdown for Neumann BCs completely disappears if the discretization is carried out first on the corresponding functional (11.1b) as described in Section 3.1.3. No special treatment of BC is required, and the problem handled in [33] is not an issue! See, for example, [7].

11.1.3 Optimal control

Many applications give rise to *optimal control* problems. An instance is given by the planning of a route for a vehicle traveling between two points, satisfying equations of motion, such that fuel consumption is minimized or the travel time is as short as possible. Typically, the m state variables $\mathbf{y}(t)$ satisfy an ODE system which involves a control function $\mathbf{u}(t)$,

$$\mathbf{y}' = \mathbf{f}(t, \mathbf{y}, \mathbf{u}), \qquad 0 \leq t \leq b. \tag{11.3a}$$

This system may be subject to some side conditions,

$$\mathbf{y}(0) = \mathbf{c}, \tag{11.3b}$$

although it is possible that $\mathbf{y}(b)$ is prescribed as well, or that there are no side conditions at all. Inequality constraints are also popular, but let's leave them out of the present quick discussion; see Bryson and Ho [39] for full details on models and analytical methods. The control $\mathbf{u}(t)$ must be chosen so as to optimize some cost function, say,

$$\text{minimize} \quad J = \phi(\mathbf{y}(b), b) + \int_0^b K(t, \mathbf{y}(t), \mathbf{u}(t)) dt \tag{11.4}$$

subject to (11.3).

The necessary conditions for an optimum in this problem are found by considering the Hamiltonian function

$$H(t, \mathbf{y}, \mathbf{u}, \boldsymbol{\lambda}) = \sum_{i=1}^m \lambda_i f_i(t, \mathbf{y}, \mathbf{u}) + K(t, \mathbf{y}, \mathbf{u}),$$

where $\lambda_i(t)$ are *adjoint variables*, $i = 1, \ldots, m$. The conditions

$$y_i' = \frac{\partial H}{\partial \lambda_i}, \qquad i = 1, \ldots, m,$$

yield the state equations (11.3a), and in addition we have ODEs for the adjoint variables,

$$\lambda_i' = -\frac{\partial H}{\partial y_i} = -\sum_{j=1}^m \lambda_j \frac{\partial f_j}{\partial y_i} - \frac{\partial K}{\partial y_i}, \qquad i = 1, \ldots, m, \tag{11.5a}$$

and

$$0 = \frac{\partial H}{\partial u_i}, \qquad i = 1, \ldots, m_u. \tag{11.5b}$$

This gives a DAE in general; however, $\mathbf{u}(t)$ can often be eliminated from (11.5b) in terms of $\mathbf{y}$ and λ, yielding an ODE system. Additional side conditions are

$$\lambda_i(b) = \frac{\partial \phi}{\partial y_i}(b), \qquad i = 1, \ldots, m. \tag{11.5c}$$

The system (11.3), (11.5) comprises a boundary value ODE (or DAE).

An *indirect approach* for solving this optimal control problem involves the numerical solution of the boundary value ODE just prescribed. This is the optimize-first approach. In contrast, a *direct approach* involves the discretization of (11.4), (11.3), and the subsequent numerical solution of the resulting large, sparse, finite dimensional, constrained optimization problem. Each of these two approaches has its advantages (and advocates). We have already mentioned above that there are some phenomena that the direct approach could fail to reproduce, but it often leads to significantly easier simulations as well [28, 27].

Finally, let us mention the issue of personal preference for mathematical analysis. The discretization first approach, while often bringing more structured and conservative discretizations to the fore, may seem on other occasions to bypass rigorous analysis of PDE and ODE methods in appropriate function spaces. Indeed, there are papers in the literature which describe successful practical methods, starting from some discrete optimization or variational formulation, with the differential system serving as mere background. The latter may cause strong reactions among certain mathematicians.

11.2 Nonuniform meshes

This discussion starts where Sections 3.1.5 and 2.4 have left off, and familiarity with that material is assumed below. We have briefly surveyed in Section 3.1.5 methods for handling nonuniform meshes and some discretization issues associated with them, but we have not said much outside Section 2.4 about how such meshes are selected.

Let us state right away that the departure from uniform meshes in a PDE context can be a costly adventure. Bookkeeping is harder, orders of accuracy can get reduced, basic parallelism may be lost, and programming becomes more delicate. Thus, there has to be a strong reason for choosing nonuniform meshes in practice! A spatial mesh changing smoothly and pleasantly as a function of time may yield a pretty mesh motion animation, but this does not necessarily translate to a more efficient PDE algorithm than one on a boring uniform mesh that is only twice as dense in each direction, say.

However, there are often situations in practice where the solution is so much steeper in certain narrow spatial locations than elsewhere, that a mesh which is dense only where the solution derivatives are large can really be advantageous in terms of computational efficiency. A solution profile of this sort is depicted in Figure 11.1. It was obtained, truth to be told, by solving (9.45c), not (9.45a), using a basic discretization scheme on a uniform 200×200 mesh—this after all is a chapter on *additional* topics. Nonetheless, the point is that solutions like this one with very sharp profiles do arise in practice, and a simplifying transformation such as (9.45b) is often not available.

The concept of automatic **mesh refinement** is typically intertwined with some sort of **error control**. An error indicator is calculated and used to decide how to select the mesh, usually in an iterative, dynamic fashion as described below. The error indicator

11.2. Nonuniform Meshes

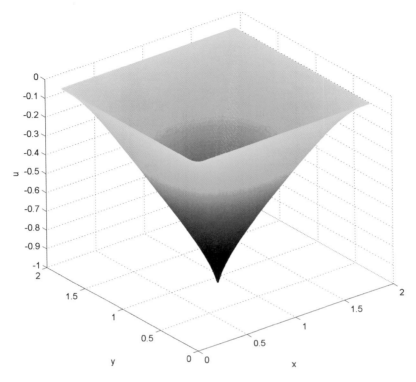

Figure 11.1. *Solution $u(T_*, x, y)$ of the problem defined by (9.45) for $\lambda = 2$ and $T_* = .1975$.*

most often relates to the pointwise solution accuracy, if not directly, then indirectly through local error measures. This practice in itself is not beyond question either, as we have repeatedly seen in Chapters 6 and 7: in many cases pointwise accuracy is of less interest than reconstruction of other, problem-specific quantities. However, despite the recent surge of interest in geometric integration (GI) methods, controlling the solution error is still of importance in many practical situations, and software packages often aim at satisfying such desires in a soft sense. We now proceed to briefly describe mesh refinement methods.

11.2.1 Adaptive meshes for steady state problems

Let us say a few words first about adaptive meshes for steady state problems before considering time-dependent ones. The first distinction that must be made is between *static* and *dynamic* mesh selection. In the static case we know ahead of time where a finer mesh is needed, and the criteria for the mesh selection do not depend on estimates of the sought solution. For example, the Poisson problem $-\Delta u = 1$ on the unit square with homogeneous Dirichlet BC has a jump discontinuity in the second derivatives of the solution at the square's corners, because along the boundaries $u_{xx} = 0$ and $u_{yy} = 0$. Hence it makes sense to have a finer mesh near these corners. This is even more important at the inner (reentrant) corner of an L-shaped domain, because the solution there is already discontinuous in its first derivative. See, e.g., [171]; there are many other references for this. Another static situation is where an approximate solution is computed in advance on a coarse mesh and is then used for selecting the fine mesh. As usual with such ideas the coarse mesh must be both fine enough to capture "danger zones" and coarse enough so that subsequent mesh refinement in such zones is actually needed.

But now let us concentrate on the fully adaptive, dynamic case. Thus we envision a sequence of meshes on which approximate solutions are computed. Each such solution is then used in a **monitor function** $M(u)$, according to which the algorithm decides where to refine the next mesh. This is done until sufficient accuracy is achieved globally.

Typically the local truncation error for a method of order p behaves like $h_j^p |u^{(p)}(x_j)|$, where h_j is the mesh width at mesh point x_j and $u^{(p)}$ is the pth derivative of the exact solution evaluated locally. The error in the solution relates to the local truncation error through an integral involving Green's function, and this has the effect of producing an error estimate of the same order but *global*: $|v - u| = O\left(h^p \|u^{(p)}\|\right)$, where $h = \max h_j$. Although this notation suggests a one-dimensional problem, the above lines extend to the elliptic PDE case.

It may therefore appear that although we really want to control the global error, it is practically better to control the local truncation error and to select the mesh accordingly, as is approximately done for initial value ODEs. Many people do precisely this. The local truncation error may be estimated for instance using more than one mesh by local extrapolation [171].

However, this logic, although valid, does not always outweigh other considerations. For boundary value problems the solution is evaluated globally anyway, i.e., an approximate solution is found on a current mesh covering the entire domain Ω, and this solution is used to decide if and where to refine the mesh to obtain a better solution in any case. Moreover, for boundary value ODEs it is often possible to isolate a local leading term of the global error and to use that to select the next mesh. We refer to Chapter 9 of [10] for much more on this topic. The upshot is that for boundary value ODEs it makes practical sense to use the approximate solution on the current mesh in order to estimate high derivatives of the solution and then attempt to select an *equi-distributing* mesh where a quantity such as $h_j^p |u^{(p)}(x_j)|$ is kept roughly independent of j. This implies that where the solution derivatives are large the mesh width is proportionately small.

For boundary value PDEs, typical expressions for the error are much more involved for several reasons, including that the Green's function is unbounded and that even the local truncation error contains more than one contribution. Again estimating the local truncation error based on previous meshes is an option. Also popular are equi-distribution methods according to simple monitor functions such as the first derivative $h_{i,j} |\nabla u_{i,j}|$ and the *arclength* $h_{i,j}\sqrt{1 + |\nabla u_{i,j}|^2}$. In both of these the approximate solution is used to estimate the gradient locally and the step size is chosen so as to keep these quantities roughly equal for all i, j.

For time-dependent problems the situation can become significantly more complex. If the locations in space which require mesh refinement are stationary in time, then the previous considerations apply, except of course that there is another source of error embodied in the time discretization. One cause for concern is that a discretization scheme which is stable

11.2. Nonuniform Meshes

only if the time step k is restricted by a finite stability function $S(h)$ can be severely limited. This is because it is the smallest spatial mesh width which determines k. For parabolic type problems this gives the nod to implicit schemes with unconditional stability, because explicit schemes do not scale up well for fine meshes. Moreover, if there is a sharp front moving as a function of time then the locations where the mesh must be fine vary in time as well. These concerns gives rise to the methods briefly discussed next.

11.2.2 Adaptive mesh refinement

This method, or approach, was developed by Berger, Oliger, and Colella [24, 23]. It is aimed at solving fluid flow problems such as those arising in hydrodynamics, and in particular systems of hyperbolic conservation laws, in several space dimensions. An underlying coarse grid covering the spatial domain allows finding a rough solution using a relatively liberal time step even with an explicit method such as those considered in Chapter 10. This allows rough error estimation which is used to determine where more resolution is needed. Finer, higher resolution meshes in *both space and time* are then locally superimposed on the coarse one, and the process is repeated until the desired accuracy is reached. Since the fine meshes have smaller time steps associated with them we have a multi time-stepping method. The fluxes between the fine and coarse meshes are manipulated so that conservation is enforced.

There are software packages associated with this approach available at Berger's website,

```
http://cs.nyu.edu/berger
```

One of these is an AMR version of the package CLAWPACK mentioned in Section 10.5.

AMR is very popular among physicists and other scientists and engineers; see, e.g., [64]. A library toolkit that allows AMR use in a parallel setting is provided in [124]. It uses an Octree data structure and allows adjacent refinement levels to differ by a factor of two.

While the AMR approach has proved to be of great practical importance, it targets a specific class of problems. A more general approach would be to allow the spatial mesh to vary as a function of time for a given time-dependent PDE system. This is considered next.

11.2.3 Moving meshes

Much of the discussion in this book has concentrated on methods that attempt to separate the treatment in time and space: semi-discretizations are devised in space or in time first, and this is followed by a discretization in the remaining variable(s). This approach is often convenient, but occasionally it is essential, as for ENO and WENO methods and for methods that impose inequality algebraic constraints at the end of each time step.

With this in mind, a natural general first approach for adaptive mesh selection for time-dependent problems is to apply a steady state algorithm as described in Section 11.2.1. Such a method is sometimes called **rezoning**: we integrate a few steps in time using the current spatial mesh, then pause to redistribute the mesh points in order to achieve a better error equi-distribution according to some simple monitor function. A local interpolation scheme must be applied to obtain missing solution values on the new spatial mesh, and this may introduce some smearing of sharp solution fronts, reminiscent of semi-Lagrangian methods. Some procedure for estimating the time discretization error has to be incorporated as well. Typically, an implicit time discretization suitable for stiff problems is used to avoid multi time-stepping. The resulting procedure has seen use in practice.

A more conceptually elegant approach is to allow the mesh to move continuously in time. We have considered allowing spatial mesh points to move in a prespecified way in Example 7.5, but now we want this done automatically, as part of the solution process. Thus, the mesh nodes may be considered as particles, and they are allowed to move in time, obeying a **moving mesh PDE** (MMPDE) that describes a coordinate transformation that in one dimension is written as $x(\xi, t)$. In Section 3.1.5 we have described such a transformation (see in particular (3.25) and Example 3.7), but here the stakes are much higher as the transformation depends on time and is specified indirectly, in order to equi-distribute the error according to some monitor function. There are several papers, particularly by Huang, Russell, and their collaborators, which describe theoretical and practical details of these ideas; see [40, 114] and references therein.

An essential concern with these methods is regarding the relative importance given to the PDE problem to be solved vs. the work associated with the moving mesh. It is practically important to ensure that the MMPDE plus the mesh transformation do not take over and produce an overall much harder numerical problem to solve. Occasionally researchers attempt to decouple the two tasks, obtaining methods that look more like rezoning but with the time variation better taken into account in the mesh selection process.

For problems in one space variable, satisfactory results have been reported in the sense of having a general purpose method. However, in more than one space variable, where finite elements with varying, unstructured meshes are typically used in space, the success record of such methods is more mixed. Several of these efforts have been devoted to parabolic *blowup problems*, i.e., problems whose solution blows up in finite time [40].

11.3 Level set methods

Many applications involve tracking interfaces between different media as they move in time. Classical instances are the simulation of crystal growth and the visualization of water waves and other fluid motion. The latter is required for many purposes, including special effects in large-budget movies.

Other applications do not originally involve motion, but an artificial motion may be introduced. For instance, in the design of an object such as a turbine wing an optimal shape is desired based on some criteria. An iterative algorithm aimed to achieve that optimal shape may be viewed as dynamically altering the shape at each iteration. Thus, the interface defining the shape is moving as a function of artificial time [9].

Let us restrict discussion, for simplicity, to two spatial variables. A propagating interface is then a parametric curve in motion. However, tracking such a curve may not be simple. For one thing, it represents a moving discontinuity. Imagine, for instance, a shape S in two dimensions defined by its characteristic function $\chi = \chi_S$, such that $\chi(x, y) = 1$ if $(x, y) \in S$ and $\chi(x, y) = 0$ otherwise. The characteristic function of course has a discontinuity at the interface, which is where everything happens. Moreover, a method that attempts to track interfaces directly may hit a snag, or at least become very complicated,

11.3. Level Set Methods

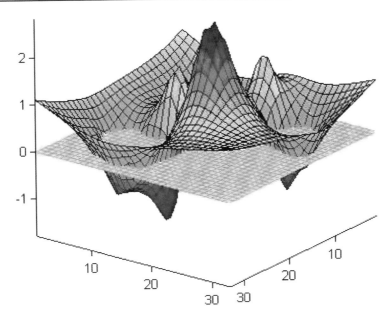

Figure 11.2. *A level set function ϕ defining shapes in two dimensions. The points (x, y) where $\phi(x, y) = 0$ define the interface between the regions inside and outside the shapes.*

if the topological nature of the interface changes, e.g., one body splits into two or more separate ones, or some separate bodies merge into one.

An incredible amount of recent attention has been devoted to an alternative approach that views an interface as the zero-*level set* of a smoother function $\phi(x, y) \in \mathcal{H}_1$; see Figure 11.2. Thus, the function ϕ whose derivative exists and is bounded describes the less smooth characteristic function χ. The seminal paper in this area is Osher and Sethian [141], although some similar ideas have been around before. These two authors have subsequently written many papers as well as a book each [152, 140], but the area is still far from having settled down. The devil is often in the detail here. A MATLAB toolbox written by Mitchell thus greatly helps:

http://www.cs.ubc.ca/~mitchell/ToolboxLS/index.html

The actual motion of the interface depends on the application in question, but it is generally governed by a *Hamilton–Jacobi equation*,

$$\phi_t + H(\nabla \phi) = 0. \tag{11.6}$$

For instance, in case of a known external velocity field $\mathbf{w}(t, x, y)$ we have $H(\nabla \phi) = \mathbf{w} \cdot \nabla \phi$, and the equation for ϕ is advection with variable coefficients,

$$\phi_t + \mathbf{w} \cdot \nabla \phi = 0. \tag{11.7}$$

The equation (11.6) is generally of hyperbolic type. But unlike in Chapter 10 we expect ϕ to vary continuously. In fact, letting $\mathbf{u} = \nabla \phi$ and applying ∇ to (11.6) we have

$$\mathbf{u}_t + \nabla H(\mathbf{u}) = \mathbf{0}.$$

In one dimesion this is a conservation law for u. Anticipating discontinuities in $\mathbf{u}$ therefore leads us to expect ϕ to be differentiable but perhaps not smoother than that.

For the external motion equation (11.7) we can apply the techniques from Chapter 10, especially the upwind methods discussed in Section 10.5. ENO, WENO, and SSP techniques can also be applied, one dimension at a time [140], with the saving grace that the judicious polynomial interpolation described in Section 10.3.2 is applied directly to ϕ, which indeed is expected to be as smooth as the primitive function w of (10.25).

Other methods for solving (11.6) for different situations, as well as a *particle level set method*, are discussed in Osher and Fedkiw [140], where several applications are given.

Bibliography

[1] P. Amestoy, I. Duff, J.-Y. L'Excellent, and J. Koster. A fully asynchronous multifrontal solver using distributed dynamic scheduling. *SIAM J. Matrix Anal. Appl.*, 23:15–41, 2001.

[2] M. Anitescu and F. Potra. Formulating dynamic multi rigid-body contact problems with friction as solvable linear complementarity problems. *ASME Nonlinear Dynamics*, 14:231–247, 1997.

[3] V. I. Arnold. *Mathematical Methods of Classical Mechanics*. Springer, New York, 1978.

[4] D. Aruliah and U. Ascher. Multigrid preconditioning for Krylov methods for time-harmonic Maxwell's equations in three dimensions. *SIAM J. Sci. Comput.*, 24:702–718, 2002.

[5] U. Ascher. Stabilization of invariants of discretized differential systems. *Numer. Algorithms*, 14:1–23, 1997.

[6] U. Ascher and E. Boxerman. On the modified conjugate gradient method in cloth simulation. *The Visual Computer*, 19:526–531, 2003.

[7] U. Ascher and P. Carter. A multigrid method for shape from shading. *SIAM J. Numer. Anal.*, 30:102–115, 1993.

[8] U. Ascher, E. Haber, and H. Huang. On effective methods for implicit piecewise smooth surface recovery. *SIAM J. Sci. Comput.*, 28:339–358, 2006.

[9] U. Ascher, H. Huang, and K. van den Doel. Artificial time integration. *BIT*, 47:3–25, 2007.

[10] U. Ascher, R. Mattheij, and R. Russell. *Numerical Solution of Boundary Value Problems for Ordinary Differential Equations*. SIAM, Philadelphia, 1995.

[11] U. Ascher and R. I. McLachlan. Multisymplectic box schemes and the Korteweg-de Vries equation. *Appl. Numer. Algorithms*, 48:255–269, 2004.

[12] U. Ascher and R. I. McLachlan. On symplectic and multisymplectic schemes for the KdV equation. *J. Scient. Comput.*, 25:83–104, 2005.

[13] U. Ascher and L. Petzold. Projected implicit Runge–Kutta methods for differential-algebraic equations. *SIAM J. Numer. Anal.*, 28:1097–1120, 1991.

[14] U. Ascher and L. Petzold. *Computer Methods for Ordinary Differential Equations and Differential-Algebraic Equations*. SIAM, Philadelphia, 1998.

[15] U. Ascher and S. Reich. On difficulties in integrating highly oscillatory Hamiltonian systems. In P. Deuflhard, J. Hermans, B. Leimkuhler, A. Mark, S. Reich, and R.D. Skeel, editors, *Algorithms for Macromolecular Modelling*. Springer, New York, 1998.

[16] U. Ascher and S. Reich. The midpoint scheme and variants for Hamiltonian systems: Advantages and pitfalls. *SIAM J. Sci. Comput.*, 21:1045–1065, 1999.

[17] U. Ascher, S. Ruuth, and R. Spiteri. Implicit-explicit Runge–Kutta methods for time-dependent partial differential equations. *Appl. Numer. Math.*, 25:151–167, 1997.

[18] U. Ascher, S. Ruuth, and B. Wetton. Implicit-explicit methods for time-dependent partial differential equations. *SIAM J. Numer. Anal.*, 32:797–823, 1995.

[19] O. Axelsson and V. A. Barker. *Finite Element Solution of Boundary Value Problems*. Academic Press, New York, 1984.

[20] D. Baraff and A. Witkin. Large steps in cloth simulation. In *SIGGRAPH*, ACM, 1998, pp. 43–54.

[21] R. Barrett, M. Berry, T. F. Chan, J. Demmel, J. Donato, J. Dongarra, V. Eijkhout, R. Pozo, C. Romine, and H. van der Vorst. *Templates for the Solution of Linear Systems: Building Blocks for Iterative Methods*. SIAM, Philadelphia, 1993.

[22] A. Bellen and M. Zennaro. *Numerical Methods for Delay Differential Equations*. Oxford Science Publications, London, 2003.

[23] M. Berger and P. Collela. Local adaptive mesh refinement for shock hydrodynamics. *J. Comp. Phys.*, 82:64–84, 1989.

[24] M. Berger and J. Oliger. Adaptive methods for hyperbolic partial differential equations. *J. Comp. Phys.*, 53:484–512, 1984.

[25] R. Bermejo. An analysis of an algorithm for the Galerkin-characteristic method. *Numer. Math.*, 60:163–194, 1991.

[26] F. Black and M. Scholes. The pricing of options and corporate liabilities. *J. Pol. Econ.*, 81:637–659, 1973.

[27] G. Bock. Recent advances in parameter identification techniques for ODE. In P. Deuflhard and E. Hairer, editors, *Numerical Treatment of Inverse Problems*, Birkhäuser, Boston, 1983, pp. 95–121.

[28] G. Bock and K. Plitt. A multiple shooting algorithm for direct solution of optimal control problems. In *Proc. IFAC 9th World Congress*, Budapest, 1984, pp. 242–247.

[29] J. L. Bona, M. Chen, and J.-C. Saut. Boussinesq equations and other systems for small-amplitude long waves in nonlinear dispersive media I. Derivation and linear theory. *J. Nonlinear Sci.*, 12:283–318, 2002.

[30] A. Bossavit. *Computational Electromagnetism. Variational Formulation, Complementarity, Edge Elements.* Academic Press, New York, 1998.

[31] E. Boxerman and U. Ascher. Decomposing cloth. In *Proc. ACM SIGGRAPH/Eurographics Symp. Computer Animation*, 2004, pp. 153–161.

[32] J. P. Boyd. *Chebyshev & Fourier Spectral Methods.* Springer, New York, 1989.

[33] A. Brandt. Multigrid Techniques: 1984 Guide with Applications to Fluid Dynamics. The Weizmann Institute of Science, Rehovot, Israel, 1984.

[34] K. Brenan, S. Campbell, and L. Petzold. *Numerical Solution of Initial-Value Problems in Differential-Algebraic Equations.* North-Holland, Amsterdam, 1989.

[35] F. Brezzi and M. Fortin. *Mixed and Hybrid Finite Element Methods.* Springer, New York, 1991.

[36] T. J. Bridges. Multi-symplectic structures and wave propagation. *Math. Proc. Camb. Phil. Soc.*, 121:147–190, 1997.

[37] T. J. Bridges and S. Reich. Multi-symplectic integrators: Numerical schemes for Hamiltonian PDEs that conserve symplecticity. *Phys. Lett. A*, 284 (4-5):184–193, 2001.

[38] T. J. Bridges and S. Reich. Numerical methods for Hamiltonian PDEs. *J. Physics A*, 39:5287–5320, 2006.

[39] A. Bryson and Y. C. Ho. *Applied Optimal Control.* Ginn and Co., Waltham, MA, 1969.

[40] C. Budd, W. Huang, and R. D. Russell. Moving mesh methods for problems with blow-up. *SIAM J. Sci. Comput.*, 17:305–327, 1996.

[41] C. J. Budd and M. D. Piggott. Geometric integration and its applications. In *Handbook of Numerical Analysis Vol. XI*, P. G. Ciarlet and F. Cucker, editors. North-Holland, Amsterdam, 2003, pp. 35–139.

[42] R. L. Burden and J. D. Faires. *Numerical Analysis.* Brooks/Cole, Pacific Grove, CA, 2001.

[43] K. Burrage and J. C. Butcher. Stability criteria for implicit Runge–Kutta methods. *SIAM J. Numer. Anal.*, 16:46–57, 1979.

[44] J. C. Butcher. *The Numerical Analysis of Ordinary Differential Equations.* John Wiley, New York, 1987.

[45] C. Canuto, M.Y. Hussaini, A. Quarteroni, and T.A. Zang. *Spectral Methods in Fluid Dynamics.* Springer, New York, 1987.

[46] M. Carpenter, D. Gottlieb, S. Abarbanel, and W.-S. Don. The theoretical accuracy of Runge–Kutta time discretizations for the initial bounday value problem: A study of the boundary error. *SIAM J. Sci. Comput.*, 16:1241–1252, 1995.

[47] G. Carrier and C. Pearson. *Partial Differential Equations: Theory and Technique*. Academic Press, New York, 1976.

[48] K.-J. Choi and H.-S. Ko. Stable but responsive cloth. *ACM Trans. Graphics (SIGGRAPH)*, 21(3):604–611, 2002.

[49] A. J. Chorin. Numerical solution of the Navier–Stokes equations. *Math. Comp.*, 22:745–762, 1968.

[50] B. Cockburn, G. E. Karniadakis, and C.-W. Shu, editors. *Discontinuous Galerkin Methods: Theory, Computation and Applications*. Lecture Notes in Computer Science 11, Springer, New York, 2000.

[51] B. Cockburn and C.-W. Shu. The local discontinuous Galerkin method for time-dependent convection-diffusion systems. *SIAM J. Numer. Anal.*, 35:2440–2463, 1998.

[52] R. Courant, K. O. Friedrichs, and H. Lewy. Über die partiellen differenzengleichungen der mathematischen physik. *Phys. Math. Anal.*, 100:32–74, 1928.

[53] S. Cox and P. Matthews. Exponential time differencing for stiff systems. *J. Comp. Phys.*, 176:430–455, 2002.

[54] J. Crank and P. Nicholson. A practical method for numerical evaluation of solutions of partial differential equations of the heat-conduction type. *Proc. Camb. Philo. Soc.*, 43:50–67, 1947.

[55] G. Dahlquist. Error analysis for a class of methods for stiff nonlinear initial value problems. In G. Watson, editor, *Numerical Analysis*, Springer, Berlin, 1975.

[56] T. A. Davis. *Direct Methods for Sparse Linear Systems*. SIAM, Philadelphia, 2006.

[57] J. Demmel. *Applied Numerical Linear Algebra*. SIAM, Philadelphia, 1997.

[58] P. G. Drazin and R. S. Johnson. *Solitons: An Introduction*. Cambridge University Press, London, 1989.

[59] H. Elman, D. Silvester, and A. Wathen. *Finite Elements and Fast Iterative Solvers*. Oxford University Press, London, 2005.

[60] H. W. Engl, M. Hanke, and A. Neubauer. *Regularization of Inverse Problems*. Kluwer, Dordrecht, 1996.

[61] B. Engquist and S. Osher. Stable and entropy satisfying approximations for transonic flow calculations. *Math. Comp.*, 34:45–75, 1980.

[62] B. Feldman, J. O'Brien, and B. Klingner. Animating gases with hybrid meshes. *ACM Trans. Graphics (SIGGRAPH)*, 24(3):904–909, 2005.

[63] B. Fornberg. *A Practical Guide to Pseudospectral Methods*. Cambridge University Press, London, 1998.

[64] B. Fryxell, K. Olson, P. Ricker, F. Timmes, M. Zingale, D. Lamb, P. MacNeice, R. Rosner, J. Turan, and H. Tufo. FLASH: An adaptive mesh hydrodynamics code for modeling astrophysical nuclear flashes. *Astrophysical Journal Supplement*, 131:273–334, 2000.

[65] Z. Ge and J. Marsden. Lie-Poisson Hamilton-Jacobi theory and Lie-Poisson integrators. *Phys. Lett. A*, 133:134–139, 1988.

[66] C. W. Gear. *Numerical Initial Value Problems in Ordinary Differential Equations*. Prentice-Hall, Englewood Cliffs, NJ, 1973.

[67] V. Girault and P. A. Raviart. *Finite Element Methods for Navier-Stokes Equations*. Springer, Berlin, 1986.

[68] G. H. Golub and C. F. van Loan. *Matrix Computations*. Johns Hopkins University Press, Baltimore, 1988.

[69] S. Gomez, A. Perez, and R. Alvarez. Multiscale optimization for aquifer parameter identification with noisy data. In *Computer Methods in Water Resources XII, Vol. 2*, WIT Press, 1998.

[70] J. B. Goodman and R. J. LeVeque. On the accuracy of stable schemes for two dimensional conservation laws. *Math. Comp.*, 45:15–21, 1985.

[71] D. Gottlieb and S. A. Orszag. *Numerical Analysis of Sprectral Methods: Theory and Applications*. SIAM, Philadelphia, 1977.

[72] S. Gottlieb, C.-W. Shu, and E. Tadmor. Strong stability-perserving high-order time discretization methods. *SIAM Review*, 43:89–112, 2001.

[73] P. M. Gresho and R. L. Sani. *Incompressible Flow and the Finite Element Method*. John Wiley, New York, 1998.

[74] E. Guendelman, A. Selle, F. Losasso, and R. Fedkiw. Coupling water and smoke to thin deformable and rigid shells. *ACM Trans. Graphics (SIGGRAPH)*, 24(3):973–981, 2005.

[75] Y. Guo, Z. Pan, and M. J. Ward. Touchdown and pull-in voltage behavior of a MEMS device with varying dielectric properties. *SIAM J. Appl. Math.*, 66:309–338, 2005.

[76] B. Gustafsson, H.-O. Kreiss, and J. Oliger. *Time Dependent Problems and Difference Methods*. John Wiley, New York, 1995.

[77] B. Gustafsson, H.-O. Kreiss, and A. Sunstrom. Stability theory of difference approximations for mixed initial boundary value problems II. *Math. Comp.*, 24:649–686, 1972.

[78] E. Haber, U. Ascher, and D. Oldenburg. Inversion of 3D electromagnetic data in frequency and time domain using an inexact all-at-once approach. *Geophysics*, 69:1216–1228, 2004.

[79] E. Haber, S. Heldmann, and U. Ascher. Adaptive finite volume method for distributed non-smooth parameter identification. *Inverse Problems*, 23:1659–1676, 2007.

[80] T. Hagstrom. Radiation boundary conditions for the numerical simulation of waves. *Acta Numerica*, 8:47–106, 1999.

[81] E. Hairer, C. Lubich, and G. Wanner. *Geometric Numerical Integration*. Springer, New York, 2002.

[82] E. Hairer, S. P. Norsett, and G. Wanner. *Solving Ordinary Differential Equations I: Nonstiff Problems*. Springer, New York, 1993.

[83] E. Hairer and G. Söderlind. Explicit, time reversible, adaptive step size control. *SIAM J. Sci. Comput.*, 26:1838–1851, 2005.

[84] E. Hairer and D. Stoffer. Reversible long-term integration with variable stepsizes. *SIAM J. Sci. Comput.*, 18:257–269, 1997.

[85] E. Hairer and G. Wanner. *Solving Ordinary Differential Equations II: Stiff and Differential-Algebraic Problems*. Springer, New York, 1996.

[86] J. Hale. *Theory of Functional Differential Equations*. Springer, New York, 1977.

[87] F. H. Harlow and J. E. Welch. Numerical calculation of time-dependent viscous incompressible flow of fluids with free surface. *Phys. Fluids*, 8:2182, 1965.

[88] A. Harten, B. Engquist, S. Osher, and S. Chakravarthy. Uniformly high order accurate essentially non-oscillatory schemes, III. *J. Comp. Phys.*, 71:231–303, 1987.

[89] M. T. Heath. *Scientific Computing: An Introductory Survey*. McGraw-Hill, New York, 2002.

[90] I. Higueras. On strong stability preserving time discretization methods. *J. Scient. Comput.*, 21:193–223, 2004.

[91] R. Hiptmair. Multigrid method for Maxwell's equations. *SIAM J. Numer. Anal.*, 36:204–225, 1998.

[92] C. Hirsch. *Numerical Computation of Internal and External Flows*. John Wiley, New York, 1988.

[93] G. R. Hjaltason and H. Samet. Speeding up construction of QuadTrees for spatial indexing. *VLDB Journal*, 11:109–137, 2002.

[94] M. Hochbruck, C. Lubich, and H. Selhofer. Exponential integrators for large systems of differential equations. *SIAM J. Sci. Comput.*, 19:1552–1574, 1998.

[95] M. Hochbruck and A. Ostermann. Explicit exponential Runge–Kutta methods for semilinear parabolic problems. *SIAM J. Numer. Anal.*, 43:1069–1090, 2005.

[96] H. Holden and N. Risebro. *Front Tracking for Hyperbolic Conservation Laws*. Springer, New York, 2002.

[97] T. Y. Hou and P. D. Lax. Dispersive approximation in fluid dynamics. *Comm. Pure Appl. Math.*, 44:1–40, 1991.

[98] W. Hundsdorfer and J. G. Verwer. *Numerical Solution of Time-Dependent Advection-Diffusion-Reaction Equations*. Springer, New York, 2003.

[99] J. M. Hyman and B. Nicolaenko. The Kuramoto-Sivashinsky equation: a bridge between PDE's and dynamical systems. *Phys. D*, 18:113–126, 1986.

[100] A. Iserles. *A First Course in the Numerical Analysis of Differential Equations*. Cambridge University Press, Cambridge, UK, 1996.

[101] E. X. Jiang. *Numerical Simulations of Semiconductor Devices by Streamline-Diffusion Methods*. Ph.D. thesis. Institute of Applied Mathematics, University of British Columbia, 1995.

[102] G. Jiang and C. Shu. Efficient implementation of of weighted ENO schemes. *J. Comp. Phys.*, 126:202–228, 1996.

[103] J. Jin. *The Finite Element Method in Electromagnetics*. John Wiley, New York, 1993.

[104] C. Johnson. *Numerical Solution of Partial Differential Equations by the Finite Element Method*. Cambridge University Press, Cambridge, UK, 1990.

[105] A.-K. Kassam and L. N. Trefethen. Fourth-order time-stepping for stiff PDEs. *SIAM J. Sci. Comput.*, 26:1214–1233, 2005.

[106] D. Kaufman, T. Edmunds, and D. Pai. Fast frictional dynamics for rigid bodies. *ACM Trans. Graphics (SIGGRAPH)*, 24(3):904–909, 2005.

[107] H. B. Keller. A new difference scheme for parabolic problems. In *Numerical Solution of Partial Differential Equations, II (SYNSPADE 1970)*. Academic Press, New York, 1971, pp. 327–350.

[108] B. L. N. Kennett. *Seismic wave propagation in stratified media*. Cambridge University Press, Cambridge, UK, 1983.

[109] S. Knapek. Matrix-dependent multigrid homogenization for diffusion problems. *SIAM J. Sci. Comput.*, 20:515–533, 1998.

[110] H.-O. Kreiss. Numerical Methods for Solving Time-Dependent Problems for Partial Differential Equations. Lecture Notes, Universite de Montreal, Montreal, Canada, 1978.

[111] H.-O. Kreiss and J. Lorenz. *Initial-Boundary Value Problems and the Navier–Stokes Equations*. Academic Press, New York, 1989.

[112] Y. Kuramoto and T. Tsuzuki. Persistent propagation of concentration waves in dissipative media far from thermal equlibrium. *Prog. Theor. Physics*, 55:356–369, 1976.

[113] J. D. Lambert. *Numerical Methods for Ordinary Differential Systems*. John Wiley, New York, 1991.

[114] J. Lang, W. Cao, W. Huang, and R. D. Russell. A two-dimensional moving finite element method with local refinement based on a posteriori error estimates. *Applied Numer. Math.*, 46:75–94, 2003.

[115] B. Leimkuhler and S. Reich. *Simulating Hamiltonian Dynamics*. Cambridge University Press, Cambridge, UK, 2004.

[116] R. J. LeVeque. *Finite Volume Methods for Hyperbolic Problems*. Cambridge University Press, Cambridge, UK, 2002.

[117] R. J. LeVeque. *Finite Difference Methods for Ordinary and Partial Differential Equations*. SIAM, Philadelphia, 2007.

[118] D. Levy and E. Tadmor. From semidiscrete to fully discrete: Stability of Runge–Kutta schemes by the energy method. *SIAM Review*, 40:40–73, 1998.

[119] S. Leyendecker, S. Ober-Blöbaum, J. Marsden, and M. Ortiz. Discrete mechanics and optimal control for constrained multibody dynamics. In *Proc. 6th Intl. Conf. Multibody Systems, Nonlinear Dynamics and Control*. ASME, 2007.

[120] A. I. Liapis and D. W. T. Rippin. A general model for the simulation of multicomponent adsorption from a finite bath. *Chem. Eng. Science*, 32:619–627, 1977.

[121] X. Liu, S. Osher, and T. Chan. Weighted essentially non-oscillatory schemes. *J. Comp. Phys.*, 115:200–212, 1994.

[122] F. Losasso, F. Gibou, and R. Fedkiw. Simulating water and smoke with an octree data structure. *ACM Trans. Graphics (SIGGRAPH)*, 23(3):457–462, 2004.

[123] S. MacLachlan and D. Moulton. Multilevel upscaling through variational coarsening. *Water Resources Res.*, 42:W02418, 2006.

[124] P. MacNeice, K. M. Olson, C. Mobarry, R. deFainchtein, and C. Packer. PARAMESH: A parallel adaptive mesh refinement community toolkit. *Comput. Phys. Comm.*, 126:330–354, 2000.

[125] R. Manson, S. Wallis, and D. Hope. A conservative semi-Lagrangian transport model for rivers with transient storage zones. *Water Resources Res.*, 37:3321–3329, 2001.

[126] J. Marsden and M. West. Discrete mechanics and variational integrators. *Acta Numerica*, 10:1–155, 2001.

[127] R. Mattheij, S. Rienstra, and J. ten Thije Boonkkamp. *Partial Differential Equations*. SIAM, Philadelphia, 2005.

Bibliography

[128] R. I. McLachlan. On the numerical integration of ordinary differential equations by symmetric composition methods. *SIAM J. Sci. Comput.*, 16:151–168, 1994.

[129] R. I. McLachlan. Symplectic integration of hamiltonian wave equations. *Numer. Math.*, 66:465–492, 1994.

[130] R. I. McLachlan and G. R. W. Quispel. Geometric integrators for ODEs. *J. Physics A*, 39:5251–5285, 2006.

[131] B. Merriman, J. Bence, and S. Osher. Motion of multiple junctions: A level set approach. *J. Comp. Phys.*, 112(2):334–363, 1994.

[132] O. Meshar, D. Irony, and S. Toledo. An out-of-core sparse symmetric indefinite factorization method. *ACM Trans. Math. Software*, 32:445–471, 2006.

[133] A. R. Mitchell and D. F. Griffiths. *The Finite Difference Method in Partial Differential Equations.* John Wiley, New York, 1980.

[134] S. L. Mitchell, K. W. Morton, and A. Spence. Analysis of box schemes to reactive flow problems. *SIAM J. Sci. Comput.*, 27:1202–1223, 2006.

[135] C. Moler and C. Van Loan. Nineteen dubious ways to compute the exponential of a matrix. *SIAM Review*, 20:801–836, 1978.

[136] J. J. Monaghan. Smoothed particle hydrodynamics. *Rep. Prog. Phys.*, 68:1703–1759, 2005.

[137] K. W. Morton and D. F. Mayers. *Numerical Solution of Partial Differential Equations*, 2nd ed. Cambridge University Press, Cambridge, UK, 2005.

[138] A. Nachbin. A terrain-following Boussinesq system. *SIAM J. Appl. Math.*, 63:905–922, 2003.

[139] J. Nocedal and S. Wright. *Numerical Optimization.* Springer, New York, 1999.

[140] S. Osher and R. Fedkiw. *Level Set Methods and Dynamic Implicit Surfaces.* Springer, New York, 2003.

[141] S. Osher and J. Sethian. Fronts propagating with curvature dependent speed: Algorithms based on Hamilton-Jacobi formulations. *J. Comp. Phys.*, 79:12–49, 1988.

[142] P. Perona and J. Malik. Scale-space and edge detection using anisotropic diffusion. *IEEE Transactions on Pattern Analysis and Machine Intelligence*, 12(7):629–639, 1990.

[143] A. Preissmann. Propagation des intumescences dans les canaux et rivières. In *Proc. First Congress French Association for Computation*, Grenoble, 1961.

[144] R. D. Richtmyer and K. W. Morton. *Difference Methods for Initial-Value Problems.* John Wiley, New York, 1967.

[145] L. Rudin, S. Osher, and E. Fatemi. Nonlinear total variation based noise removal algorithms. *Physica D*, 60:259–268, 1992.

[146] S. Ruuth. Implicit-explicit methods for reaction–diffusion problems in pattern formation. *J. Math. Biol.*, 34:148–176, 1995.

[147] S. Ruuth. Efficient algorithms for diffusion-generated motion by mean curvature. *J. Comp. Phys.*, 144:603–625, 1998.

[148] S. Ruuth and R. Spiteri. Two barriers on strong-stability-preserving time discretization methods. *J. Scient. Comput.*, 17:211–220, 2002.

[149] Y. Saad. *Iterative Methods for Sparse Linear Systems*. PWS Publishing Company, Boston, 1996.

[150] J. M. Sanz-Serna and M. P. Calvo. *Numerical Hamiltonian Problems*. Chapman and Hall, London, 1994.

[151] G. Sapiro. *Geometric Partial Differential Equations and Image Analysis*. Cambridge University Press, Cambridge, UK, 2001.

[152] J. A. Sethian. *Level Set Methods and Fast Marching Methods: Evolving Interfaces in Geometry, Fluid Mechanics, Computer Vision, and Materials Science*. Cambridge University Press, Cambridge, UK, 1999.

[153] L. F. Shampine. Conservation laws and the numerical solution of ODEs. *Comp. Maths. Appls. B*, 12:1287–1296, 1986.

[154] L. F. Shampine. *Numerical Solution of Ordinary Differential Equations*. Chapman and Hall, London, 1994.

[155] L. F. Shampine, I. Gladwell, and S. Thompson. *Solving ODEs with MATLAB*. Cambridge University Press, London, 2003.

[156] C.-W. Shu. Essentially non-oscillatory and weighted essentially nonoscillatory schemes for hyperbolic conservation laws. In *Advances in Numerical Approximation of Nonlinear Hyperbolic Equations*. Springer Lecture Notes in Math 1697, 1998, pp. 325–432.

[157] R. Spiteri and S. Ruuth. A new class of optimal high-order strong-stability-preserving time discretization methods. *SIAM J. Numer. Anal.*, 40:469–491, 2002.

[158] G. Strang and G. Fix. *An Analysis of the Finite Element Method*. Prentice-Hall, Engelwood Cliffs, NJ, 1973.

[159] J. C. Strikwerda. *Finite Difference Schemes and Partial Differential Equations*, 2nd ed. SIAM, Philadelphia, 2004.

[160] S. H. Strogatz. *Nonlinear Dynamics and Chaos*. Addison-Wesley, Reading, MA, 1994.

[161] A. M. Stuart and A. R. Humphries. *Dynamical Systems and Numerical Analysis*. Cambridge University Press, Cambridge, UK, 1996.

[162] D. Sulsky, H. Schreyer, K. Peterson, R. Kwok, and M. Coon. Using the material-point method to model sea ice dynamics. *J. Geophysical Res.*, 112:C02S90, 2007.

[163] A. Taflove. *Computational Electrodynamics: The Finite-Difference Time-Domain Method*. Artech House Publishers, New York, 1995.

[164] J. W. Thomas. *Numerical Partial Differential Equations: Finite Difference Methods*. Springer, New York, 1995.

[165] V. Thomée. A stable difference scheme for the mixed boundary problem for a hyperbolic first-order system in two dimensions. *J. Soc. Indust. Appl. Math.*, 10:229–245, 1962.

[166] A. N. Tikhonov and V. Ya. Arsenin. *Methods for Solving Ill-Posed Problems*. John Wiley, New York, 1977.

[167] L. N. Trefethen. Group velocity in finite difference schemes. *SIAM Review*, 24:113–136, 1982.

[168] L. N. Trefethen. Pseudospectra of linear operators. *SIAM Review*, 39:383–406, 1997.

[169] L. N. Trefethen. *Spectral Methods in MATLAB*. SIAM, Philadelphia, 2000.

[170] L. N. Trefethen and D. Bau, III. *Numerical Linear Algebra*. SIAM, Philadelphia, 1997.

[171] U. Trottenberg, C. Oosterlee, and A. Schuller. *Multigrid*. Academic Press, New York, 2001.

[172] S. Turek. *Efficient Solvers For Incompressable Flow Problems*. MacMillan, New York, 1999.

[173] C. V. Turner and R. R. Rosales. The small dispersion limit for a nonlinear semidiscrete system of equation. *Studies Appl. Math.*, 99:205–254, 1997.

[174] K. van den Doel and U. Ascher. Real-time numerical solution of Webster's equation on a nonuniform grid. *IEEE Trans. Audio, Speech, Language Processing* 2008, to appear.

[175] S. Vandewalle. *Parallel Multigrid Waveform Relaxation for Parabolic Problems*. Teubner, Leipzig, 1993.

[176] J. M. Varah. Stability restrictions on second order, three level finite difference schemes for parabolic equations. *SIAM J. Numer. Anal.*, 17:300–309, 1980.

[177] R. Viera and E. Biscaia, Jr. Direct methods for consistent initialization of DAE systems. *Computers Chem. Eng.*, 25:1299–1311, 2001.

[178] C. Vogel. *Computational Methods for Inverse Problem*. SIAM, Philadelphia, 2002.

[179] R. Wang and R. Spiteri. Linear instability of the fifth-order WENO method. *SIAM J. Numer. Anal.*, 45:1871–1901, 2007.

[180] G. Wei and J. Kirby. Time-dependent numerical code for extended Boussinesq equations. *J. Waterway, Port, Coastal Ocean Eng.*, 121:251–261, 1995.

[181] J. Weickert. *Anisotropic Diffusion in Image Processing*. Teubner, Stuttgart, 1998.

[182] G. B. Whitham. *Linear and Nonlinear Waves*. John Wiley, New York, 1974.

[183] P. Wilmott, S. Howison, and J. Dewynne. *The Mathematics of Financial Derivatives*. Cambridge University Press, Cambridge, UK, 1995.

[184] I. Yavneh. Why multigrid methods are so efficient. *Comput. Sci. Eng.*, 8:12–22, 2006.

[185] K. S. Yee. Numerical solution of initial boundary value problems involving Maxwell's equations in isotropic media. *IEEE Trans. Antennas Propagation*, 14:302–307, 1966.

[186] H. Yoshida. Construction of higher order symplectic integrators. *Phys. Lett. A*, 150:262–268, 1990.

[187] N. J. Zabusky and M. D. Kruskal. Interaction of 'solitons' in a collisionless plasma and the recurrence of initial states. *Phys. Rev.*, 15:240–243, 1965.

[188] P. F. Zhao and M. Z. Qin. Multisymplectic geometry and multisymplectic Preissmann scheme for the KdV equation. *J. Phys. A*, 33(18):3613–3626, 2000.

Index

*Boldfaced type indicates terms of particular importance
and page numbers on which terms are defined.*

accuracy, 22, 123
accuracy order, 13, 38, 94
Adams methods
    absolute stability, 54
adaptive mesh refinement (AMR), 371
adiabatic invariant, 200, 202
advection, 1
advection equation, 7, 8, 13, 147
advection-diffusion equation, 1, 327
algebraic multigrid, 320
algorithm
    CG, 315
    Gaussian elimination, 73
    multigrid, 319
    PCG, 316
    tridiagonal system, 74
alternating direction implicit (ADI), 290
amplification factor, 18, 136
amplification matrix, 136
amplitude, 3
animation, 1, 305
anisotropic diffusion, 102
ansatz, 19
approximate inverse, 317
artificial diffusion, 164
artificial time, 372
artificial viscosity, 164, 358
astrophysics, 246

backward differentiation formula (BDF), 40
backward error analysis, 196
basis function, 112
biharmonic equation, 274
Black–Scholes model, 32
boundary conditions (BC), 1, 5, 58, 92, 253
    absorbing, 263, 272
    Dirichlet, 2, 11, 107, 253
    essential, 109, 274
    extrapolation, 254, 262, 272
    inflow, 261
    mixed, 107
    natural, 104, 109, 274
    Neumann, 107, 253
    nonreflecting, 262, 272
    outflow, 261, 346
    periodic, 2, 11, 20, 253
    radiating, 272, 274
boundary value ordinary differential equation (BVODE), 71
Boussinesq system, **250**, 311
box scheme, 165, **167**
Burgers equation, **157**, 170, 260, 272

Cauchy problem, 5, 6
cell average, 338, 342
cell Reynolds number, 358
chaos, 44, 191, 252
characteristic curve, 8, 16, 17
characteristic variable, 259, 352

Chebyshev points, 55, 78, 245
Chebyshev polynomial, 78, 245
classical wave equation, 217
cloth simulation, **68**, 69, **308**
collocation
    Gauss, 47, 174, 188
    Radau, 47
compact
    discretization, 94
    set, 331
    support, 331
complexity
    flop, 139
compressible fluid flow, 356
computational fluid dynamics (CFD), 58, 66
computer graphics, 1, 67
computer vision, 1
condition number, **26**, 31, 282
conforming finite elements, 110
conjugate gradient algorithm (CG), **315**
conjugate gradient method (CG), 281
    preconditioned (PCG), 281, 307, **316**
conservation
    $\mathcal{L}_2$-norm, 169
    energy, 181
    form, 157
    mass, 181
    momentum, 181
conservation law, 156, 225, 329
conservative system, 23, 182
consistency, 22, 38, 123
convergence, 22, 94, 120
corner-transport upwind, 358
coupling, 204
Courant number, 19, 237
Courant–Friedrichs–Lewy condition (CFL), **15, 16**, 18, 220
crystal growth, 372

dangling node, 118
decoupling, 204
delay differential equation, 88

diagonally implicit Runge–Kutta (DIRK), 61
difference
    backward, 96
    centered, 24, 96
    compact, 95, 242
    forward, 24, 95, 96
    implicit, 97, 243, 269
differential algebraic equation (DAE), 56, 64–65, 108, 181, 204, 256, 269, 288
    index, 65
    index reduction, 65
    inequality constraint, 288
    limit, 204
    semi-explicit, 65
differentiation matrix, 83, 243
diffusion, 1
diffusion-generated motion, 303
dimensional splitting, 276, 356
discontinuous Galerkin (DG), 113
discrete Fourier transform (DFT), 82, 300
discretization
    full discretization, 92
    semi-discretization, 37, 91
dispersion, 23, 211, 212, 231
dispersion relation, 212
dissipation, 23, 153, 211
dissipativity, 153, 278
    strict, 153
divergence, 183
divided differences, 74
domain of dependence, 16, 220
donor-cell upwind, 358
double well, 303
Dufort–Frankel scheme, 148

earth sciences, 1
elasticity, 67
elliptic
    differential operator, 93
    problem, 9
energy
    conservation, 181
    norm, 315

energy method, 153, 169–177
entropy condition, 333
    Lax, 333
    Oleinik, 333
error
    bound, 128
    boundary layer, 270
    control, 368
    local, 53
    local truncation error, 40, 43
essentially nonoscillatory scheme (ENO), 328, 345, 362
Euler equations, 327, 356
Euler–Lagrange equation, 104, 109, 194, 365
Eulerian method, 246
exponential Runge–Kutta, 311
exponential time-differencing, 310

fast Fourier transform (FFT), 82, 236, 300
Fermi–Pasta–Ulam (FPU), 199
finite difference
    centered, 14
    one-sided, 13, 19
finite difference method, 92, 93
finite difference scheme
    explicit, 14
    leapfrog, 14, 19
finite element method (FEM), 67, 93, 94, 110, 195
    discontinuous Galerkin, 113
finite volume, 100, 342
finite volume method, 92, 93, 110, 195
fixed point iteration, 80, 312
fixed point iteration and Newton's method, review, **80**
floating point operation (flop), 73
fluid flow, 1, 305
flux, 329
    Buckley–Leverett, 362
    fluctuation, 355
    limiter scheme, 342
    numerical, 336
    splitting, 337, 351

forward problem, 284, 287
Fourier
    analysis, 135
    mode, 18
Fourier transform, 6, 18, 29, 31
    continuous, review, **29**
    discrete and fast, review, **82**
    periodic, review, **31**
fractional step method, 276
frequency, 6, 212
front tracking, 329
Function spaces, review, **28**

Gauss divergence formula, 102, 168, 227, 335
Gauss–Newton method, 289
Gauss–Seidel, 317
Gaussian
    points, 47, 78, 80
    quadrature, 80
Gaussian elimination, **73**
    without pivoting, 73
geometric integration, 181
Gershgorin's theorem, 27
ghost
    differential equation, 202
    Hamiltonian system, 196
    unknown, 108
Gibbs phenomenon, 160
Godunov's scheme, 338–340
Godunov–Ryabenkii condition, 264
gradient, 182
grid, 318
groundwater flow, 304
group velocity, 212

Hamilton–Jacobi equation, 373
Hamiltonian, 182
    error, 192
    separable, 189
Hamiltonian PDE, 298
Hamiltonian system, 70, 181, 222
harmonic average, 100
harmonic oscillator, 66, 183
hat function, 132
heat equation, 7, 9

Henon–Heiles Hamiltonian system, 191
Hessian matrix, 185
high resolution, 328
highly oscillatory problem, 66
hyperbolic
    differential operator, 91
    problem, 7, 8
    system, 11

image processing, 1
implicit difference scheme, 97, 243, 269
implicit-explicit method (IMEX), 173, 306
incomplete Cholesky decomposition, 317
incomplete LU-decomposition (ILU), 317
incompressible Euler equations, 305
incompressible Navier–Stokes, 157, 305
inexact Newton, 292
initial condition, 1
initial value problem, **4**, 38
initial-boundary value problem, 2, 5, 107
inner product, 26, 31, 171
inner-outer iteration, 292
integration by parts, **170**
interpolation
    error, 76
    Lagrange form, 75
    piecewise polynomial, 77
    polynomial, 94
invariant, 181
inverse problem, 71, 284, 287
    regularization, 289
inverting electromagnetic data, **285**
iterative methods for linear systems, review, **312**

Jacobi, 317
Jacobian matrix, 49

Korteweg–de Vries equation (KdV), 148, 151, 170, 211, 230–242, 276, 311
Krylov space, 315
Krylov space method, 281, **315**
Kuramoto–Sivashinsky equation (KS), **252**, 311

Lagrange polynomial, 75
Lagrangian, 194
Lagrangian method, 34, 246
Laplace equation, 9
Lax–Friedrichs scheme, 147
Lax–Richtmyer theorem, 126
layer
    initial, 57, 58
    transition, 327
leapfrog, 14, 218
Legendre polynomial, 78, 174
level set, 373
level set method, 372–374
line search, 314
linear multistep method, 39
Lipschitz
    constant, 49
    continuity, 49, 154
local error, 52
local truncation error, 38, **40**, 43, 120, 122, 128
locally one dimensional (LOD), 294
logarithmic norm, 35
Lorenz equations, 44

marker-and-cell (MAC), 305
mass-spring model, 67
material derivative, 157
material point method, 246
mathematical finance, 1
MATLAB
    bvp4c, 89
    cholinc, 319
    dde23, 88
    delsq, 319
    fft, 83
    ode23s, 63
    ode45, 53, 192

Index 391

pcg, 319
MATLAB script
    leapfrog for advection, **21**
    Lorenz butterfly, 44
    Robertson's equations, 61
matrix
    banded, 74
    block structure, 136
    block-tridiagonal, 122
    circulant, 236
    commutativity, 275
    condition number, **26**, 31, 282
    defective, 140
    diagonalizable, 27
    diagonally dominant, 27
    eigenvalue, 25, 26
    eigenvector, 26
    exponential, 31
    fill-in, 281
    Hermitian, 25, 141
    Jordan canonical form, 27
    LU-decomposition, **73**
    normal, 25, 139, 141
    orthogonal, 25
    positive definite, 26
    simultaneously diagonalizable, 275
    singular value, 28
    skew-Hermitian, 25
    skew-symmetric, 25, 184
    sparse, 73
    spectral radius, 26, 139, 277
    symmetric, 25
    symmetric positive definite, 10, 73
    tridiagonal, 74, 94
    unitary, 25
matrix power and exponential, review, **30**
Maxwell's equations, 106, 134, 224, 249, 287
    quasi-static approximation, 105
mean curvature, 104
mechanical systems, 1
mesh, 12
    adaptive mesh refinement (AMR), 371
    coordinate transformation, 114
    dual, 99
    equi-distribution, 370
    finite elements, 116
    function, 13, 50
    monitor function, 370
    nonuniform, 115
    Octree, 117
    primal, 99
    Quadtree, 117
    ratio, 117
    refinement, 368
    rezoning, 371
    selection, 368
    stretching, 114
    tensor product, 115
    uniform, 12
method
    A-stable, 54, 124
    accuracy, 123
    Adams, 40
    Adams–Bashforth, 40
    Adams–Moulton, 41
    additive, 276
    backward differentiation formula (BDF), 42, 108, 120, 256
    backward Euler, 41, 125
    box, 165, 167, 336
    classical fourth order Runge–Kutta (RK4), 44, 159
    collocation, 47, 174, 188
    collocation at Radau points, 108
    compact, 211, 219
    consistency, 40, 43, 123
    contractivity, 349
    convergence, 120
    Crank–Nicolson, 124, 165
    diagonally implicit Runge–Kutta (DIRK), 61
    dispersion, 211
    dissipativity, 153, 278
    Dufort–Frankel, 148

energy, 153
energy conserving, 148, 159
essentially nonoscillatory
    (ENO), 345
Euler, 123
Eulerian, 246
explicit, 39, 120
explicit Runge–Kutta, 43
explicit trapezoidal, 42
exponential integration, 276
finite element, 110
finite volume, 100, 342
fixed point iteration, 80
flux limiter, 342
forward Euler, 37, 40, 43
four step Adams–Moulton
    (AM4), 41
Gauss–Newton, 289
Gauss–Seidel, 313
Godunov, 338
implicit, 39, 121, 124
implicit Runge–Kutta, 47
implicit-explicit (IMEX), 276,
    306
Jacobi, 313
L-stable, 56
Lagrangian, 34, 212, 246
Lax–Friedrichs, 159, 352
Lax–Wendroff, 156, 213, 279,
    294
leapfrog, 121, 140, 158, 218
level set, 303
MacCormack, 158, 279, 294,
    355
meshless, 246
midpoint, 47
monotone, 159, 328
monotonicity, 86
monotonicity preserving, 341
multiple shooting, 72
multistep, 39
multisymplectic, 211
Newmark, 87
Newton, 80
nodal, 113
nonnegative, 86

one-sided, 66
order, 40, 43, 48
particle, 212, 246
partitioned Runge–Kutta, 189
predictor-corrector, 60
Radau collocation, 256
Richtmyer, 158, 355
Rosenbrock, 60
Runge–Kutta (RK), 42, 120, 268
Runge–Kutta–Chebyshev, 55,
    282
semi-implicit, 60, 70
semi-Lagrangian, 252, 305, 359
shooting, 72
spectral, 212, 242, 280
splitting, 359
Störmer–Verlet, 187, 191
stability, 120, 125
staggered midpoint, 191
stiff decay, 125
streamline diffusion, 366
strict stability, 126
strictly dissipative, 153
strong stability preserving
    (SSP), 347
successive overrelaxation
    (SOR), 313
symmetric, 66
symmetric successive
    overrelaxation (SSOR), 313
symplectic, 211, 222, 276, 298
symplectic Euler, 187
theta, 84, 165
time reversible, 189, 195
total variation bounded, 341
total variation diminishing, 341
trapezoidal, 41
underrelaxation, 313
upstream, 34
upwind, 34, 359
variational, 194
weighted essentially
    nonoscillatory (WENO),
    350
method of lines, 71, 91, 108, 144
midpoint method

Index

dynamic equivalence to
trapezoid method, 207
mode, 3
modified Hamiltonian system, 196
modified PDE, 164–165, 178, 335
molecular dynamics, 205
monotone scheme, 159, 328, 340
monotonicity, 347
monotonicity preserving, 341
moving mesh, 229, 371–372
multigrid algorithm, **319**
multigrid method, 281, 307, 316, **317**, 367
multistep method, 39–42
    Adams, 40
    Adams–Bashforth, 40
    Adams–Moulton, 41
    backward differentiation formula (BDF), 42
multisymplectic structure, 225, 231, 249, 298

Navier–Stokes equations, 157, 327
Newton's method, 80
nodal representation, 75
nonlinear Schrödinger equation (NLS), 34, 170, 298, 311
norm
    logarithmic, 35
    matrix, 25
    vector, 25

Octree, 117, 371
operator splitting, 189, 234, 276
optimal control, 365, 367
    adjoint variables, 367
    Hamiltonian function, 367
optimization, 104
order of accuracy, 38, **40**, 43, 48, 94
order reduction, 268
ordinary differential equation (ODE), 3, 6, **38**
    autonomous, 44
    boundary value problem, 71
    initial value problem, 71
orthogonal and trigonometric polynomials, review, **77**
orthogonal projection, 310

Padé approximation, 294
parabolic
    differential operator, 93
    problem, 7, 8
    system, 10
parasitic wave, 213
Parseval's equality, 29, 31
partial differential algebraic equation (PDAE), 180
partial differential equation (PDE), 1
    constant coefficient, 6
    hyperbolic, 3
    parabolic, 3
particle method, 67, 93, 246
phase velocity, 212
piecewise polynomial, 77
polynomial
    Chebyshev, 55, 78
    Legendre, 78
    trigonometric, 79
polynomial interpolation, review, **74**
preconditioned conjugate gradient algorithm (PCG), **316**
preconditioning, 281
predictor-corrector approach, 60
primitive function, 343
properly posed problem, 5

quadrature
    Gaussian, 80
    midpoint rule, 80
    Simpson's rule, 80
    trapezoidal rule, 80
quadrature rule, review, **79**
Quadtree, 117
quasi-linearization, 292
quasilinearization, 233

Radau collocation, 48
Radau points, 47
Rankine–Hugoniot jump condition, 332
rarefaction wave, 333

region of absolute stability, 53, 144, 145
regularization, 289
relaxation, 317
relaxation method, 312–314
reviews
    discrete and fast Fourier transforms, 82–83
    fixed point iteration and Newton's method, 80–81
    Fourier transform, the continuous case, 29–30
    Fourier transform, the periodic case, 31–32
    function spaces, 28–29
    Gaussian elimination and matrix decomposition, 73–74
    iterative methods for linear systems, 312–320
    matrix norms and eigenvalues, 25–28
    matrix power and exponential, 30–31
    orthogonal and trigonometric polynomials, 77–79
    polynomial interpolation, 74–77
    quadrature rule, 79–80
    Taylor's theorem, 24–25
Reynolds number, 327
Riemann problem, 332, 333, 339, 352
    approximate solver, 355
Rosenbrock method, 60, 311
roundoff error, 22
Runge–Kutta method, 42–48

semi-discretization, 255, 342
semi-group, 12
semi-implicit backward Euler, 70
semi-Lagrangian method, 252, 305
shallow water equations, 250, 301
shock
    capturing, 329
    jump condition, 332
similarity solution, 333, 334
similarity transformation, 27
simultaneously diagonalizable, 275

sine-Gordon equation, 221
singular value, 28
smooth particle hydrodynamics (SPH), 246
smoother, 7, 22
Sobolev space, **28**
solution operator, **12**
sonic point, 340, 350
sound synthesis, 225
space variable, 1
specral method, 247
spectral
    accuracy, 243
    collocation method, 243
    Galerkin method, 243
spectral derivative, 83
spectral method, 93, 242, 300, 304, 325
spectral radius, 26, 139
speech synthesis, **225**
splitting, 312
splitting method, 189, 234, 276, 292, 356
spurious mode, 215
stability, 17, 120, 125
    0-stability, 215
    A-stability, 54, 124
    AN-stability, 207
    condition, 18, 19
    conditional, 125
    function, 125
    Godunov–Ryabenkii condition, 264
    L-stability, 56
    ODE methods, 53
    region, 53, 144
    stiff decay, 125
    unconditional, 125
    unconditionally unstable, 18
    von Neumann, 139, 143
staggered grid, 94
staggered mesh, 94, 101, 117
stationary iterative method, 312
steady state, 12
steepest descent, 314, 323
step size, 39

stiff decay, 56, 125, 256
stiffness, 23, 55, 296
    transient, 57
Strang splitting, 191
strict dissipativity, 278
strong stability preserving (SSP), 347
summation by parts, 171
superposition, 352
symbol, 6, 154
symmetric discretization, 94
symplectic
    map, 183
    method, 70, **185**
symplectic discretization, 222

Taylor's theorem, review, **24**
tensor product mesh, 115, 116
test equation, 53
theorem
    convergence-stability equivalence, 126
    monotonicity and TVD, 341
    ODE convergence, 50
    stability for perturbed problem, 129
    stability of dissipative schemes, 154
    symplectic perturbation, 196
    well-posed ODE, 49
time variable, 1
total variation (TV), 104, 341
    bounded (TVB), 341
    diminishing (TVD), 341
trapezoid method
    dynamic equivalence to midpoint method, 207
triangle mesh, 111
trigonometric
    basis function, 31
    polynomial, 79
truncation error, 94
two time scales, 297, 301

upstream difference, 34
upwind scheme, 34

variational
    approach, 94
    form, 111
    principle, 109, 365
variational method, 194–195
viscosity, 1
von Neumann stability condition, 139, 143, 276

warm start, 283
wave equation, 8, 217, 259
wave number, 3, 6, 212
weak solution, 331
weighted essentially nonoscillatory scheme (WENO), 328, 350, 362
well-posed problem, **4**, 49